T0344347

Construction Process Planning and Management

An Owner's Guide to Successful Projects

Sidney M. Levy

AMSTERDAM • BOSTON • HEIDELBERG • LONDON
NEW YORK • OXFORD • PARIS • SAN DIEGO
SAN FRANCISCO • SINGAPORE • SYDNEY • TOKYO

Butterworth-Heinemann is an imprint of Elsevier

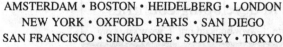
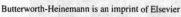

Butterworth-Heinemann is an imprint of Elsevier
30 Corporate Drive, Suite 400, Burlington, MA 01803, USA
Linacre House, Jordan Hill, Oxford OX2 8DP, UK

© 2010 ELSEVIER Inc. All rights reserved.
Except Appendix A: Copyright Construction Management Association of America, Inc., 2005.
All rights reserved. Appendix B: AIA Document A101™ – 2007. Copyright © 1915, 1918, 1925, 1937,
1951, 1958, 1961, 1963, 1967, 1974, 1977, 1987, 1991, 1997 and 2007 by The American Institute of
Architects. All rights reserved. Appendix C: AIA Document A102™ – 2007 (formerly A111™ – 1997).
Copyright © 1920, 1925, 1951, 1958, 1961, 1963, 1967, 1974, 1978, 1987, 1997 and 2007 by
The American Institute of Architects. All rights reserved.

No part of this publication may be reproduced or transmitted in any form or by any means, electronic
or mechanical, including photocopying, recording, or any information storage and retrieval system,
without permission in writing from the publisher. Details on how to seek permission, further
information about the Publisher's permissions policies and our arrangements with organizations such
as the Copyright Clearance Center and the Copyright Licensing Agency, can be found at our
website: www.elsevier.com/permissions.
This book and the individual contributions contained in it are protected under copyright
by the Publisher (other than as may be noted herein).

Notices
Knowledge and best practice in this field are constantly changing. As new research and experience
broaden our understanding, changes in research methods, professional practices, or medical
treatment may become necessary.
Practitioners and researchers must always rely on their own experience and knowledge in evaluating
and using any information, methods, compounds, or experiments described herein. In using such
information or methods they should be mindful of their own safety and the safety of others, including
parties for whom they have a professional responsibility.
To the fullest extent of the law, neither the Publisher nor the authors, contributors, or editors,
assume any liability for any injury and/or damage to persons or property as a matter of products
liability, negligence or otherwise, or from any use or operation of any methods, products,
instructions, or ideas contained in the material herein.

Library of Congress Cataloging-in-Publication Data
Application submitted

British Library Cataloguing-in-Publication Data
A catalogue record for this book is available from the British Library.

ISBN: 978-1-85617-548-7

For information on all Butterworth–Heinemann publications
visit our Web site at www.elsevierdirect.com

Printed in the United States of America
Transferred to Digital Printing, 2011

Working together to grow
libraries in developing countries

www.elsevier.com | www.bookaid.org | www.sabre.org

ELSEVIER   BOOK AID
            International   Sabre Foundation

Contents

Preface

Traversing the design and construction process, even for an experienced hand, is sometimes daunting. When dealing with unfamiliar terminology and professionals in the field of architecture, engineering, and construction, project owners may need some guidance along the way and, at times, wish they had a relative in the building business to provide them with some helpful tips.

That is the purpose of this book: to offer project owners who are new to the design and construction process some firsthand experience from someone who has been in this business for 40 years, and, for owners who have been involved with many projects, perhaps shed some new light on problems they may have encountered previously and wish to avoid in the future.

Of fundamental importance in this process is the establishment of a good working relationship with the architect's team and the contractor. In the heat of a stressful moment during either design or construction, these strong relationships will prevail, resulting in a reasonable approach to a reasonable solution. Being able to view that difficult situation from the perspective of others is another trait that takes on added importance when hundreds or tens of thousands of dollars are at stake.

An owner has a right to expect professional performance from the architect and contractor. After all, they have committed or will shortly be committing significant funds to each of these professionals. The architect and engineers will in turn be committing their staff and their reputation to the design of your building, and in today's complex building systems, incorporating compliance with a multitude of local, state, and federal rules and regulations, and that is a mighty task.

Selecting an architectural firm specializing in the type of project being considered and interviewing their past clients are two ways to move toward a well-designed project. Visiting some of those recently completed projects can give an owner an opportunity to look at the architect's work and talk to those former clients to learn a little bit more about their experience working with that architect. Because the relationship between architect and owner is a close one, the client must feel comfortable with the design team.

A contractor can be selected first by offering bid documents to a selected group of builders, and, second, upon selection, negotiating the final terms, conditions, and contract sum. Contractors work on slim profit margins, and most try to control their project by monitoring their costs closely. And most reputable contractors will assume some responsibility for minor contract obligation interpretations. Owners driving a particularly hard bargain with the contractor may find that there is little "wiggle" room left for the builder.

An owner should consider selecting a contractor on the basis of reputation and not solely on the bottom line. Integrity and excellent past client relationships are the hallmarks of a successful, competitive contractor. A good working relationship between the design team and the contractor can be promoted and encouraged by the owner, since this is an essential element in a successful project.

Owners must be as fair in their dealings with their design and construction team, as they expect those members to be fair to them. The three tenets of a well-executed construction project can be summed up in three words: fair, responsible, and reasonable. An engaged and knowledgeable owner is a prime requisite for a successful construction project, and hopefully this book will make those tasks somewhat easier.

Sidney M. Levy

The genesis of a construction project

1

The design and construction industry represents a huge chunk of the American economy. There are 2.53 million construction companies in the United States, and the total value of construction in place in 2007 was $1.137 trillion. Of this total, about $499 billion was residential housing and $637 billion was nonresidential construction: commercial buildings, schools, factories, roads, and bridges. Architectural and related engineering services included 107,386 establishments employing 1,266 million people and generating revenue of $161 billion.

ARCHITECTURAL INDUSTRY SNAPSHOT

The practice of architecture is centuries old, but in this country the profession did not become recognized until the mid-1800s as the Industrial Revolution unfolded. Before that time, and in the decades that followed, carpenters and masons not only built structures for their clients but served as quasi-designers as well. The era of the Master Builder flourished in the early twentieth century; skilled contractors employing crews of laborers, carpenters, bricklayers, plumbers, and other trades offered clients the benefit of not only their construction experience but their ability to customize past projects to fit the new owner's requirements.

The American Institute of Architects (AIA) was founded in 1857; at that time, anyone could claim to be an architect. The AIA was responsible for establishing schools of architecture—the first at the Massachusetts Institute of Technology in Cambridge, Massachusetts, in 1865. The first graduate from an architectural program was Nathan C. Ricker, who graduated from the University of Illinois architecture program in 1873. This school also had the distinction of conferring the first architect's degree on a woman.

One of the driving forces behind the use of an architect was the proliferation of government regulations and the resultant increase in the complexity of building structures and the types of equipment being offered by various manufacturers. The legal system also began to weigh in on liability issues and to pursue the principle of due diligence with respect to the adequacy of a structure's design.

© 2010 by Elsevier, Inc. All rights reserved.
Doi_No = 10.1016/B978-1-85617-548-7.00001-X

The practice of architecture has changed dramatically since those early days. Seventy-five years ago, a visit to an architect's office would reveal rows of aspiring architects hunched over their drawing boards preparing hand-drawn designs, erasers at hand, moving T-squares and plastic triangles to form the shape of their structure. Today, the pens and pencils, T-squares, triangles, and erasers have been replaced by powerful computers and specialized software programs. The only noise emanating from those work stations are the mouse clicks and printers from which these intricate designs flow.

Increasingly advanced software allows an architect to produce a complete list of all of the materials required for the project simultaneously with the progression of the design itself. If you design a 30-foot-long, 10-foot-high drywall partition, the computer automatically generates a materials list: 15×10 steel studs and 14 sheets of gypsum drywall. If the architect has a database of costs, an estimate for both labor and materials will also be created.

Architects using other types of computer software can produce three-dimensional images to be viewed by their engineers and clients. By adding a time sequence—the fourth dimension—the client can actually see the virtual building being constructed from the ground up before the first shovel of earth is uncovered. This innovation, called building information modeling (BIM), which is now offered by large architectural firms, will undoubtedly become mainstream, and new innovations will continue to amaze potential clients.

THE ARCHITECT AND THE CLIENT

The Architect's Handbook of Professional Practice defines the term *client* as "at least one person with whom one architect will deal with on what is most often a remarkably intimate basis." There is an intimacy between owner and architect similar to that of contractor and owner, as these professionals consume so much of the owner's time and money to produce a product with a potential 100-year lifespan. The structure that results will either please or displease its owner-occupant for years to come, so it had better be done right. The owner places those responsibilities on the architect and the builder.

In the *Architect's Handbook*, Gordon Chong states, "Unlike architects, who view the design and construction of a building as 'an end,' the majority of our clients see buildings as 'a means' to satisfying a wider set of requirements.... One of the most important challenges facing us as architects is to ensure that we fully understand our clients and their motivations."

THE CONSTRUCTION INDUSTRY

The three basic segments of the construction industry are building contractors, also known as general contractors, who construct residential, commercial, industrial, and institutional buildings; heavy and civil-engineering contractors, who build roads, highways, bridges, tunnels, and other similar projects; and specialty contractors, more familiarly known as subcontractors, who perform specific trade work such as carpentry, roofing, electrical, plumbing, heating-ventilating-air-conditioning (HVAC) work, and a host of other tasks, The 2002 U.S. Census Bureau statistics revealed that the construction industry had a total of 8.9 million wage and salary employees.

Technology in the construction industry has not proceeded at the pace enjoyed by the manufacturing sector. Today there are some robot-driven, software-guided bull-dozers, but the full impact of technological advances has not reached down to the average general contractor. Although most contractors have computers with sophisticated scheduling, estimating, and cost-control software programs, they still lay up one brick after another to build a wall and nail studs and sheetrock to construct partitions.

AN OWNER'S MAJOR COMMITMENT

To some businesspeople, their company's new construction project may represent one of the largest corporate investments they will ever make, and it is one in which it is wise to proceed carefully and systematically. The process of design and construction is not that complex for professionals in that field, but it is an environment that requires expanding an owner's knowledge and experience. It requires a well-thought-out plan of what the company hopes to achieve in the design and construction of their new or renovated office building, corporate headquarters, or manufacturing facility.

Lessons learned from problems with operations in the old building and pitfalls to avoid in the new design should be carefully annotated when a new project is being considered. Without a careful plan of what you as an owner wish to achieve, it may be difficult to convey those needs and requirements to an architect whose responsibility it is to convert them into a plan that a builder can follow. But before we get ahead of ourselves, as an owner there are some strategic decisions that you must consider as the plan for the new building begins to take form.

Project delivery

The "project delivery system"—the method by which an owner gets from point A to point B—has several different options. Selecting an architect is usually the first step in this process but not necessarily an absolute. There are several ways to proceed with the design and construction of the project. Each of these project-delivery systems is discussed in much detail in the following chapters in this book, but for now let's look at the basics. The project's genesis can take several forms: design-bid-build, design-build, construction management, and program management.

Design-bid-build

The most prevalent project-delivery system in the public sector, also employed by a large number of private-sector clients, is the design-bid-build process. It is a rather straightforward approach: An owner engages an architectural or engineering firm to produce a complete set of plans, specifications, and specific project requirements. These documents, referred to as bid documents, are distributed to a selected list of general contractors, who are prequalified as far as construction experience and financial strength is concerned. These contractors will submit their price to complete the work as outlined in the bid documents. It seems like a simple approach, but, as we shall see later, there can be a lot of twists and turns in the process. Since the contractor bidders will be estimating the costs for all of the work stated in the bid documents—the plans and specifications—the quality of the bids will depend on the quality of the plans and specifications. If something is missing from the plans, the bidders may not include that missing item in their price because they are concerned that their competition will not.

Some refer to design-bid-build as design-bid-redesign-rebid. An owner's budget may not reflect the actual cost of the construction project, and upon receiving bids, the lowest bidder may have submitted a price in excess of that budget. The owner must go back to the drawing board to redesign (at additional cost) and rebid, sometimes in a time frame of rapidly increasing costs, with the result that the project requirements may have diminished but the cost of work increased.

In many instances, the bids received by an owner during the conventional design-bid-build process exceed the owner's budget, as just noted. This can occur for one of many reasons. The delay between the completion of the plans and specifications and the date when bids are solicited may be subjected to inflationary forces in the marketplace. Historically, inflation in the construction industry has outpaced the Consumer Price Index (CPI), and a time lag of 12 months, for example, may generate increased labor and material costs of 5 percent or more. Some owners may not have allowed for that adjustment. Alternatively, the owner's budget may have been assembled with unrealistic prices to begin with, and the market will return the more realistic costs.

For whatever reason, when the design-bid-build process results in bids that exceed the owner's budget, an architect and owner may decide to work with the lowest qualified bidder, review costs, and make changes that are acceptable to all parties to reduce the price of the work to fit the owner's budget. This may require some design changes, and an owner and his or her design consultants must carefully consider all of those costs and also ascertain that neither the program nor the quality will be impacted by the changes. The negotiated scope of work and resultant price can then be incorporated into the negotiated construction contract.

Another approach to a negotiated contract is to select a general contractor with whom the owner and/or architect has had previous successful dealings and ask that builder to work with the architect to develop a cost-effective design that meets the owner's program. In this process, the contractor can share current estimating experience with the architect and advise on constructability issues, material and labor costs, and availability. The architect can then review these comments and incorporate the accepted changes in the design, and the owner can negotiate a contract agreement with the general contractor.

Design-build

Engaging an architect at the conceptual stage of a project is not the only way to proceed down the path to design and construction. An increasingly popular process called design-build is being employed in both private- and public-sector work. The essence of design-build is to place both activities in the hands of one firm: a design-builder. Some design-build firms were created when a general contractor employed architects and/or engineers on staff to provide a full-service organization. Other general contractors offering design-build services form a joint venture with an architectural firm or hire an architect much as they hire subcontractors to perform the design work.

Architects can also be the lead team member in a design-build situation, inviting a contractor with whom they have worked successfully on previous projects to join with them. This process of placing design and construction in the hands of one entity has the advantage of being able to monitor real-time costs as the design progresses to keep the owner's budget on track. The contractor employs the current database of costs in parallel with the progression of design so if changes need to be made to remain on budget, these changes can be reviewed quickly by the owner, who may elect other cost-saving options or increase their budget. At least there are fewer surprises.

According to advocates of design-build, the entire schedule for the project is significantly shortened because the "build" side of the team can begin to order materials and equipment, engage subcontractors more quickly, and get a jump on construction. A reduced schedule means less construction financing, which is more expensive than permanent financing. Owners using design-build report fewer change orders—another plus to this project-delivery system.

But design-build is not for everyone; it requires an owner to have a specific detailed plan in place and experienced staff on hand to manage the process from the owner's standpoint; in some instances, state laws do not permit design-build projects. There are a number of firms that specialize in design-build support, and the Design Build Association of America (DBIA) is a source for more information and a list of design-build firms.

The construction management approach

The Construction Management Association of America (CMAA) considers construction management a *service* as opposed to the hiring of a contractor who delivers a *product:* the building designed by the architect. But the construction-management approach is also a project-delivery system. Unlike the arm's-length contract transaction between an owner and a general contractor in a design-bid-build or design-build system, the construction manager (CM) is the owner's agent and acts, as such, on his or her behalf.

The construction manager provides the owner with sufficient professional office and field staff to complete the construction project. These services can be provided during the design stage, the construction stage, or both. Construction management can be viewed in much the same way as a situation in which an owner has experienced construction professionals on staff to handle the upcoming construction project.

As an owner's agent, the CM will serve as a representative of the owner when engaged to assist in the design, the construction, or both, stages of the project. Some consider the hiring of a CM during the design stage to be most important because this is where these professionals can bring their knowledge of costs and means and methods of construction to bear as they work with the designers to produce the most cost-effective project.

CM services are divided into two basic phases: design and construction. An owner employing a CM during design is able to tap that professional's knowledge of local labor pools, material and equipment vendors, and a current estimating database and then advise on scheduling and value-engineering procedures. A CM during construction will provide the owner with a seasoned project manager, project superintendent, and other professional staff to meet the owner's needs and interests throughout the project. A list of CM firms is available through the Construction Management Association of America website: *www.cmaanet.org*.

The program manager

Taking the role of construction manager a little further, a program manager's responsibilities are wider and more varied. The CMAA defines the role of program manager as one that includes not only assistance in the design and construction process but also development, planning, environmental study, and interaction with local, state, and federal government regulatory agencies. The program manager can also be engaged to oversee multiple owner projects, each of which may be in various stages of development.

The construction consultant

If an owner does not have experienced staff in either discipline, another approach to design and construction is possible. In this case, an owner can hire a consultant to represent him or her. These consulting firms have experience in all phases of the construction process and can be hired for specific phases. They can work with the architect during design development to comment on costs versus design, and they can work with the owner during the bidding process to interview prospective bidders. These consultants will review bids, offer advice on contractor selection, and work with the owner's attorney during the preparation of the contract for construction to ensure that ample protection is included in that contract. The consultant can be engaged to review change orders and assist in resolving disputes and claims from the contractor. These consultants generally work on an hourly rate and are available on an as-needed basis.

There are a number of different types of consultants that an owner may consider as he or she begins to firm up the project plans. Estimating consultants can provide cost information of a general or detailed nature. This could prove helpful in the project's planning stage to establish a budget and determine whether available financing sources are adequate.

Cost manuals, such as those published by R.S. Means or McGraw-Hill's Sweets Division, provide component, unit-cost, and square-foot pricing for many different types of construction and can be ordered over the Internet.

Building-code consultants are available to discuss compliance with local building officials or federal regulators when significant renovations are being considered and where upgrades may result in the need to comply with such regulations as the Americans with Disabilities Act (ADA).

Scheduling consultants can be hired to produce a detailed construction schedule and update that schedule as construction proceeds. They can also review a project in crisis to establish responsibility for delays that may have occurred along the way.

The fast-track approach

We often hear about "fast-track" projects, but what exactly does this mean? A conventional project-delivery system can be expedited via the "fast-track" method. It involves assigning priority to the development of specific design drawings and accompanying specifications that will allow for ordering those essential components early on instead of waiting for the normal progression of design development. All of this is done with an eye to either accelerating the start of the project or certain phases of construction.

Using a building's structural-steel framework as an example, this is how the fast-track process works. In the normal process of design, the structural drawings for both foundation and superstructure will be the first ones produced. They will be followed by the production of the architectural drawings and the design of the building's electrical and mechanical systems. Upon the production of a complete set of drawings (with the exception of the design-build process), a contractor will be selected and a contract awarded, allowing the builder to begin ordering materials and equipment.

Under a conventional schedule, it is not until a general contractor is brought on board that a structural steel subcontractor is engaged. And only after detailed drawings have been approved by the subcontractor will an order for steel be placed. This process will produce structural steel on the job site about 12 to 16 weeks *after* the contract for construction is signed.

Utilizing the fast-track method, the owner can award the structural steel job to a subcontractor as soon as the steel design has been completed. If a contractor has been selected but a contract sum has not been negotiated because all of the other drawings necessary for a complete estimate have not been produced, the owner can authorize that contractor to place an order for the structural steel immediately. On the advice of the architect and engineer, the owner can award a contract to a structural steel subcontractor and "assign" this contract to the selected general contractor, who will fold the scope and cost of that work into the contract for construction.

By either means, the fast-track approach allows the entire steel production cycle to be triggered and delivered to those waiting foundations much sooner. The fast-track process can also be used when other long-lead-time equipment is required— for example, a specialized piece of machinery from an overseas manufacturer or a complex HVAC component. The engineer can complete the design for this equipment out of sequence with the normal progression of design documents so an advance order can be placed.

Fast tracking is more complicated than this brief explanation indicates, but it is a concept that can be pursued by an owner as the design phase progresses when it is advisable to accelerate a project's completion date.

Considerations in selecting a project-delivery system

Some practitioners refer to design-bid-build more accurately as design-bid-redesign-rebid, which is a costly process. Tracking design development with a realistic data-base of costs from either an architect who is well versed in that type of construction, a builder, or an estimating service is key to avoiding that recycling process of rede-signing and rebidding.

The design-build delivery system appears to be best suited to an owner with past experience in similar projects that can be accurately conveyed to the design-builder. But this does not rule out the first-time owner who has a very clear picture of the company's requirements. Discussions with a construction-management firm as a new project is under consideration may either enforce an owner's opinion that CM is the way to go or direct him or her to look for other options.

Many problems that arise with all delivery systems can be traced back to whether a realistic budget has been established in the first place. Consultation with an archi-tect, builder, or estimator early on is an important step to take.

Green and sustainable building

Green building is based on designs that are more environmentally sensitive—those that tend to lessen the impact on our environment. Commercial and institutional buildings (schools, hospitals, public buildings) have an enormous impact on our environment. Studies have shown that buildings in the United States consume about 65 percent of all electrical consumption; they generate about 30 percent of all greenhouse-gas emissions; and they consume 30 percent of our raw materials. And buildings produce one-third of our total waste output, which amounts to nearly 136 million tons per year. These statistics have driven the green and sustainable build-ing movement. The green-building concept attempts to reduce our dependence on energy and make those buildings more energy-efficient.

Considering going green?

If an owner is considering "green," he or she should investigate all aspects of this approach and consult with architects, engineers, and contractors who have some experience in this process. Since the green-building concept is relatively new, there may be a dearth of experienced professionals in certain geographic areas. Such professional organizations as the American Institute of Architects (AIA) and the Associated General Contractors of America (AGC) may be able to provide some assis-tance. And, of course, the United States Green Building Council (USBGC) should be contacted. It has developed a rating system known as LEED (Leadership in Energy and Environmental Design) to certify new and renovation design based on compli-ance with the achievement of certain goals.

The green-building movement involves not only environmentally friendly components but a whole-building design process that takes into account the site on which the structure is to be built. This whole-building approach focuses on the following features:

- Reducing energy costs
- Reducing maintenance costs
- Reducing the impact of the structure on the environment
- Providing building occupants with a healthier and safer environment
- Creating a more productive environment

Sustainability

The definition of *sustainability* can best be explained by Paul Hawkins in his book *The Ecology of Commerce: A Declaration of Sustainability*: "Sustainability is an economic state where the demands placed upon the environment by people and commerce can be met without reducing the capacity of the environment to provide for future generations."

The sustainable-building movement tries to encourage the use of renewable resources instead of depleting the ones we have been consuming in our construction components. We are all familiar with a material known called Masonite, which is made from waste and recycled-wood products and classified as a medium-density fiberboard (MDF), a sustainable material. That plywood-like panel with visible flakes embedded in it that we see fastened to the exterior of residential projects is a sustainable material, since it is made of reconstituted wood chips. These products and many more are referred to as engineered-wood products. Hay is another sustainable product and can be used as an insulating material; there are many more examples.

CONTRACTS AND THE CONTRACTOR

A construction project is basically a continuum of contracts: contracts between an owner and the architect, between an owner and a contractor, and between the general contractor and the subcontractors. The owner-general contractor agreement is a trickle-down contract. The term *general contract* refers to the contract between the owner and the general contractor. Just read the standard clause in a general contractor's subcontract agreement, which goes something like this:

The foregoing incorporation of the General Contract, by reference, shall pose upon the SUBCONTRACTOR, the same obligations and responsibilities with respect to the work to be performed by it under this subcontract as are imposed upon the CONTRACTOR under said general Contract. The General Contract shall be made available to the SUBCONTRACTOR at his request for inspection at CONTRACTOR'S office at any mutually agreed upon reasonable time.

An owner should be aware that any terms and conditions in the agreement with the general contractor may be shared with any subcontractor engaged by that GC.

Contractors deal with contracts on a daily basis and are much more familiar with their content than the average owner. They are acutely aware of the terms and conditions of a contract that may be favorable to them and are very much aware of those provisions they must avoid. An attorney with a general practice may not be as experienced as one whose firm deals primarily in construction law, not only to assist in the preparation of a construction contract but to offer guidance and advice if a dispute or claim arises. In all fields of endeavor, there are honest businesspeople and those who try to push the limits of creditability—and beyond.

ETHICS IN THE CONSTRUCTION INDUSTRY

In 2004, FMI, the nation's largest management-consulting firm for the construction industry, teamed up with CMAA to survey project owners, architects, engineers, construction managers, and contractors to gauge their concerns about ethics in the industry. The results, culled from 270 responses, might be kept in mind as we traverse the design and construction industries in the chapters that follow.

The key concerns expressed by the respondents to the survey were fourfold:

1. There appeared to be a breakdown in trust and integrity.
2. There was a perceived loss of reputation for the industry.
3. There was a need to provide a code of ethics and standards.
4. There was a need to create a more equitable bidding process.

Concerns were voiced by owners, architects, engineers, and contractors; they all seem to point to a need for fairness on the part of each party to the construction process.

Concerns about architects and engineers included the following:

- Owners stated that architects and engineers do whatever makes the owner happy, often at the expense of the contractor.

- Architects and engineers need to express fairness when dealing with contractors or making decisions that affect the owner.

- Design professionals knowingly issue plans and specifications that are deficient.

Concerns about contractors included the following:

- Bid shopping, a practice where contractors use one subcontractor's price to drive down the price of another to achieve the lowest cost, often an unrealistically low price

- Change-order games, played by a general contractor who knowingly submits a low bid in the hope of gaining more profit by issuing questionable change orders as construction proceeds

- Payment games, the receipt of payment from one owner, which should be used to pay for labor, materials, and equipment for that project, commingled with funds to pay for other projects

- Instituting claims that are vague or specious

- Engaging subcontractors whose past performance has been unreliable

Concerns about owners included the following:

- Owners who authorize work but argue about paying for it

- Owners who are very late in their payment of contractor requisitions

- Owners who pass off responsibility to others when they are the party that should assume responsibility and resolve problems promptly and equitably

- Owners who lack ethical behavior, such as advertising bogus low bids to drive down the price of bidding contractors

- Little dialogue between owners and contractors about the expectations of both parties

It appears from this study that there is plenty of blame to go around, indicating the need to maintain and enforce ethical business practices by owner, architect, and contractor alike. So with that in mind, we will now begin the design and construction process.

Selecting and working with an architect

2

The architectural and related engineering services industry occupies a significant position in the U.S. economy. According to the U.S. Census Bureau 2002 Economic Census, there are 107,386 architectural and engineering establishments in this country, with a total annual payroll of $68 billion and a total annual sales revenue of $161.8 billion.

PSMJ Resources, headquartered in Newton, Massachusetts, with branch offices in Great Britain and Australia, conducts more that 200 architecture/engineering/construction seminars and conferences annually. It produces the industry's premier annual survey, which includes a review of the architecture/engineering/construction community's management salary, financial performance, and fees and pricing status.

The Big Picture Results of the *PSMJ 2008 Benchmark Survey's Executive Summary* sums up the key overall indicators for this group of design professionals:

- Financial performance of project activities remained excellent.

- Operating profits before incentive bonuses and taxes as a percentage of net revenue leveled off at 15.2 percent, the same as the 2007 survey.

- Operating profit margins remain high but appeared to be poised for a downturn.

- Gross revenues increased by only 9 percent as compared with a 14 percent increase in 2007—a decrease of 36 percent.

- Backlog of work continued to build but at a 25 percent lower rate than the 12 percent growth rate in the 2007 survey.

- Staff growth was 5.5 percent, a 23 percent reduction from the 7.1 percent growth reported in 2007.

- Overhead rates increased 3 percent, a reversal of previous trends of declining overhead rates.

- Direct labor costs increased by 4 percent.

© 2010 by Elsevier, Inc. All rights reserved.
Doi_No = 10.1016/B978-1-85617-548-7.00002-1

THE CHANGING WORLD OF THE ARCHITECT

Today the process of turning a building concept into a reality is a rather complex procedure; the architect must deal with code compliance with any number of local, state, and federal public agencies; rising costs; and the quest to find experienced personnel. The list of building codes and regulatory agencies that are part of the design and construction process is an alphabet soup: BOCA (Building Official and Code Administration); ADA (Americans with Disabilities Act); ASHRAE (American Society of Heating, Refrigeration and Air-Conditioning Engineers); ASTM (American Society of Testing and Materials); NFPA (National Fire Protection Association); and EPA (Environmental Protection Agency), to name just a few. When public funds are involved, the Davis-Bacon Act establishes minimum labor rates for construction workers.

The practice of architecture and the tools of its trade have changed dramatically over the last half-century. A visit to an architect's office in the 1950s would reveal designers working at drafting tables with pens, pencils, T-squares, and plastic triangles, creating their designs by hand. Computer-assisted design replaced many of those drafting tables as computer hardware became more readily accessible and more affordable and software programs for architects and engineers proliferated. Nowadays, you see architects and engineers staring at a computer screen.

THE ARCHITECT AND ENGINEER SELECTION PROCESS

For those who have had no prior dealings with an architect, selecting the right one for your project has been made somewhat easier by the American Institute of Architects (AIA) headquartered in Washington, D.C., with branches throughout the country. The AIA website—*http://architectfinder.aia.org*—is a good place to start. There is a pulldown menu from which one can select the type of building under consideration and a list of architects experienced in that particular type within a specific geographic area. After one or more architectural firms have been selected, a request-for-qualifications form can be purchased from the AIA and sent to the selected architect or architects. Coupled with an interview or two and a look at some of the firms' recent projects, an owner can quickly find a suitable architect for the project.

The architect's qualification statement

The relationship between owner and prospective architect will be a close one, and compatibility with concepts and personnel assigned to the design process is important. By issuing a request for qualifications, an owner can start the process of becoming familiar with the firm that will occupy a great deal of time, money, and energy when the final selection is made.

The qualification statement will include some basic business information from the architectural firm, qualifications for the project, the types of services provided, a list of references, and information about the team that will be working with the owner.

The following basic information is included:

1. Name and address of the firm and any branch offices

2. The type of organization: sole proprietorship, partnership, corporation, limited-liability corporation (LLC); if the project is a joint venture with another firm or other design consultants; a description of the full nature of the collaboration

3. Prime contact at the architect's firm

The general statement of qualifications can be as simple as a brochure of the company's history, years in business, types of projects completed, and photographs of the interiors and exteriors of recent work. General information typically includes the following:

1. Names of organization's principals
2. Professional history
3. Registration status; most states require licensing, and some firms have multistate licenses
4. Professional affiliations
5. Key personnel
6. Total number of staff
7. Number of registered architects
8. Honors and awards received
9. Professional and civic involvement

Some firms employ the types of engineers required to complete the design, whereas other architectural firms may subcontract one or more of these consultants:

1. Civil engineers
2. Structural engineers
3. Mechanical engineers
4. Electrical engineers
5. Interior designers
6. Landscape architects
7. Others

A list of projects representative of the architect's recent work (past five years) with a brief description of each and accompanying photographs are usually provided. If the firm has been associated with other architectural firms in the design of other buildings, a list of those projects should be included along with the nature of that participation. This list should include the following information:

1. Project name and location
2. Project size in square feet and number of floors if a multistoried building
3. Cost

4. Project owner
5. Completion date
6. Contractor or construction manager
7. Brief description

The references section should include a representative sample from recent past projects and possibly current projects. A resume of key personnel who will participate in the project will be provided. Each person's educational background, licensing, or registration information; number of years with the firm; previous employment if applicable; and any experience, awards, and particular expertise should be included.

Upon receipt of these qualification statements, the next step is to interview the firm and those individuals from both the owner's and the architect's firms who will be working together. This is important because often a firm's principal(s) will be present during the initial interview with the owner, but you as an owner will be devoting considerable time to the particular architect or architects assigned to the project throughout the design and possibly construction period, and you need to feel comfortable with those designers.

Interviewing an architect

When interviewing an architect to determine whether he or she is the one for your project, an owner must be prepared to discuss the nature, scope, and special requirements of the proposed project. Of course, a timetable for construction, availability of the proposed site, and financing arrangements should also be addressed. To assist in helping prospective clients to better understand the role of client and architect during this initial interview, the AIA prepared a list of 20 questions to ask during the interview process:

1. What are the important issues the architect sees in the program you have described, and what challenges does it present?

2. How will the architect approach this project?

3. How will the architect extract enough information from the owner's team to meet the requirements and their goals?

4. How will the architect establish priorities and make decisions, particularly when it comes to the owner's budget?

5. Who from the architect's firm will be dealing with the owner directly, and will this person remain during design and into construction?

6. Does the architect appear to be very interested in this project, or does the owner detect an attitude of indifference?

7. How busy is the architect? If he or she is too busy at the time the owner wishes to commence design, will this impact the quality and/or delivery of the design?

8. If prior interviews have been conducted with other firms, what sets this firm apart from the others?

9. How does the architect establish fees, and what method does he or she suggest for this project?

10. What would the architect expect the actual fee could be by using one or more of the preceding methods?

11. Has the architect explained the steps in the design process?

12. How would the architect organize this process?

13. What does the architect expect you, the owner, to provide?

14. What is the architect's design philosophy? Can you see some projects for a similar use that they recently completed?

15. What is the architect's experience and track record of cost estimating? Have previous estimates previously followed the results of the contractor's bids?

16. What will the architect provide in the way of sketches, drawings, models, and renderings along the way to define the project?

17. If the scope of the projects changes, will there be additional fees? What constitutes a change requiring additional fees, and what would these fees be?

18. What services can the architect provide during construction?

19. How long does the architect expect the design and construction phases to take?

20. Can the architect provide a list of past clients with projects similar in function to the one now being proposed?

Once the selection of an architectural firm has been made, the owner must now decide what services will be required of the design team, and this depends on a number of variables, such as the number and types of other design consultants on the firm's staff.

THE ARCHITECT'S TEAM

In most projects, the architect is the team captain; he or she requires the services of other design consultants, each one of whom will work together in a coordinated design effort:

■ The structural engineer's job is to design the building's foundation and superstructure after consultation with the architect, whose visual scheme for the building's exterior may impact the structural design.

■ The civil engineer's responsibility is to investigate soil conditions for foundations; design driveways, roadways, and parking areas; and, working with other engineers, provide the size and location of all underground utilities (gas, water, electric power, storm sewer) required for the project.

■ The electrical engineer will establish the building's total power requirements and design the circuitry within the building proper. The electrical engineer may also consult with a lighting designer to provide the most efficient and cost-effective interior and exterior (site) lighting for the building. A "low-voltage" consultant may also be employed to design the building's security and data and telecommunications systems. The electrical engineer will consult with the civil engineer on the design of the underground incoming electrical conduits and cables.

■ The mechanical engineer has the responsibility of designing the plumbing; heating, ventilating, and air conditioning (HVAC); and fire-protection systems within the building, and also consults with the civil engineer to design all incoming underground utilities: water, sanitary sewer, storm sewer, and fire-protection mains.

■ The landscape architect will work with the civil engineer to design the contours of the property for both practical and aesthetic purposes. The landscape architect will set aside grass and planting areas and select regional plantings for beauty and sturdiness.

■ The interior designer can interface with the architect to select materials and colors for various finishes—floors, ceilings, walls—and can assist in the selection of artwork within the structure.

Building information modeling

The latest application of computer imagery in the design process is referred to as building information modeling (BIM), the creation of 3D and 4D models of the owner's building, sometimes referred to as virtual design. By creating a 3D image of the proposed project and passing it by all of the related consulting engineers for their reviews and comments, the end product avoids some of the problems associated with the more conventional method of computer-assisted design (CAD) and ensures that everything fits in its allotted space.

One of the major problems that can occur in a complex, multistoried design is when all of the various design consultants—the structural engineer, the electrical engineer, the mechanical engineer, and other involved in the design development—work independently and are not thoroughly coordinating their work with other members of the team. This is essential in ensuring that all work is being reviewed by other designers so electrical and mechanical elements, for example, can fit within the confines allotted for them by the architect, who, in turn, is working within the confines dictated by the structural engineer.

With BIM, the structural "skeleton" is transmitted to all other consultants electronically; they must fit their work into the system. This 3D image can be rotated on its axis, and when the mechanical engineer designs a 3-foot by 2-foot heating and air conditioning duct for the third floor and finds that there is not enough room to pass under a steel beam, he or she can reduce the depth and increase the width so a 4-foot by 18-inch duct can fit. When such conflicts occur and items of work don't fit in their allotted space, all parties are aware of these conflicts and adjust their systems accordingly.

By adding the fourth dimension—time—BIM can create a virtual construction schedule that diagrammatically shows the building evolving from foundation through superstructure to completion. This "virtual," visual construction schedule is very effective when used as a tool to review actual progress during the weekly or biweekly project construction meetings. While standing outside the construction field office, the status of construction can be compared to the contract schedule as represented by the BIM imagery to dramatically reveal the actual progress as opposed to the planned schedule.

Today, the cost of BIM is such that it is cost-effective only on very large projects, but just like the computer that replaced the T-square and pencil, as software and hardware costs decline, this system will become mainstream and prove to be a valuable tool for owners, design consultants, and builders.

DEFINING THE SERVICES OF THE DESIGN TEAM

Selecting an architect is the first of many decisions an owner must make. The architect's team can provide a number of services beyond the basic preparation of the project plans and specifications, and as the owner begins to discuss the project requirements, he or she should review in-house capabilities to determine if the basic "plans and specs" scope of work should be expanded to include other services that extend through construction. Although not all architectural firms can furnish all of these services, they may be able to provide sources for an owner to tap into. The AIA defines architect services in six basic categories:

- Design services—architectural, civil (sitework), structural, mechanical, electrical—the basics

- Landscape and interior-design services

- Evaluation and planning services such as surveying an owner's existing facilities to provide spatial comparisons with the new facility

- Project financing assistance, site analysis, and development planning

- Bidding or negotiation services—furnishing bid documents, addenda, and responses to contractor questions

- Analysis of alternates and value-engineering proposals submitted by bidders, bid evaluations, and contract selection and award

- Contract administration services, including periodic onsite visits during construction, full-time onsite project representation, contractor payment requests, testing and inspection services, quotation requests, and change order reviews

- Contract cost-accounting services

- Furniture and equipment purchases and installation administration

- Interpretation and assistance in contractor disputes and claims

- Facility administration services, including monitoring equipment startup procedures, maintenance, and operations program assistance.

- Warranty issues, postconstruction review and analysis, preparation of record drawings—"as-builts"

During the design stage, it may be advisable to have the architect prepare the bid documents, receive the bids, and advise the owner on contractor selection. This process starts with the detailed preparation of the necessary documents, forms, and other instructions, which will be sent to the prospective contractor bidders. Other architectural services may be selected after a contractor award is imminent, and these are referred to as "construction services" or "contract administration" services.

Contract administration

Contract administration, also referred to as construction services, affords the owner the assurance that what is being built is what has been designed. The architect will act as a go-between for owner and contractor in the administration of the construction contract throughout the entire construction process and even into postconstruction activities. The following are some typical contract administration services:

- Evaluating work in place to ensure compliance with the plans and specifications

- Ensuring that all applicable building codes are being observed

- Reviewing and approving shop drawings—those detailed material and equipment specification sheets submitted by the general contractor—to indicate compliance with the requirements of the plans and specifications

- Approving samples required by the contract documents

- Reviewing and commenting on the results of the various testing and inspection procedures required by the project

- Reviewing and approving or amending the contractor's monthly requisition for payment

- Reviewing and approving or rejecting claims for work-change orders

- Conducting periodic progress meetings with the contractor, vendors, and subcontractors

- Reviewing and commenting on contractor-proposed changes to the contract documents, which may increase or decrease the contract sum

- Supervising the project's completion to include a review of as-built drawings; preparation, inspection, and sign-off on all punch list items; and all other contract closeout requirements

- Ensuring compliance with all warranty and guarantee requirements

- Serving as an initial arbiter if and when contract disputes arise

Architects view contract-administration services as spending a dollar to save a hundred dollars. And without experienced staff in the owner's organization, these added costs are generally well worth the money spent.

An owner may also engage a construction consultant to assist not only in the design process but all through the construction phase, acting as the owner's representation with the design team and the contractor. These consultants often work on an hourly rate basis and can be brought on board as needed.

Owner's responsibility for services

One of the first services required of an owner embarking on a construction project and a new building site will be a civil engineer, who will act as a consultant to the architect in the preparation of the project's various site plans and conduct the following tests required by the structural engineer:

- Performing a series of test borings with an auguring machine at selected areas throughout the site to determine the capacity of the soil so foundations of the proper type and size can be sent to the structural engineer (Typically for a multistoried commercial building, one boring for every 50 to 100 lineal feet of structure is required; one boring for every 100 to 150 lineal feet is required for other commercial buildings.)

- Determine the classification of the soils to determine if they are suitable as a subgrade for use by the contractor as a base for paved areas, both asphalt and concrete

- Test the permeability of the soil

- Determine the presence of any underground water or springs (Depending on the time of year these tests are performed, these tests may not be so reliable. The absence of springs in summer may not be typical of conditions during the early spring season in the same location.)

- Determine the presence of rock (another imperfect observation, since no rock may be present in the exact location of one test boring but may exist in significant amounts in areas between borings)

- Determine if other underground obstructions exist, such as an old foundation from a previously abandoned building

- In addition to test borings, which are typically 6 inches in diameter or slightly smaller, a test pit can be dug with an excavator or backhoe to uncover a wide area—say, 10 feet deep and 10 feet wide—if some underground obstruction is suspected.

Inspection services

These types of investigations are usually included in the architect's fee, but other tests and inspections may be required. Certain inspections and tests are contracted for directly by the owner to ensure that the testing and inspection firm is impartial and is working solely in the owner's interest. Both the civil engineer and the architect can be helpful in providing a list of companies that offer these inspection and testing services. The following are some of the types of inspections and testing required during the course of construction:

- Inspect soils prior to placement of foundations, inspect any soil requiring compaction, or, in the case of special foundations like pilings, provide load tests to verify the load-bearing capacity of the piles

- Concrete sampling and testing, including inspecting the concrete as delivered by the transit mix company to ensure compliance with a slump test; provide concrete compression testing to ensure that the concrete meets the specifications

- Posttensioning of concrete slabs if required

- Vibration monitoring of equipment if required

- Permeability of soil testing

- Steel inspections at the site and at the steel subcontractor's shop

- Moisture testing to determine if certain flooring materials can be installed over concrete slabs

- Masonry inspections

- Air infiltration testing

Using the same basic procedure for selecting an architect and a contractor, the request for qualification can be used to solicit bids from testing and inspection companies.

After an architect has been selected, he or she may have had previous experience with a testing and inspection service or be able to provide a list of these types of companies that may be contacted. In the event that a request for qualification for testing and inspections is to be prepared, the format of the RFQ will include those inspections and tests dictated by the project's design; for example, a cast-in-place concrete structure will require significant concrete testing but possibly little or no structural-steel testing.

For concrete foundations, buildings, or sites, testing will include the following:

- Verification of mix designs and certificate of compliance

- Compaction of aggregate base course under concrete or other paved areas

- Material samples, including aggregate-base materials

- Concrete batch plant inspections

- Verification of size, type, and placement of reinforcement in cast-in-place concrete, including splicing inspection if required

- Concrete cylinder and slump tests, including handling, delivery, and testing of cylinders

- Verification of concrete-slab flatness and finish tolerance. (This is important when an automated warehouse is being built and extremely flat floors are required for the operation of the robotic or automated equipment.)

- If posttensioned slabs are designed, recording and verifying stressing operations and sampling and inspection of posttensioning strands

For structural steel and miscellaneous metals, tests will include the following:

- Steel-fabrication plant inspections

- Field inspections—qualifying welders, inspecting all steel erection and connections, including magnetic-particle testing and ultrasonic testing of field welds

- Metal decking when used as the structural form for concrete floors; welding procedures for decking and shear studs

- Shop and field welds for metal fabrications such as steel railings and stairs

- Visual inspections and torque tests for some structural-steel bolt connections

- Pull-test to assure that hangers and rods encased in concrete are secure. (Remember those connections that failed to hold the concrete slabs above the travel lanes in Boston's Big Dig?)

- Measure the thickness, density, and bond strength of sprayed-on fireproofing applied to structural-steel members.

Other forms of testing and inspection that may be required include the following:

- Hydrostatic or air testing of pipes and fittings

- Inspection of formwork and shoring operations

- Roofing and waterproofing inspections

- Inspection of sheet metal flashings and trim

- Inspection of fire-stopping (material used to plug openings to prevent the spread of fire)

- Façade mock-up testing

- Concrete moisture testing (needed when flooring materials are to be installed over freshly poured concrete floors)

- Testing of acoustical ceilings for compliance with sound ratings

- Elevator testing

- Vibration isolation testing (large HVAC equipment is often mounted in a vibration-isolating device to minimize vibration caused during the machine's operating cycle)

- Testing of fuel storage tanks

A testing and inspection request for qualifications would follow this format:

Project notice: Owner (state name) seeks statements of qualifications to perform special testing and inspection services as described below for the (name the project and its location). The project delivery system is a (type of construction contract).

General description of work: The architect can provide a general description of the project; its size, function, type of structure, and other identifying information that will provide the testing and inspection company with a basic description of the scope of work required; and the anticipated construction schedule.

Schedule: A statement such as "The project is currently in the preparatory stage of construction. Expected start of construction is (date) with an anticipated completion date of (date).

Proposed scope of services: In consultation with the architect, the owner will include a list of services incorporating some of the items listed above.

Proposed cost: The architect may supply an estimated range of services so the testing and inspection company can discern an order of magnitude of work that will impact pricing. A large project with many repetitive types of testing and inspection requirements, for example, will general receive better rates. In this section of the RFQ, some form of rate schedule should be furnished by the testing and inspection company: hourly or daily rate, lump sum based on a certain quantity, and so on. This can be spelled out as follows:

1. Hourly rate schedule: Submit a summary of hourly rates for each staff member who would or could be billed to the project. Include regular and overtime rates. If subcontractors are to be hired, include their hourly rates. Travel time will not be included; only actual hours spent on the site will be reimbursed.

2. Testing/inspection: Submit proposed unit costs based on estimated quantities to perform testing and inspection services for the description of scope provided in the RFQ, including associated drawings and specifications.

3. Travel costs and other costs if required outside the general metropolitan area of the proposed project.

4. Identify and offer any potential savings associated with the required testing and inspection services stated to be provided.

The proposal includes the following:

1. Provide a cover letter (two-page limit) that should address the manner in which the testing and inspections company will perform these services, dedicated personnel anticipated for this project, daily reports to be generated, and other communication requirements.

2. Document the proposal with the following: statement of qualifications, resumes of proposed personnel, and samples of project testing and inspection forms.

PREPARATION OF THE BID DOCUMENTS

The process of selecting a contractor begins with the preparation of the bid documents, which is more involved than just the preparation of the project's plans and specifications. The architect can provide valuable assistance in the preparation of these documents, which will include the following:

- Advertising for bids if a list of potential bidders is not available; assistance in preparing the advertisement and offering advice on which publications will be most effective

- Preparing instructions to bidders to include the location where copies of plans and specs are available, how much they cost (generally a refundable fee; if the contractor is not selected, upon return of the documents in good condition, the fee is refunded), when the bid is due, special requirements such as payment and performance bonds, insurance requirements, where bids are to be submitted, and the deadline for the submission

- Preparation of a bid form so all bids can be adequately evaluated

- Type of construction contract to be awarded to the successful bidder (i.e., lump sum, cost-plus, construction manager, etc.)

- A complete list of each drawing in the set and any addenda (changes to the initial set of drawings)

- The complete specification manual, or manuals if more than one has been prepared

- Notification of a prebid conference if advisable

Chapter 5 discusses the preparation of a bid document in more detail.

The prebid conference

The architect can play a significant role in a prebid conference. Depending on the size and complexity of the project and the resultant plans and specifications, a prebid conference with all prospective bidders may be called, and the architect will assume a meaningful role in this process. An explanation of the process will point out the positive effects of such a meeting.

Questions about the plans and specifications may require further explanation by the architect and engineer, particularly specific owner requirements, as well as general familiarization with the project. Just meeting with the bidding contractors and engaging in idle talk can reveal much to both owner and contractor. An owner who seems to be reasonable will appeal to a contractor and make him or her possibly "sharpen the pencil" when preparing a bid; conversely, an owner who comes on too strong will have a negative effect. And the same can be said of an owner's opinion of a contractor or two. But that aside, the prebid conference will accomplish the following:

- Provide background information on the project, its nature, and the architect's and owner's expectations

- A question and answer opportunity for the bidders to discuss conflicts, inconsistencies, omissions, and unclear items in the plans and specifications so the architect/engineer can respond and clarify. (The A/E will usually respond after the meeting in writing to all attendees.)

- Discussion of any alternates, allowances, or unit pricing included in the bid package

- Indication that the architect and owner are open to any cost savings proposed by the bidders

If the prebid conference or meeting includes a question and answer session, as most do, some of the questions will be answered with a clarification or the issuance of a small 8½ × 11 drawing referred to as an "SK" (sketch). Other questions may require a revision to the plans and/or specifications. When changes are required to be made to the drawings before a contract is awarded, these changes are referred to as "addenda," and the date of the change to a specific drawing is made in the title box of the drawing(s) affected (Figure 2-1). When several different addenda are issued, they will be numerically dated—for example, Add#1—09/09/09, Add#2—10/10/09, and so on.

Changes made to the plans and specifications after a contract for construction has been awarded to the builder are referred to as "bulletins." Multiple bulletins are also numbered sequentially and dated in the title box of the drawing(s) affected by that change. When looking at a set of plans, one can determine when the changes occurred: addenda before contract award, bulletin after a construction contract has been issued.

Once bids are received after the prebid conference, they may require some analysis by the A/E team and the owner. The comparison of bids can be made much

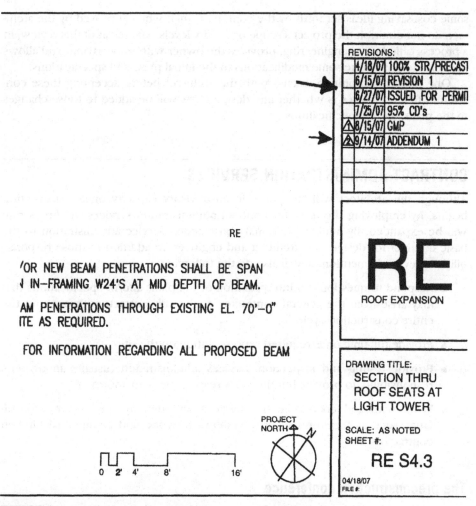

REVISIONS:

4/18/07	100% STR/PRECAST
6/15/07	REVISION 1
6/27/07	ISSUED FOR PERMIT
7/25/07	95% CD's
⚠8/15/07	GMP
⚠9/14/07	ADDENDUM 1

RE

RE
ROOF EXPANSION

DRAWING TITLE:
SECTION THRU
ROOF SEATS AT
LIGHT TOWER

SCALE: AS NOTED
SHEET #:

RE S4.3

04/18/07
FILE #:

'OR NEW BEAM PENETRATIONS SHALL BE SPAN
↲ IN–FRAMING W24'S AT MID DEPTH OF BEAM.

AM PENETRATIONS THROUGH EXISTING EL. 70'–0"
ITE AS REQUIRED.

FOR INFORMATION REGARDING ALL PROPOSED BEAM

PROJECT
NORTH

N

0   2'   4'   8'   16'

FIGURE 2-1

Title block from an architectural drawing showing revisions and addendas.

easier if the bid documents are explicit in how the contractors are to prepare their bids. Chapter 5 presents an in-depth discussion of bid document preparation and analysis.

At the owner's direction, the architect will gather, evaluate, and analyze the bids and, in conjunction with the owner, select the most qualified bidder, or bidders. If two bidders with similar scopes, prices, and qualifications participate, another interview with both may be required prior to the final selection.

Some owners may decide to select the best-qualified bidder, arrange a meeting with the architect, and negotiate a final price with the contractor after reviewing some cost-saving ideas set forth by the contractor that, when approved by the architect, do not diminish the project's scope or quality levels. The result of this review, in a process called value engineering, provides the owner with some savings and allows the contractor to make some modifications to the initial plans and specifications.

One important point to discuss with the architect before accepting these contractor-offered changes is whether any design costs will be added to those changes to the plans and/or specifications.

CONTRACT-ADMINISTRATION SERVICES

Although the architect will still serve in an advisory capacity once construction begins, by employing the team for contract administration services the firm's role will be expanded. By adding additional construction service administration to the basic contract for design, the architect and engineer, in addition to those responsibilities previously mentioned, will also do the following:

- Respond to questions on interpretation of the plans and/or specifications that may arise from the general contractor or the subcontractors throughout the entire construction cycle

- Provide full-time or as required representation on the site

- Provide testing and inspection services administration (usually an owner's responsibility to provide but the A/E's responsibility to approve)

- Work with and coordinate the services provided by the owner's consultants, such as furniture, security systems, artwork, and commercial kitchen contractors

The preconstruction conference

Somewhat similar to the prebid conference that all prospective bidders attend, once a contractor selection has been made, a preconstruction conference will be conducted to establish several ground rules during construction and highlight some of the owner's requirements and procedures to establish a working relationship among the owner, architect, engineer team, and contractor. Although many of the items discussed at the preconstruction conference have been included in the bid documents, a rundown of key items will act as a checklist for all parties. The architect should take the lead in addressing many of the issues included in the contract documents that require amplification or definition. A typical preconstruction conference checklist includes the following:

1. Some projects include an official notice to proceed, which establishes the date of the start of construction and will be used to determine the "contract" time for completion of the project. If a notice to proceed is not indicated,

the official date of the start of the project should be established at the pre-construction conference.

2. An explanation and introduction to the architect, engineer, and owner's team and the members who will be involved in the day-to-day operations.

3. Channels of communication between all parties will be established: mailing addresses, e-mail addresses, telephone numbers.

4. Establish a schedule and protocol for all future job meetings, and assign responsibility to prepare minutes of those meetings. Some architects prefer to have the general contractor conduct those meetings, while other architects would rather take the lead.

5. Verification that insurance certificates meeting contract requirements have been submitted and have been accepted by the owner prior to the start of construction. The same holds for verification of insurance certificates for all subcontractors as the general contractor awards such subcontract agreements.

6. Procedures for the submission of questions to the architect, the submittal of proposed change orders (PCO), number of copies required, and when additional copies are to be sent to the various design engineers.

7. The method for submission and approval of the contractor's progress payment will be reviewed. Will a "pencil copy" (draft) be required for architect review prior to submission of final copy? If so, when is this to be submitted to the architect? When will final requisition submission be required, and how long will it take to review prior to issuance of payment to the contractor?

8. If a list of proposed vendors and subcontractors is required by contract, it should be submitted at this time, with a date for review and approval/comment established and updated as more vendors and subcontractors are added. The contractor must also include any sub-subcontractors, known as second- and third-tier subcontractors.

9. Establish responsibility to apply for temporary utilities, which in some cases are the owner's responsibility and in other cases the contractor's.

10. State the required time for a baseline construction schedule to be submitted and procedures for updating the schedule at specified intervals.

11. Advise the owner of contractor-provided job site security procedures to be taken to protect the owner's site and work.

12. Review parking requirements, particularly if this is an urban site with restricted onsite parking available.

13. Advise the contractor of acceptable storage requirements for materials stored on- and offsite.

14. Request the contractor to post all required permits in the field office for review by the architect.

15. Define the extent of the owner's testing responsibilities and the contractor's obligations; also establish notification procedures for the contractor to advise the A/E of the need for testing in sufficient time to bring the testing company to the site.

16. Explain the nature and extent of any owner-furnished equipment, proposed dates of delivery, and owner's responsibility to unload, store, and install.

17. Review with the contractor the responsibility for job site safety and cleaning to ensure that the owner's site is safe and clean at all times.

18. Advise the contractor of the nature of any separate contracts anticipated by the owner and how the general contractor is expected to interface with them.

19. Review all closeout documents, including the preparation, updating, and final submission of as-built drawings so all parties are aware early on of their responsibilities.

20. Review the warranty, guarantee requirements, and extra materials to be furnished by the subcontractors prior to project closeout.

THE OWNER-ARCHITECT CONTRACT FOR SERVICES

Several considerations should be kept in mind when negotiating the cost of architectural services. As we already discussed, the extent of services required is one thing to consider: "Do I want to contract for design only, or do I want to include some or all construction-related services?" The standard AIA Architect-Owner contract for design is AIA Document B141 Standard Form of Agreement Between Owner and Architect with Standard Form of Architect's Services, and the standard contract for construction services is AIA Document B163 Standard Form of Agreement Between Owner and Architect with Description of Designated Services.

THE DESIGN PROCESS

The design process consists of many phases, beginning with the initial meeting between the project owner and the architect where concepts were exchanged and continuing on to the production of what is often referred to as "working drawings." Drawings at this point represent the final stage of design and are required to obtain a building permit and thereby commence construction. The documents produced at each of these stages are referred to as "deliverables."

A project owner may not always want to proceed directly to "working drawings" for a number of reasons:

1. Full funding for the project many not yet be in hand, but some design documents may be necessary to obtain the funding required.

2. Conceptual drawings may be necessary to start the budgeting process, not only for construction purposes but for the interior fit-up of the building and costs associated with the relocation from an existing facility.

3. Approval of the project by the owner's board of directors.

4. The client's spatial needs may not be fully defined, and a preliminary set of basic floor plans may be required to do so.

There are a number of options open to an owner to obtain less-than-complete drawings, and the architect can be helpful in selecting the proper phase or stage of design to meet the owner's needs. The architect may also be able to provide a cost associated with each stage of design.

The initial schematic phase of design will require the architect to obtain some basic information from the owner: What is the function of the building—commercial, retail, corporate office? What square footage will be required, and how many people will occupy the space? Are there any special needs that must be addressed? What does the *owner* consider the project's goal? The architect will then begin to produce a "schematic design," which will generally include the following drawings:

- A site plan locating the building on the owner's construction site

- A floor plan, dimensioned, possibly with a lobby floor if multistoried

- A section through the building (as though one took a knife and sliced the building in half, much like a layer cake); this allows an owner to see the spatial dimensions of a typical floor and ceiling heights.

- An elevation—the view one would see standing outside and looking at the completed building, with floor to floor and overall height dimensions

- Possibly a computer image of the building in 3D, a colored rendering, or a model

At the schematic level, the architect will also include an estimate based on this design, a proposed construction start, and the length of both design and construction. Unlike the Consumer Price Index (CPI), inflation in the construction industry tends to be higher. As an example, in the September 2008 issue of *Engineering News Record,* it was reported that construction costs were up 6.3 percent as of the third quarter. An anticipated start of construction would take any inflationary forces into account.

The design-development stage picks up where schematic design leaves off and includes some structural, MEP (mechanical-electrical-plumbing), and architectural details. Windows will be sized and located accurately in the building's exterior elevation, and partitions and rooms will be defined, along with some descriptions of basic materials.

The owner will receive more detailed floor plans, sections, and elevations with full dimensions. Components such as door types, sizes, and materials will be indicated. The architect will also provide "outline specifications," a descriptive narrative of basic building components.

It is important for both the architect and the owner to review each phase of design as it is produced to be assured that they have included the essential program requirements. This is no easy task for an owner who is not versed in design and construction terminology or has little experience in reading and interpreting construction drawings. An architect will welcome questions, no matter how basic, because the designers will need to explain what they have created and confirm that their design meets the needs of the owner's program.

The drawings produced at this stage, known as construction documents, will include sufficient information and detail for the contractor to estimate the cost of the project. They will also include a complete specification manual. These are also the drawings and specifications that will be presented to the local building department to obtain a building permit, so they are also referred to as "the permitting set."

The specification book or manual contains detailed information about each material, piece of equipment, and component of construction to be installed in the project. The Construction Specifications Institute (CSI) developed a numbering system for specification manuals in which each section is divided into "divisions." This 40-year-old system is known as MasterFormat®; it was upgraded in 2004 and now contains 49 divisions, starting with Division 1—General Requirements on through Division 04—Masonry, Division 05—Metals, Division 22—Plumbing, and so forth. Figure 2-2 shows a two-page portion of a typical specification manual.

The specifications manual is assembled according to the MasterFormat® numbering system, so when the architect says, "Would you please go to Division 5?," he or she may want to review some structural steel requirements or inspection procedures.

The specifications manual also includes installation instructions, quality-control measures, and product warranty information. Although it may be difficult to understand all of the information contained in this spec manual, the owner's representative assigned to the project should read various sections of the manual, preferably with the architect, and highlight or flag the important sections. Don't be afraid to ask the architect lots of questions about what's inside this thick book.

Architects use a standard lettering and numbering system for the construction drawings they produce, primarily to act as an index to separate the various design disciplines: civil, architectural, structural, mechanical, electrical, plumbing, and so forth. Each series of drawings is numbered sequentially:

- C drawings designate civil drawings—sitework, underground utilities.

- D drawings refer to partial or complete demolition of existing structures on the site if required.

TABLE OF CONTENTS

INTRODUCTORY INFORMATION
Document 00001 - Project Title
Document 00002 - Project Directory
Document 00010 - Table of Contents

CONTRACTING REQUIREMENTS
Agreement forms issued by Construction Manager

CONSTRUCTION PRODUCTS AND ACTIVITIES

DIVISION 1   GENERAL REQUIREMENTS
Section 01100   -   Summary
Section 01210   -   Allowances
Section 01310   -   Project Management and Coordination
Section 01320   -   Construction Progress Documentation
Section 01330   -   Submittal Procedures
Section 01410   -   Regulatory Requirements
Section 01420   -   References
Section 01450   -   Quality Control
Section 01500   -   Temporary Facilities and Controls
Section 01600   -   Product Requirements
Section 01630   -   Product Substitution Procedures
Section 01700   -   Execution Requirements
Section 01732   -   Cutting and Patching
Section 01739   -   Selective Demolition
Section 01750   -   Starting and Adjusting
Section 01770   -   Closeout Procedures
Section 01780   -   Closeout Submittals

DIVISION 2   SITE CONSTRUCTION
Section 02200   -   Excavation and Backfilling
                    Attachment – Utility Trench Typical Section
Section 02310   -   Drilled Mini Piles
Section 02753   -   Plain Cement Concrete Paving
Section 02821   -   Chain Link Fences and Gates

DIVISION 3   CONCRETE
Section 03310   -   Concrete Work
Section 03361   -   Concrete Sealers
Section 03410   -   Structural Precast Concrete
Section 03600   -   Grout
Section 03700   -   Concrete Repairs
Section 03921   -   Concrete Resurfacing and Patching

DIVISION 4   MASONRY
Section 04230   -   Reinforced Unit Masonry
Section 04810   -   Unit Masonry Assemblies
Section 04910   -   Unit Masonry Restoration
Section 04930   -   Unit Masonry Cleaning

Table of Contents
00010-1

FIGURE 2-2

Partial table of contents from a specifications manual showing the contents of Division 1—General Requirements to Division 9—Finishes.

Table of Contents
00010-2

FIGURE 2-2 *Continued*

- A drawings are architectural.

- S drawings are structural.

- M drawings are mechanical (HVAC).

- P drawings are plumbing.

- E drawings are electrical.

- FP drawings are for fire protection.

- Other designations to fit a particular project, such as GI (general information), FS (food service), and AV (audiovisual), may be added.

As the drawings pass through their various stages from schematic to construction, each revision and its date are shown in a title block (Figure 2-3), which also includes the drawing's title, the architect's project number, and the scale to be used to measure any portions of the work portrayed on that drawing.

There are two scales (similar to the rulers we used in school) required to extrapolate drawing dimensions into actual construction dimensions: the architect's scale and the engineer's scale. The architect's scale is used to measure all plans except the civil C drawings, for which an engineer's scale is required.

Each scale usually takes the form of a triangular-shaped ruler. The architect's scale uses divisions within the standard inch. There is a 1/8-inch scale, which means that 1/8 inch as measured on the drawing equals one foot; another side of the triangle will have a 1/4-inch scale, where 1/4 inch equals one foot. Other scales used are 1/2 inch to the foot, 1/8 inch to the foot, and 3/4 inch to the foot.

The engineer's scale is divided into decimal fractions of an inch, so a scale of 10 means that one increment equals 1 foot; a 20 scale means that one increment on this scale equals one foot. Engineers' scales usually come in 10, 20, 30, 40, 50, and 60 increments. The civil drawings will indicate a scale, such as 1:10 or 1:50, meaning one increment in the scale is equal to one foot.

NEGOTIATING THE ARCHITECT'S FEE AND OTHER FORMS OF COMPENSATION

After determining the scope of work to be performed by the architect, there are various accepted practices for compensation:

- In time-based compensation, the architect will bill for hours spent working on the project and will request reimbursement based on salaries of staff plus expenses, which are usually listed separately. The firm then uses a "multiplier" that includes personnel fringe benefits and the company's overhead and profit. If this method is employed, it is best to have the architect prepare a list of billable hourly rates and anticipated expenses so there is no disagreement when the invoices begin to arrive in the mail.

)ENTIFIED WITHIN THE CONTRACT
 REVIEW AND APPROVAL BY THE
ION AND NATIONAL PARK SERVICE
IITED TO:

R WALL REPAIR AND RESTORATION,
\R COLOR AND TEXTURE, MASONRY
C.

P DRAWINGS.

: WORK SO THAT AN APPROPRIATE
S IN ADDITION TO THE TIME
\LL BE ALLOWED FOR REVIEW AND
/IEW IS REQUIRED.

REVISIONS:

06/27/07	ISSUED FOR PERMIT
07/25/07	PROGRESS SET
08/15/07	ISSUED FOR GMP

GI
GENERAL INFORMATION

PROJECT #: 0202.123
DRAWING TITLE:

CODE & HISTORIC REFERENCE

SCALE: AS NOTED
SHEET #:

RE.G4

DATE: 2/22/07
FILE #:

FIGURE 2-3

Title block from an architectural drawing with dates of various development stages: permit application, a continuing progress set, and a pricing (GMP) set.

- A stipulated sum is a total dollar amount for all provided services.
- A percentage of the cost of work is based on a percentage of the estimated or actual cost of the work (generally between 4 and 12 percent, depending on the scope of services under contract).

- In the square footage of the structure being designed method, the owner needs to know how the square footage is determined. Is it the inside dimension or the outside dimension of the structure? Is it measured from the centerline of the exterior wall to the centerline of the opposite wall(s)?

- The unit cost method is often used in structures with repetitive designs, such as a hotel, motel, or apartment house, where a unit cost is determined based on the number of rooms or similarities of space. However, an owner needs clarification as to whether these "unit prices" include elements other than room designs, such as the lobby, mechanical storage rooms, meeting rooms, and so forth.

- The royalty method is based on a share of the income or profit created by the project.

Some of these means of compensation may require some clarification:

- The hourly rate approach may be appropriate during the period when the project scope is being investigated, after which a lump sum or percentage-of-cost method may be more practical.

- A stipulated sum contract may be considered. Generally, this approach includes all personnel costs, including overhead and profit, but it may not cover some reimbursable expenses, so a list of inclusions and exclusions will be needed.

- The question of a payment schedule should be addressed. Does the architect intend to submit invoices on a time-based schedule, or will billings be submitted when certain design milestones have been achieved? If the latter is selected, what percentages will apply to each milestone—for example, 10 percent at schematic design, 30 percent at design development, and the balance when construction documents are produced?

Other methods often employed are based on the services required and the nature and the complexity of the project. If a schematic design is all that is required, at the present time, an owner may wish to reimburse the design team on an hourly basis plus any costs to reproduce the schematic design, provide a rendering of the project, and so forth. Although these hourly rates usually include the salaries, benefits, overhead, and profit for those designated to perform the work, clarification of the architectural firm's billing rates beforehand is strongly advised.

Most architectural firms include services for other design professions, such as structural design, mechanical-electrical-plumbing (MEP), and civil engineering, which might include in-house staff or subcontracted work. If these firms are "subcontracted" to the architect, each bill will include not only the cost of the work but their respective overhead and profit as well.

Some design firms directly employ all engineering services under one roof, but they are the exception rather than the rule. As we discussed previously, an architect's fee, based on the cost of construction, can vary from 4 to 12 percent, depending on the nature and extent of services to be provided. Gone are the days when all architects employed the same fee structure. Nowadays the fee structure is established by negotiations between the owner and the architect.

An architect who employs other design consultants will assemble costs for the project, and the work sheets may look like this:

Architect's fee, which includes design work	$
Civil engineering	$
Structural engineering, substructure	$
Structural engineering, superstructure	$
Mechanical engineering	$
Electrical engineering	$

Depending on the nature and scope of the project, other consultants may be required to meet specific owner needs:

- Waterproofing consultant—frequently used on high-rise curtain walls or buildings with intricate window designs or stone or masonry veneer

- Green building consultant—if the project is expected to meet certain U.S. Green Building Council (USGBC) LEED ratings, some special materials and equipment may require additional research

- Low-voltage electrical or building systems management consultants for automated building systems and data and voice communication systems

- Interior design—office furniture, fixtures, and equipment (FF&E) layout; color coordination; artwork consultation

Each one of these consultants will incorporate the cost of his or her services, during and possibly after construction, to which an architect will add administration fees. Quite often an architect will include a "contingency" in the cost estimate that would be tapped if he or she encounters extra work required to satisfy the contract with the owner—in other words, if something was left out or the architect had to redesign a minor element of construction.

PSMJ Resources, Inc., headquartered in Newton, Massachusetts, is the architectural and engineering consulting firm that conducts annual surveys for those industries. Their 2008 PSMJ A/E CEO Benchmarks Survey included a comparative staff billing rate chart for the years 2007–2008 (Table 2-1). This listing of billing rates for various architect and engineering personnel employed on a project can be used as a guideline when analyzing an architect's compensation. Note the add-ons on the bottom of the chart for other consultant fees and reimbursable costs. Table 2-2 contains additional material taken from this PSMJ study, revealing management compensation results, including base salary, bonus, and total direct compensation figures for management positions for the years 2007 and 2008.

Other information gathered by PSMJ indicates that as of the time of this survey the design community was in fairly good economic health. A comparison of key

Table 2-1 Architect Comparative Staff Billing Rates for 2007 and 2008

Hourly Billing Rates	2007		2008	
	Median	Mean	Median	Mean
Principal	$170	$173	$170	$171
Associate	140	141	135	139
Project Manager	125	124	125	124
Project Architect	110	114	115	115
Project Engineer	110	114	110	111
Architect	100	102	99	102
Senior Engineer	120	126	125	122
Intern Architect	79	76	80	78
Engineer	95	100	96	98
Junior Engineer	81	86	82	80
Designer	88	90	90	90
Senior Drafter/CAD	84	83	81	84
Drafter/CAD	70	71	70	69
Spec Writer	100	103	86	93
Estimator	100	104	85	95
Job Site Inspector	75	81	85	89
Clerical	55	57	55	55
Markups				
Outside Consultants	10%	12%	10%	9%
Reimbursables	10	9	10	9

Source: By permission of PSMJ Resources, Inc., Newton, MA.

indicators for the years 2007 and 2008 (Table 2-3) reflects increases in revenue but also increases in direct labor costs. Reductions in the backlog of work have also resulted in staff cuts. For the year 2008, financial indicators varied according to the size of the A/E firm:

- Small firms with a staff size of 21 to 50 reported the highest operating profit margin.

- Firms reporting the lowest overhead rates were those of mid- to large size, with staffs between 101 to 750 members.

Table 2-2 Architect Management Compensation Results for 2007 and 2008

Position	Base Salary		Bonus		Total Direct Compensation	
	2007	2008	2007	2008	2007	2008
Chairman of the Board	$172,989	$194,045	$100,000	$69,383	$247,219	$290,338
Chief Executive Officer	167,111	175,000	63,045	66,000	250,000	250,000
Executive Vice President	148,900	171,571	90,703	50,000	241,000	238,664
Senior Vice President	137,280	145,000	55,000	44,000	195,242	195,000
Other Principals	109,000	130,000	28,253	23,486	146,000	161,962
Director of Finance	126,535	147,500	30,970	40,952	165,000	204,800
Controller	80,266	84,772	6,000	10,000	88,033	99,000
Business Manager	69,123	68,798	5,239	4,000	72,651	76,500
Director of Administration	67,000	94,126	6,465	13,016	73,465	101,101
Director of Operations	110,000	144,858	11,500	29,264	125,000	173,310
Director of Quality Control	108,200	125,800	15,302	17,500	162,000	152,000
Director of Business Development	92,853	104,202	6,000	10,000	100,900	121,732
Director of Human Resources	84,889	82,000	5,000	5,940	89,616	88,158
Director of Computer Operations	87,907	91,520	6,000	6,750	95,802	98,043
Branch Office Manager	109,200	114,400	15,000	12,000	125,000	128,049
Department Head	97,219	103,333	7,581	8,650	106,000	116,076
Senior Project Manager	85,000	90,000	7,250	7,392	95,000	101,245
Junior Project Manager	66,500	70,030	4,965	4,250	71,271	77,500

Source: By permission of PSMJ Resources, Inc., Newton, MA.

- The largest firms, with 500 or more employees, reported the largest increase in gross revenue.

- Firms with the largest backlog were the very largest ones, with staff size in excess of 750.

Table 2-3 Comparison of Key Indicators for 2007 and 2008

Medians	2007	2008	% Change
Net Revenues per Total Staff	$113,745	$118,126	4%
Net Revenues per Direct Labor Hour	$81.35	$84.95	4%
Direct Labor Costs per Direct Labor Hour	$27.97	$29.11	4%
Total Costs per Direct Labor Hour	$72.78	$78.86	8%
Equity per Total Staff	$19,013	$21,205	12%
Operating Profit (Net Revenues)	15.24%	15.19	0%
Overhead Rate (before Incentive/Bonus)	154.47%	159.71	3%
Chargeability (Payroll Dollars)	60.8%	59.72	−2%
Backlog Change	12.0%	9.0%	−25%
Gross Revenues Change	14.0%	9.0%	−36%
Staff Size Change	7.1%	5.5%	−23%
Net Multiplier	3.03	3.09	2%
Average Work-in-Process Days	21.58	22.81	6%
Average Collection Days	66.72	70.44	6%

Source: By permission of PSMJ Resources, Inc., Newton, MA.

The case for reimbursable expenses

No matter which form of architect compensation is finally decided, to avoid any misunderstanding, the architect should address the issue of reimbursable expenses, such as travel, cost of producing plans and specifications, telephone, e-mail, and so forth. For example, in a stipulated sum contract, this sum does not usually include reimbursable expenses, which can be significant. A single set of "contract" plans can be as much as $300 to $500, so it is best to have an understanding of what anticipated reimbursables will be billed and an estimate of their cost.

The standard architect agreement

The AIA Documents Committee considered certain basic concepts when they prepared the various standard agreements for design and construction services, and it is important for owners to understand the basic tenets that permeate these documents:

- These documents address "reasonable expectations" and include customs and services that prevail in the industry on a national basis. An interpretation of "reasonable" seems to imply that, due to the complexity of today's design and construction process, these documents will not reach the level of being error-free.

- Risks are shared with those parties who have the most direct control over that portion of the process. Where no party has been identified, risks are directed to the party best suited to protect against an anticipated loss. This seems to be another form of sharing risk, and it shields the architect from some liability in the design process.

- Risks are also allocated along the lines that have evolved from case law—for example, when an assignment of risk has been established by the courts, this precedent may apply to other similar cases.

- Most AIA contracts are standard forms, and most completed contracts contain "exhibits" and "schedules" added by the owner to customize the scope of work and the party responsible for that work.

Most contract negotiations, whether they be with the design consultants or the contractor, revolve around risk allocation. Neither owner, architect, nor contractor wishes to assume what is perceived as too much risk, and when level heads prevail, assignment of risk can be fairly distributed by the process of give-and-take.

THE OWNER'S ROLE IN THE DESIGN PROCESS

An owner should take an active role in the design process to provide the design professionals with the information they require to develop the project design. Some owners may not have experience in interpreting two-dimensional designs and all those lines on the drawing, and if that is the case, it is best to ask the architect to explain those arcane symbols and lines.

While avoiding the temptation to micromanage, an owner needs to provide continuity of oversight so each stage of design development is familiar to that owner group or owner-assigned individual from start to finish. An owner's senior staff should assume the responsibility to not only assist the design team in understanding the owner's requirements but also to review the design and either change or accept it as it evolves. If sufficiently knowledgeable staff is not available to fulfill this function, the owner has several options:

- Engage a construction consultant to monitor the design to ensure that it meets the owner's program

- Engage a construction manager (CM) as the project-delivery system of choice

- Select an experienced general contractor to work with the architect during design to assist in estimating, material selection, and constructability issues; negotiate some form of construction contract with the contractor to cover these services

Constructability and coordination

The definition of *constructability* may vary from design professional to design professional, but it can be as simple as providing the knowledge and experience to plan, design, procure, and execute the most cost-effective, highest-quality overall project objective.

It may be as simple as determining the type of structural system based on the anticipated start of construction. If it appears that a project in a geographic area where harsh winter weather is the norm and the start of the project puts the structural framework right in the dead of winter, it might be wise to consider a steel structure rather than cast-in-place concrete, which could require costly "winter protection." The design of a structure located in an area far from a precast concrete manufacturer and where transportation costs from factory to job site can be substantial should also be considered when selecting a structure or façade.

As structural and architectural drawings are transmitted to the various engineering disciplines, the complex problem of designing the plumbing, heating-cooling-air-conditioning systems, electrical work, and sprinkler systems to fit in their allotted space is a challenging task. Some back-and-forth coordination is required to allow ductwork to fit above the ceilings and not conflict with plumbing, sprinkler piping, HVAC equipment, or electrical components. And as these systems rise from floor to floor in a midrise or high-rise building, space must be provided in partitions and other vertical areas to encapsulate these "risers."

This is just another of those daunting tasks facing an architect. The project specifications manual usually includes a provision requiring the contractor to perform a series of coordination efforts by producing a "coordination drawing" that is passed among those subcontractors required to install the various components just listed. During this process, hopefully all conflicts or disparities will be uncovered and everything will fit—but that doesn't always happen.

Lack of adequate coordination during the design stage can have serious consequences if problems are brought to light by the general contractor as they and their subcontractors work through the coordination-drawing process. There have been instances where the length and width of the dimensions of a floor designed by the structural engineer for a multistoried building are inconsistent with the dimensions established by the architect because they have not communicated with each other adequately. Now suppose that an electrical engineer has followed the structural drawings and the mechanical engineer has followed the dimensions on the architectural floor plans. It is rather easy to discern the problems—and added costs—awaiting an owner.

When owners force the architect to ready the drawings for bid purposes before they have been fully reviewed and coordinated, they are looking for trouble. Owners should avoid making that mistake. Surveys of general contractors have highlighted some concerns that cause strained relationships between owner and builder:

- Specifications that are either incomplete or are ill-suited for the job at hand. (I remember a school project that required wall-mounted TV brackets in each classroom; the specifications said that the brackets should be able to swivel so the "patients can view the television screen.")

- Unrealistic schedules set by the owner, architect, or contractor. When an owner demands a schedule for completion of a design that leaves no time for review and coordination, he or she is asking for trouble. An architect who provides an unrealistic schedule may impact the start of construction, hence completion, hence the length of time the owner must carry that more costly construction financing. A contractor who agrees to a unrealistically short construction schedule just to please an owner does everyone a disservice.

- Enough cannot be written about the need for a final review of the design documents by each member of the design team. Floor openings designed to carry ductwork, waste and water mains, and electrical conduits from floor to floor in a multistory building must show the same openings on the structural, architectural, plumbing, HVAC, and electrical drawings. This may sound elementary, but many projects have soared in costs because these "elementary" reviews were not performed during design but were exposed during construction.

- Then there are weather-related problems. As we just mentioned, building a cast-in-place structure during winter months in the northern part of the country needs to take into account the costs to protect the building when freezing temperatures occur. And these costs can be in the high six figures for large buildings.

Now that we have presented the architectural process, it is time to review general contracting and then move on to the construction process.

The architectural, engineering, and contracting industries

3

THE ARCHITECTURAL PROFESSION

An architect is a licensed individual tasked with the responsibility to plan and design a wide range of structures from single-family houses to factories, schools, hospitals, and commercial buildings that reach ever skyward.

In the United States, aspiring architects must meet three requirements relating to education, experience, and licensing. More than 25 states require a bachelor or master's degree in architecture from a college or university accredited by the National Architectural Accrediting Board (NAAB) in order to attain the requisite educational requirement. As far as experience is concerned, the intern architect must earn 700 training units, which are divided into 16 categories. Two of these categories involve immersion in design and construction documents and management skills along with their design training.

Design and construction training involves the following activities:

- Site and environmental analysis
- Schematic design
- Engineering-systems coordination
- Building cost analysis
- Code research
- Design development
- Construction documents
- Specifications and material research
- Document checking and coordination

Management training covers project-management techniques along with office-management skills.

© 2010 by Elsevier, Inc. All rights reserved.
Doi_No = 10.1016/B978-1-85617-548-7.00003-3

45

The architect's intern program

The National Council of Architectural Registration Boards (NCARB), in conjunction with the American Institute of Architects, created the Intern Development Program, which requires participants to work under the direct supervision of a registered, licensed architect so they can put their "book learning" to work in a real-time situation.

For each eight hours of experience gained by working with a licensed architect, the intern receives one training unit (TU). Training can be pursued outside the United States or Canada as long as the trainee is working with a foreign firm under the supervision of a licensed architect.

State licensing requirements

Although a formal degree in architecture can be waived by some states, those candidates seeking licensing may be required to have ten years of experience working in the field of architecture and intensive knowledge gained from the Intern Development Program.

All state architect-licensing jurisdictions require certification after the intern passes a series of nine computerized exams administered by the NCARB. Each of the 50 states and 5 U.S. territories have NCARB registration boards to ensure parity among individual state rules and regulation, which may vary to some degree.

Licensing requirements, as reported by the Bureau of Labor Statistics, include a professional degree in architecture, with the preceding exceptions; at least three years of practical training; and a passing grade in all divisions of the Architect Registration Examination.

Scope of design firms

As reported by the Labor Department in the latest (2006) statistics, there were 132,000 architects in the United States. Most of these professionals work in architectural firms with fewer than five employees. The federal government, using a National Employment Matrix, projected that by 2016, more than 155,000 architects will be practicing, representing an increase of about 18 percent. About one in five is self-employed, more than twice the proportion for all occupations, according to the Bureau of Labor Statistics.

There are several giant international architectural and engineering design firms. In 2007, the top 500 design firms, as reported by McGraw-Hill's periodical *Engineering News-Record*, generated total revenues exceeding $80 billion, an increase of 15.8 percent from the previous year. The largest design firm in the United States was URS Corporation, based in San Francisco, with revenues in 2007 exceeding $4.8 billion. It generated most of its income from power plant design. In second place was Jacobs of Pasadena, California, with an annual revenue of $4.3 billion, mainly derived from the industrial and petrochemical fields. The general building design category, based on 2007 revenue, was led by AECOM Technology Corporation in Los Angeles, California, with $1.127 billion. The leading commercial office building design firm was Gensler, headquartered in San Francisco, with $652 million in 2007 revenue. The smallest architectural design firm in terms of revenue among *ENR*'s top 500 for 2007 had an annual revenue of $24 million.

LANDSCAPE ARCHITECTS

Landscape architects specialize in landscape design for residential areas, public parks and public buildings, college campuses, shopping centers, and recreational facilities such as golf courses. When working with public works agencies, architects, and engineers, landscape architects assist in the planning, location, and development of new highways and the upgrading of existing ones. They have become intimately involved in environmental issues, working with scientists and government agencies to determine the optimal method of dealing with conservation and restoration of natural resources, particularly in the construction of infrastructure including highways, tunnels, and bridges.

Landscape architects work closely with building architects and civil engineers in private construction projects. Their knowledge of plant and tree species for each particular geographic area is essential in developing aesthetic and cost-effective landscape plans for building owners. Landscape architects are often employed by developers of large residential projects as they work with development planners.

Education, training, and certification

A bachelor's degree, as a minimum, or preferably a master's degree in landscape architecture is held by most landscape architects. The undergraduate degree can be either a BLA (bachelor of landscape architecture) or a BSLA (bachelor of science in landscape architecture). Courses in landscape architecture are offered by 61 colleges and universities in the United States. They include surveying, landscape design, plant and soil science, geology, professional practices, and general management techniques.

As of January 2008, 49 states required landscape architects to be licensed, a procedure based on the Landscape Architect Registration Examination (LARE), sponsored by the Council of Landscape Architectural Registration Boards and administered in two parts: graphic and multiple-choice. Fifteen states require passing grades on an examination in addition to the LARE. There are currently 28,000 landscape architects, and the Bureau of Labor Statistics projects a 16 percent increase to 32,000 by the year 2016.

THE CONSTRUCTION-RELATED ENGINEERING PROFESSION

The range of engineering specializations ranges across a wide spectrum of activities from mining engineering to outer-space engineering, but in the design and construction industries, six categories prevail: civil, structural, mechanical, electrical, environmental, and materials engineering. These various engineering disciplines, each contributing to specific components of the building, interact with one another and the architect throughout the design and into the construction phase of the project.

Aspiring engineers pursue bachelor of science degrees, which are almost mandatory to obtain an entry-level job, and specialize in a specific type of engineering, with a strong emphasis on mathematics. The United States has 1,830 accredited colleges offering degrees in various engineering disciplines. To become a professional engineer (PE), a candidate generally is required to have a BS degree, four years of experience in relevant work, and a passing grade on a state examination. All 50 states and the District of Columbia offer state engineering licensing exams. The U.S. Department of Labor categorized engineers as those professionals who "apply the principles of science and mathematics to develop solutions to technical problems. Their work is the link between scientific discovery and the economic solution to commercial applications that meet societal needs."

As of 2006, the United States has 256,000 civil engineers, 227,000 mechanical engineers, and 153,000 electrical engineers. Thirty-one percent are employed in the manufacturing sector, 28 percent are in the building-trades sector, 5 percent are self-employed, and the rest are in various government agencies and other types of business.

Civil engineers

Civil engineering is the second oldest engineering discipline, the first being military engineering. Civil engineers deal with site and soils analysis and are engaged by the structural engineer to design a building's foundation and underground utilities such as storm, water, sewer, and electrical services from their existing location to the structure and any paved areas on the building site.

This profession has subdisciplines, such as environmental engineering, geotechnical engineering, water resources engineering, coastal engineering, materials engineering, and surveying. An owner will have occasions to require the services of at least one and possibly two or more of these engineering specialists on her project.

The geotechnical engineer will be engaged for most new construction projects. Through a series of site investigations and extrapolation from those investigations, he or she will develop an analysis of a building site's underground soil conditions and the conditions most likely to be encountered as construction gets underway. The geotechnical investigations provide the structural engineer with soil-conditions data required for building foundation design. And these geotechnical investigations and observations will become part of the bid documents and subsequently a part of the contract for construction, aiding the contractor in the preparation of the site work estimate. Geotechnical reports are often couched in a degree of vague language, which is understandable, since the soil samples taken by the civil engineer are limited in nature to the specific area or areas from which they are taken and are meant to be representative of conditions rather than a guarantee of existing underground conditions.

In consultation with the mechanical and electrical engineers, the civil engineer will design the size, location, and composition of all incoming underground utilities: water, storm sewer, sanitary sewer, and gas and electric and their connections to existing offsite mains.

Structural engineers

The structural engineer is responsible for the design of the building's foundation after reviewing the soil analysis supplied by the civil engineer. The type and size of foundation will be determined by this engineer. Will conventional concrete footings suffice, or are soil conditions of such a nature that special foundations such as piles or caisson will be required? After foundation design has been completed, the engineer will design the superstructure in collaboration with the architect and the engineers assigned to design other specific components. Aesthetic as well as structural considerations may at times dictate the design of the building's structure, requiring the structural engineer to work very closely with the architect.

Mechanical engineers

The mechanical engineer designs the building's heating, ventilating, and air-conditioning systems as well as the plumbing and fire-protection systems. He or she furnishes the civil engineer with pipe sizes and types for the layout of underground incoming utilities. The mechanical engineer confers with the structural engineer to ensure that the building, site, or roof will have sufficient structural integrity to support the weight and size of the equipment being designed. The mechanical engineer will confer with the architect to ensure that sufficient space is allotted for equipment installation and access for the routine maintenance that will occur after the building becomes operational.

In a multistory building design, the mechanical engineer will provide the structural engineer with the size and location of all floor penetrations required for HVAC, plumbing and electrical ductwork, piping, and conduit that extends from floor to floor. The structural engineer will then ensure that all such openings are structurally sound.

Electrical engineers

The electrical engineer also works very closely with the other engineering disciplines. Power requirements are generated in large part by the type and size of mechanical equipment being designed by the mechanical engineer. When these power requirements, such as cable and conduit sizing, have been established, this information is passed to the civil engineer so he or she can design the location, size, and type of underground conduits coming into the building. Lighting electrical loads are coordinated with the architect, who will be strongly influenced by the owner's requirements.

Site lighting and parking lot lighting are selected by the electrical engineer and coordinated with the civil engineer, who will design any foundations required for these site lights. These four types of engineers—civil, structural, mechanical, and electrical—who cooperate on overall building designs, operate under the supervision of the architect, who will orchestrate the process of reviewing, commenting on, and incorporating all of these seemingly diverse activities into one harmonious design. This coordination process is one of the more important aspects and responsibilities of the team captain: the architect.

The environmental engineer

The environmental engineer has come to play a bigger part in the planning, design, and construction process due to the increasing public awareness of the impact these activities can have on the community in which the structures are to be built. These concerns have manifested themselves in the many types of environmental impact studies required in both public and large private projects. Environmental engineers have the responsibility to prevent damage to the environment and alert designers to potential community health hazards on sites being considered for construction. These engineers investigate waste treatment issues, remediation of toxic sites, and potential sources of air and water pollution.

As of 2006, 54,000 environmental engineers are in this country. The projected 25 percent growth of this engineering discipline during the period 2007–2016 surpasses the second highest growth category of engineers—biomedical—which is expected to grow by 21 percent, followed by industrial engineers (20 percent) and civil engineers (18 percent).

The materials engineer

And last we have the little-publicized but very important materials engineers, who works behind the scenes testing and verifying the integrity of a wide array of materials. Turn the pages of any construction specification manual, and one can find numerous references to ASTM, the American Society for Testing and Materials, an international organization devoted to developing a source for technical standards for materials, products, systems, and services.

This organization had its beginnings in the early 1800s when a group of scientists and engineers came together to investigate the cause of frequent breaks in the steel rails on the nation's growing railroad system. Their efforts led to the development of a new type of steel for rail tracks, and today this organization has grown by leaps and bounds and has embarked on developing material and product standards on a worldwide basis. Today, ASTM provides the following services:

- Standard material specifications
- Standard testing methods
- Standard practices
- Standard reference guides
- Standardized classification of materials, products, and systems

THE CONSTRUCTION INDUSTRY

The construction industry, with 7 million wage and salary jobs and 1.9 million self-employed workers as of the 2004 Census, is one of the nation's largest industries. Nearly two out of three of those jobs represented employment by what is called specialty contractors—carpenters, plumbers, electricians, masons, and HVAC contractors—known to most as subcontractors.

Construction accounts for nearly 5 percent of the country's total nonfarm labor market, and it represents about 9 percent of America's total gross domestic product (GDP). This makes it a big business, and as the homebuilding meltdown revealed in 2007 and 2008, the impact of construction, whether high or low, affects many other related industries such as paint and flooring manufacturers, furniture and lighting fixture companies, and major appliance manufacturers.

Size and revenue of contractors

The construction industry is composed of about 710,000 businesses, mostly small companies, 91 percent of which have fewer than 20 employees. While the largest U.S.-based contractor had revenues of $22 billion in 2007, the overwhelming majority of builders had an annual volume of less than $10 million.

To provide a snapshot of the construction industry, it can be categorized as one in which building contractors range in size from a small family-owned business operating in a narrow geographic area to giant multinational firms. Finding the right one for your project is sometimes a confusing task but can be made somewhat easier by understanding how the industry works.

This is a business of high risk and relatively low profit margins. The Construction Financial Management Association (CFMA) of Princeton, New Jersey, is a nonprofit organization serving the construction financial community; every year it surveys the 7,000 members in chapters across the country to obtain financial data and the major concerns of the industry. The members include residential, nonresidential, industrial highway, and specialty contractors. The 2007 financial survey presented the following national overview as reported by the respondents:

- The year's hot topic was field personnel recruitment and the ability to retain qualified workers, a concern that will continue for the immediate future. (This can impact owners, who may see a decrease in quality levels of workers.)

- Construction jobs are good jobs, with the seasonally adjusted hourly rate of $21.08 per hour as of September 2007, a rate that reflects a 4.5 percent increase over the previous year for the same period. (Owners may find that labor increases in the construction industry exceed the overall inflation figures reported in the media.)

- Material costs are a major problem. From December 2003 to September 2007, construction material producer prices indices increased 30 percent, more than double the 13 percent rise in the Consumer Price Index (CPI). Steel, cement, diesel fuel, and other petroleum-based products were at the top of the price increase column. The construction slowdown in 2008 in the United States has had a dampening effect on price increases of some materials, while worldwide demand has increased the prices of others. The projected building cost index for 2009, as reported by McGraw-Hill in December of 2008, reflected a decrease of 0.5 percent, as opposed to an increase of 5.5 percent for the year 2007–2008.

- In 2006, shipments of construction materials exceeded $500 billion, approximately 11 percent of the total shipments by U.S. manufacturers, and shipments of construction machinery topped $36 billion, 11 percent of all U.S. machinery manufacturers. Due to the value of the dollar in relation to other world currencies, heavy equipment manufacturers like Caterpillar saw export sales rise during that period.

■ The typical construction establishment is a small company with an average employment of fewer than nine individuals.

■ Internal Revenue Service figures for 2004 show that the 700,000+ corporations in construction had a net income of $47 billion, or 3.7 percent of total receipts of $1.3 trillion, considerably below the all-industry average margin of 4.9 percent.

■ Construction is a high-turnover industry. The Small Business Administration (SBA) showed that in 2004, 77,000 companies closed shop.

CFMA reviews the member responses and prepares a Best-in-Class composite for nonresidential and industrial building contractors. Table 3-1 shows that net earnings before taxes for these Best-in-Class contractors was 8.1 percent, and net earnings were 7.5 percent, as compared to all participants, who reflected net earnings before taxes of 2.7 percent and net earnings after taxes of 2.3 percent. Table 3-2 shows a composite financial analysis of all nonresidential and industrial contractors. Table 3-3 shows a composite of Best-in-Class subcontractors.

Risky business

Construction has often been likened to an outdoor factory producing a one-off product. Subject to the vagaries of the weather, the uncertainty of price protection for long-term labor and material commitments, and the shortage of skilled laborers from time to time, the risks in the industry are fairly self-evident. As we have seen in the CFMA figures, most members reported net income before taxes and net earnings below 3 percent. It does not take much to turn these low earnings figures into a net loss and resultant financial failure. One bad project has sunk more than one construction firm.

A 2002 Dun & Bradstreet survey revealed the following failure rates for various industries in the United States:

Industry	Failure Rate
National Average	1.40%
Agriculture, forestry, fishing	0.77%
Finance, insurance, real estate	0.84% (not so in 2008!)
Mining	0.92%
Services	1.27%
Retail trade	1.36%
Wholesale trade	1.36%
Manufacturing	1.49%
Construction	**1.52%**
Transportation, communications	2.75%

Table 3-1 Composite of Best-in-Class Financial Results from All Companies Participating in the Construction Management Financial Association's 2007 Survey

Balance Sheet

	All Participants		Best in Class			All Participants		Best in Class	
	Amount	Percent	Amount	Percent		Amount	Percent	Amount	Percent
Current assets:					**Current liabilities:**				
Cash and cash equivalents	$ 8,543,794	15.3 %	$ 3,398,884	14.9 %	Current maturity on long-term debt	$ 705,386	1.3 %	$ 304,677	1.3 %
Marketable securities and short-term investments	2,436,875	4.4	811,232	3.6	Notes payable and lines of credit	1,220,433	2.2	660,780	2.9
Receivables:					Accounts payable:				
Contract receivables currently due	20,608,340	36.9	9,329,662	41.0	Trade, including currently due to subcontractors	16,405,345	29.4	3,828,893	16.8
Retainages on contracts	7,421,822	13.3	1,975,956	8.7	Subcontracts retainage	4,629,576	8.3	485,786	2.1
Unbilled work	692,231	1.2	261,083	1.1	Other	727,460	1.3	208,101	0.9
Other receivables	864,894	1.6	358,856	1.6	Total accounts payable	21,762,382	39.0	4,522,780	19.9
Less allowance for doubtful accounts	(82,553)	(0.1)	(66,099)	(0.3)					
Total receivables, net	29,544,733	52.9	11,858,427	52.1	Accrued expenses	3,758,727	6.7	2,358,966	10.4
					Billings in excess of costs and recognized earnings on uncompleted contracts	6,963,332	12.5	2,803,350	12.3
Inventories	1,084,616	1.9	538,436	2.4	Income taxes:				
Costs and recognized earnings in excess of billings on uncompleted contracts	2,810,763	5.0	1,091,441	4.8	Current	360,958	0.6	99,672	0.4
Investments in and advances to construction joint ventures	327,796	0.6	17,752	0.1	Deferred	42,806	0.1	15,493	0.1
Income taxes:					Other current liabilities	703,622	1.3	244,798	1.1
Current / refundable	17,827	0.0	6,989	0.0	Total current liabilities	35,517,646	63.6	11,010,517	48.3
Deferred	185,506	0.3	80,689	0.4	**Noncurrent liabilities**				
Other current assets	1,864,677	3.3	440,351	1.9	Long-term debt, excluding current maturities	2,196,966	3.9	1,080,664	4.8
Total current assets	45,788,679	83.9	18,252,231	80.1	Deferred income taxes	1,123,834	2.0	40,200	0.2
					Other	581,290	1.0	219,782	1.0
Property, plant and equipment	12,004,158	21.5	8,338,451	36.6	Total liabilities	39,398,737	70.6	12,360,164	54.3
Less accumulated depreciation	(6,124,529)	(11.0)	(4,873,003)	(21.4)	Minority interests	88,680	0.1	55,724	0.2
Property, plant and equipment, net	5,879,630	10.5	3,465,448	15.2	**Net worth:**				
					Common stock, par value	764,587	1.4	224,373	1.0
Noncurrent assets:					Preferred stock, stated value	90,784	0.2	29,269	0.1
Long-term investments	996,432	1.8	292,454	1.3	Additional paid-in capital	1,880,067	3.4	713,320	3.1
Deferred income taxes	387,468	0.7	25,925	0.1	Retained earnings	13,135,875	23.5	9,439,460	41.4
Other assets	1,756,975	3.1	740,157	3.2	Treasury stock	(549,836)	(1.0)	(699,721)	(3.1)
Total noncurrent assets	3,130,874	5.6	1,058,536	4.6	Excess value of marketable securities	57,027	0.1	26,959	0.1
					Other equity	963,482	1.7	626,667	2.8
					Total net worth	16,341,966	29.3	10,360,328	45.5
Total assets	$ 55,609,363	100.0 %	$ 22,776,215	100.0 %	Total liabilities and net worth	$ 55,609,363	100.0 %	$ 22,776,215	100.0 %

Statement of Earnings

	All Participants		Best in Class	
	Amount	Percent	Amount	Percent
Contract revenue	$ 154,696,293	98.0 %	$ 57,613,232	94.8 %
Other revenue	3,174,629	2.0	3,169,071	5.2
Total revenue	157,870,922	100.0	60,782,303	100.0
Contract cost	(139,283,761)	(88.2)	(48,222,898)	(79.3)
Other cost	(2,484,033)	(1.6)	(2,549,017)	(4.2)
Total cost	(141,767,794)	(89.8)	(50,771,915)	(83.5)
Gross profit	16,103,128	10.2	10,010,388	16.5
Selling, general and administrative expenses:				
Payroll	(5,442,876)	(3.4)	(2,313,411)	(3.8)
Professional fees	(226,907)	(0.1)	(150,026)	(0.2)
Sales and marketing costs	(582,300)	(0.4)	(251,134)	(0.4)
Technology costs	(184,584)	(0.1)	(145,582)	(0.2)
Administrative bonuses	(715,168)	(0.5)	(747,090)	(1.2)
Other	(4,870,600)	(3.1)	(1,494,636)	(2.5)
Total SG&A expenses	(12,022,435)	(7.6)	(5,101,878)	(8.4)
Income from operations	4,080,693	2.6	4,908,509	8.1
Interest income	416,777	0.3	97,990	0.2
Interest expense	(241,213)	(0.2)	(112,678)	(0.2)
Other income / (expense), net	31,999	0.0	21,681	0.0
Net earnings / (loss) before income taxes	4,288,256	2.7	4,915,882	8.1
Income tax (expense) / benefit	(628,886)	(0.4)	(343,803)	(0.6)
Net earnings	$ 3,659,370	2.3 %	$ 4,571,880	7.5 %
Average backlog	$ 123,121,158		$ 41,128,580	

Number of Participants

All Participants	Best in Class
Number 756	Number 189

Financial Ratios

	All Participants		Best in Class	
	Average	Median	Average	Median
Liquidity Ratios				
Current Ratio	1.3	1.4	1.7	1.8
Quick Ratio	1.1	1.2	1.5	1.6
Days of Cash	19.5	13.7	20.1	17.1
Working Capital Turnover	14.0	12.1	8.4	7.4
Profitability Ratios				
Return on Assets	7.7 %	8.1 %	21.6 %	23.5 %
Return on Equity	28.2 %	26.4 %	47.4 %	50.4 %
Times Interest Earned	18.6	11.9	44.7	37.3
Leverage Ratios				
Debt to Equity	2.4	1.6	1.2	1.1
Revenue to Equity	9.7	7.7	5.9	6.2
Asset Turnover	2.8	3.0	2.7	2.9
Fixed Asset Ratio	38.0 %	27.8 %	33.4 %	19.0 %
Equity to SG&A Expense	1.4	1.3	2.0	1.8
Underbillings to Equity	21.4 %	11.4 %	13.1 %	8.2 %
Backlog to Equity	7.6	5.7	4.0	3.3
Efficiency Ratios				
Backlog to Working Capital	10.9	8.0	5.7	4.4
Months in Backlog	9.4	7.2	8.1	6.6
Days in Accounts Receivable	46.3	$3.6	57.0	56.6
Days in Inventory	2.7	2.9	3.8	2.8
Days in Accounts Payable	43.5	35.3	28.6	25.2
Operating Cycle	27.5	38.2	52.3	55.9

Source: By permission of the Construction Financial Management Association, Princeton, NJ.

Table 3-2 Selected Financial Data from Industrial and Nonresidential Contractors Responding to the CFMA 2007 Survey

All Companies		All Industrial & Nonresidential
Number of Companies	756	269
Assets ($)	55,809	84,571
Liabilities ($)	39,399	66,205
Net Worth ($)	16,342	18,308
Net Worth to Assets	29.3 %	21.6 %
Revenue ($)	157,871	257,068
Gross Profit ($)	16,103	16,186
Gross Profit Margin	10.2 %	6.3 %
SG&A Expense ($)	12,022	11,670
SG&A Expense Margin	7.6 %	4.5 %
Net Earnings ($)*	4,288	5,201
Net Earnings Margin*	2.7 %	2.0 %
Current Ratio	1.3	1.2
Return on Assets	7.7 %	6.1 %
Return on Equity	26.2 %	28.4 %

Source: By permission of the Construction Financial Management Association, Princeton, NJ.
Note: All dollar amounts are in thousands.
*Before Taxes

Figure 3-1 tracks the failure rate for all companies for the period 1989 through 2002 and graphically displays construction company failures.

The Surety Industry Organization (SIO), the trade group for providers of construction bonds, published a study entitled "Why Do Contractors Fail?" in which it tracked failures by different types of construction companies and the reasons for business failure:

Failure Rates	2002–2004	2004–2006
Specialty trade contractors	29.0%	24.4%
Heavy and highway contractors	27.4%	21.6%
Nonresidential building contractors	25.0%	17.5%
Industrial contractors	24.6%	14.6%

Failure Rates by Age of Business

Five years or less	32%
Six to ten years	29%
Over ten years	39%

Table 3-3 Selected Financial Data from Best-in-Class Specialty Contractors Responding to the CFMA 2007 Survey

All Specialty Trade		ANNUAL REVENUE				
		Best in Class	All $0–50 M	Best in Class $0–50 M	All >$50 M	Best in Class >$50 M
Number of Companies	290	99	199	70	91	29
Assets ($)	34,375	19,321	6,338	5,777	95,688	52,016
Liabilities ($)	21,095	11,452	3,728	2,785	59,073	32,371
Net Worth ($)	13,252	7,816	2,607	2,990	36,530	19,465
Net Worth to Assets	38.6 %	40.5 %	41.1 %	51.8 %	38.2 %	37.4 %
Revenue ($)	98,307	55,920	16,331	16,604	277,574	150,819
Gross Profit ($)	18,564	10,541	2,990	3,744	52,621	26,947
Gross Profit Margin	18.9 %	18.9 %	18.3 %	22.6 %	19.0 %	17.9 %
SG&A Expense ($)	15,550	6,258	2,186	2,224	44,775	15,994
SG&A Expense Margin	15.8 %	11.2 %	13.4 %	13.4 %	16.1 %	10.6 %
Net Earnings ($)[*]	2,884	4,224	753	1,513	7,545	10,767
Net Earnings Margin[*]	2.9 %	7.6 %	4.6 %	9.1 %	2.7 %	7.1 %
Current Ratio	1.6	1.6	1.6	2.0	1.6	1.6
Return on Assets	8.4 %	21.9 %	11.9 %	26.2 %	7.9 %	20.7 %
Return on Equity	21.8 %	54.0 %	28.9 %	50.6 %	20.7 %	55.3 %

Source: By permission of the Construction Financial Management Association, Princeton, NJ.
Note: All dollar amounts are in thousands.
[*]Before Taxes

Contractor Failure Risks

1. Low profit margins
2. Slow collection of accounts receivables
3. Insufficient working capital
4. High material prices
5. Shortage of qualified, skilled workers
6. Subcontractor failure
7. Inadequate cost-tracking systems
8. Estimating problems
9. Overexpansion
10. Onerous contracts
11. Unreasonable owners

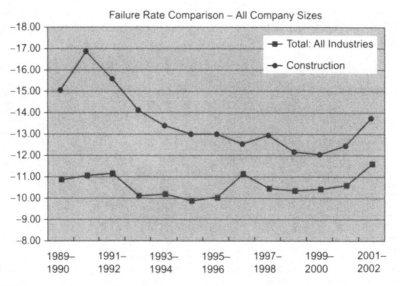

FIGURE 3-1

Failure rate comparison of all industries and the construction industry.
Source: U.S. Census, 1989–2002 Business Information Tracking Series.

A bankruptcy law firm listed the number one reason for contractor failure as a significantly underbid project; one large project improperly bid can hit the contractor so hard that he or she can never recover. The healthy builder today may not be so healthy next year. Steps taken by an owner to explore the financial health of a proposed contractor and requiring some financial security by requesting payment and performance bonds will provide less exposure in the event of contractor financial failure.

We can see by the business failure rates and the reasons for those failures that owners may find it a little easier to understand situations in which a contractor may become overly aggressive in trying to defend and protect his or her costs when interpretation of a contract provision seems to weigh heavily against the contractor with little chance for rebuttal.

The age of the master builder

In the early part of the last century, there was a class of contractors known as "master builders" who could provide the services that in today's market are provided by "design-builders." They offered a potential client their services to design and construct a project primarily using their experienced staff of managers and carpenters, laborers, plumbers, and electricians, all on staff. As a more competitive environment entered the economy, particularly after World War II, general contractors began to question their decision to maintain large crews of workers, some of whom would be idle between jobs. These large crews of skilled workers required the contractor to continually seek work, even though some jobs might not be profitable but were needed to meet a large weekly payroll. So general contractors began to divest themselves of this large weekly labor cost burden and subcontracted much of this work to specialty contractors; excavators, plumbers, electricians, drywall companies—in other words, subcontractors.

As this transition took place, those general contractors who subcontracted much of the work previously performed by their own forces became known as "brokers," a derisive term at that time. Nowadays, it is the rare general contractor who maintains a sizeable crew of skilled workers beyond a few laborers and carpenters.

A typical general contracting company organization

Although the size of the staff will vary considerably from the small, local general contractor to the international construction giant, the basic functions are the same. Project management, site supervision, estimating, purchasing, and accounting are the lifeblood of any contractor.

Control of costs, scheduling, and quality of work are the hallmarks of a successful contracting firm, and experienced staff are essential to achieving these goals. These functions are filled by the following types of staff positions.

Project management

The function of a project manager is to ensure that the requirements of the construction contract with the owner are fulfilled. The project manager is the primary interface between owner and contractor. The PM must pay attention to quality, costs, and schedule. The PM is greatly assisted in this task by the project superintendent, who is stationed at the project site and responsible for the day-to-day activity and operations on that site. Project managers are generally office based but spend considerable time at the project to confer with their project superintendent and observe firsthand how work is progressing. They deal with schedule issues, subcontractor relations, and quality issues. It is not unusual for a project manager to work on several projects at the same time, but, depending on the size and complexity of a project, they may be assigned solely to that project and work out of the field office instead of the corporate or branch office.

Project management responsibilities in each company can vary widely. Some companies rely on the project manager to purchase all materials and equipment and negotiate subcontract agreements, while other companies leave these tasks to other staff members or a purchasing department. Large and complex projects may require a project manager, a project engineer or two, a project superintendent and a few assistants, and even several foremen to oversee the operations of specific trades.

One of the key functions of project management, along with quality control and adherence to schedule, is cost control. Successful contractors have a method of updating their costs to date on a project to compare with their estimated costs; when the former exceeds the latter, the project manager must determine the reason why those costs have exceeded the estimate and take measures to control that particular cost item if possible. Not all excessive costs can be controlled, and while some costs will exceed a particular budget item, others may come in below budget and therefore provide a leveling effect. This is particularly true when any form of cost-plus-fee contract or a cost-plus with a guaranteed maximum price contract is being administered. Both of these contract formats are reviewed in detail in Chapter 4.

Project executive

In very large construction companies with multiple project managers, this expanded group of employees require close supervision. The project executive, usually an officer in the company, will assume responsibility to oversee multiple projects and the multiple project managers assigned to those projects.

The project executive is the person generally called upon when serious claims or disputes arise between contractor and owner. His or her experience is these matters will come into play in an attempt to resolve the issues quickly and amicably.

Estimator(s)

How does a general contractor prepare an estimate for the work required by the owner's plans and specifications? General contractors retain a database of costs culled from previous projects, creating a detailed unit cost for hundreds or thousands of work tasks. These costs are updated periodically for material and equipment price fluctuations and changing labor costs. General contractors also rely heavily on estimating data from specialty contractors, who maintain their own databases of costs for their particular trade. These costs are also incorporated into the general contractor's database and are updated and revised as more current costs are made available.

Many of these cost-reporting systems emanate from the field. If the contractor is performing certain tasks with her own staff, she will assign cost codes to each part of the tasks, and daily labor hours will be tracked against a particular task.

Cost-reporting techniques

As an example, let's look at a fairly simple operation: pouring a concrete footing. This task is divided into several parts: excavation, installation of wood or steel forms to define the width and length of the footing, placing concrete in those footings, and, after it cures (hardens), removing the forms. Although the excavating contractor will dig the trench for the footing, some hand labor is required; that labor may be supplied by the concrete contractor or possibly the excavating contractor or the general contractor. Each task is assigned a separate cost code within a general category assigned for concrete work:

Code	0330	Hand-trim trench prior to installing forms
Code	0340	Install footing forms
Code	0350	Place concrete
Code	0360	Strip and clean concrete forms

As laborers and carpenters complete these tasks, the hours spent daily on each task will be apportioned to the appropriate cost code. When the total length of footing is poured and the total labor expanded for that task is recorded and cost-coded tasks and other components of that operation are completed and reported, the estimator can determine unit cost: the cost per lineal foot of footing. After repetitive operations of the same task performed on many other jobs, a range of unit costs can be consolidated into an "average" cost to be used in the preparation of future estimates.

The contractor's in-house estimating department will undoubtedly have a computer software program that allows access to this database so it can prepare the estimate. Once again, large contractors will have an expanded estimating staff: chief estimator, estimator, and junior estimator. There are also a number of online estimating services, cost-estimating manuals such as those prepared by the R.S. Means Company, and local estimating services.

It is fairly evident that the quality of a contractor's estimate is one of the lifelines for survival; poorly assembled estimates representing either excessive or deficient costs will ensure that this contractor's survival rate is short term.

Purchasing

As just mentioned, some companies prefer that their project managers perform all of the purchasing functions, since this procedure also affords them an intimate look at the intricacies of the services, materials, and equipment required for their project from that particular vendor or subcontractor. Other companies prefer to have centralized purchasing department so one source can be kept current on costs, reliable suppliers, new products, and economies of scale in the purchasing function.

Accounting

The contractor's accounting department will prepare the monthly progress payment requests to the owner and perform normal accounting functions relating to accounts payable and accounts receivable. The periodic cost reports that reflect costs to date compared to estimated costs, when requested by the project manager or project executive, will be prepared by the accounting department.

The accounting department, as it prepares weekly hourly wage and salary payrolls, accumulates information that will feed into a specific project's cost report and serve as the basis for cost updating. For example, the weekly wages of both project superintendent and project manager will be tracked against the estimated cost for those functions in the project's cost estimate. If the general contractor has supplied labor from the company's own workforce, such as laborers or carpenters, the payroll of each type of worker will be cost-coded against the corresponding work task as explained previously.

Other staff members may be part of the large general contractor's workforce: a general superintendent, who oversees the activities of all project superintendents, and a director of safety, who also has multiproject responsibility.

Salary levels

Based on a few construction personnel salary surveys for the year 2007, the annual salaries of selected manager/supervisor positions are listed here. They vary depending on geographic location and the size of the company. Projects in large metropolitan areas command qualified and therefore highly compensated managers. These ranges can be viewed as guidelines.

Project manager	$89,500–$125,000
Project superintendent	$81,000–$128,000
Assistant superintendent	$35,000–$56,000
Estimator	$67,500–$88,500
Scheduler	$42,000–$81,000

Year-end bonuses also figure into some of these salaries, particularly when a successful project has been wrested from a failing one or higher than expected profits were attained by exceptional diligence by the management team. Other fringe benefits such as company cars must also be factored into management "costs."

The shift to subcontracted work

The shift from the full-service contractor with crews of laborers, carpenters, masons, and excavation crews to the "broker"-type contractor, one who subcontracted most of the work and only provided management services, has been a recurrent theme in today's economy. Specialty contractors, better known as subcontractors, provide the general contractor with an experienced group of workers when needed for a specific project. Some general contractors continue to maintain a small or even sizable crew of selected tradespeople, primarily carpenters and laborers, and advertise themselves as full-service builders who do not have to rely on subcontractors except for specialized trades such as plumbing, heating, and electrical work.

When a contractor performed work with the company's own forces, the costs would be marked up by the overhead and profit figures as stated in the contract. When work tasks are subcontracted, the subcontractor's costs will include the overhead and profit percentage, and the general contractor will add on this overhead and profit.

Union, nonunion, and merit-shop contractors

Not all general contractors are alike in the makeup of their workforce and construction crews. Some contractors hire only union-affiliated workers and subcontractors and are "union shop contractors," while some hire only nonunion workers, and subcontractors are referred to as "open shop" contractors. Others do not discriminate between union and nonunion workers but hire the most competitive qualified tradespeople; they are know as "merit shop" contractors.

Generally, large construction projects in major metropolitan areas are dominated by union contractors for good reason. Large projects require large numbers of skilled workers, and the various union trade associations can fill those needs. But in other parts of the country, large construction projects are manned by merit shop contractors and in some cases even by open shop contractors.

Union membership has been declining since from its high in 1948 when 30 percent of the labor force had collective bargaining agreements to just about 8 percent in recent years. In the construction industry there was a slight increase in the percentage of union workers between 2005 and 2006, but these figures can be misleading because the overall number of union workers increased from 8,053 in 2005 to 8,444 in 2006, primarily due to the overall expansion of construction volume from 8,053,000 to 8,444,000 workers during that period.

The double-breasted contractor

There are opportunities for union, merit, and nonunion contractors operating in the same geographic venue, and some contractors embrace all concepts. When large crews of skilled workers are required, the union can be called upon to provide them, since most unions have comprehensive craft training programs for beginning and apprentice workers; as these apprentices pass through their training programs and become skilled "journeymen," they begin to enjoy good wage rates and fringe benefits.

When contractors operate in all three jurisdictions—union, merit, and nonunion—maintaining both union and merit or nonunion contracting businesses, they are referred to as being "double-breasted." Although there may be common ownership of both entities, there are two different office locations, a separate management staff, and separate accounting departments—in effect two separate companies.

Contractor trade organizations

Two organizations represent each segment of union and nonunion contractors. The Associated Builders and Contractors (ABC; *www.abc.org*) is the national organization representing about 24,000 merit shop and nonunion contractors across the country. The Associated General Contractors of America (AGC; *www.agc.org*) represents about 33,000 firms, including 7,500 of the nation's largest general contractors, and is generally considered an organization of union contractors and subcontractors. Each organization has chapters throughout the country. Specialty contractors have two trade organizations: the American Subcontractor Association (*www.asa.online. org*) and the Association of Specialty Contractors (*www.assoc-spec-con.org*).

Differences between union and nonunion

There are some unique distinctions between classes of contractors in their trade practices and in their wage rates. Work restrictions abound in collective bargaining agreements: plumbers don't usually dig ditches for underground pipes; either laborers or operating engineers manning excavating equipment do so. Carpenters can't clean up their debris at the end of the day; laborers are designated for that work. From a cost standpoint, this makes sense because a laborer's hourly rate is less than the more skilled plumber or equipment operator. But in some cases this distinction can be a little frustrating. Unloading air-conditioning units from a delivery truck was once done by three laborers, but because the unit has an electrical connection, shouldn't union electricians also be involved in the unloading process? But wait a minute! This particular air-conditioning equipment has a compressor in it, so doesn't that require an HVAC mechanic? This practice has now stopped. Once again, the term *reasonableness* comes into play, and, depending on what the collective bargaining agreement includes, both union representative and contractor should be able to work out their problems. The collective bargaining agreements of union workers contain many of their work practices, wages, and fringe benefits, including the following:

- Wages and period through which current rate applies
- Pension
- Health and welfare
- Annuity
- Industry improvements
- Education
- Labor-management trust fund
- Travel and parking allowances
- Provision for vacation pay

The standard contract between owner and general contractor generally includes a provision that the owner may, at his or her discretion, award separate contracts for other work.

An owner must be aware of a situation that can arise when he hires a nonunion contractor who is expected to work alongside the general contractor's union workers. Jurisdictional disputes may arise, and the union representative may refuse to work when there is a nonunion worker of a similar trade working on the job site.

This can occur when a nonunion low-voltage electrical contractor, such as a private telephone and data communication installer, is hired by the owner and must work in close proximity to a union electrician on the job or when a furniture installer employing nonunion carpenters is working on the same floor as union carpenters. It is best to discuss those separate contracts with the general contractor beforehand so he can advise the appropriate union trade and avoid any potential work slowdowns.

Hourly wage rates for union workers are generally higher than those for workers without union affiliations. Union workers have collective bargaining agreements negotiated between contractor groups and local union representatives, and they usually include much higher employee benefit costs than merit or open shop organizations. As an example, a union carpenter's hourly rate exclusive of the contractor's overhead and profit but including fringe benefits (also referred to as "labor burden") in the Boston area for the year 2007 would look something like this:

Base hourly rate	$36.88
Health and welfare fund	7.78

Pension fund	4.65
Annuity fund	7.26
Education fund	1.18
Holidays	1.37
Travel	12.25
Total	$71.37 per hour

That same carpenter working for a nonunion general contractor in Baltimore, Maryland, where many nonunion and merit shop contractors flourish, would be approximately $20.00–$25.00 base hourly wage rate plus a labor "burden" of 40 percent for a total of $28.00 to $35.00 per hour.

THE CONSTRUCTION MANAGER

When a contract is executed between an owner and a general contractor, this is an "arm's-length" transaction, in effect an agreement between a buyer and a seller. A construction manager is actually the owner's agent, acting in his or her behalf during the design and/or construction process. It is as if the owner has an experienced construction professional on staff and assigned to the project. The construction management process can be employed on a project of any size and complexity but is mostly found on high-value and complex projects.

The methodology behind construction management can be seen in the provisions of the contract format. Construction Management Association of America Document A-1 is an agreement between an owner and a construction manager; various segments of this contract illustrate the relationship and services provided by the CM. A complete Document A-1 is included in Appendix A, but here are some highlights:

Working relationship:

"In providing the CM's services described in this Agreement, the CM shall endeavor to maintain, on behalf of the owner, a working relationship with the Contractor and Designer.

Basic services:

1. Prepare a construction management plan to include the owner's schedule, budget, and general design requirements for the project.

2. Assist the owner in the selection of a designer by developing lists of potential firms, developing criteria for selection, and transmitting requests for proposals.

3. Prepare preliminary budgets and budget analysis for the owner and the designers.

4. Establish a Management Information System (MIS) to facilitate communication between owner, CM, designer, contractor, and other parties involved in the project.

5. Assist in soliciting bids from contractors, conducting prebid and postbid conferences, and assisting in the selection of a contractor and the preparation of the contract for construction.

6. Review the contract documents and make recommendation to the owner with respect to constructability issues and scheduling. During construction provide project management services.

7. Provide time and cost management services during construction.

8. Review and recommend approval of progress payments and change orders.

9. Assist in an orderly and conclusive project closeout process.

Construction-manager contracts are discussed in more detail in Chapter 4.

THE DESIGN-BUILDER

Design-build, as discussed earlier, is simply a situation where an architect and a builder combine services to provide an owner with a one-source contact and contract for the design and construction of the owner's project. This is different from the conventional project approach, known as design-bid-build. The design-build team can be structured as follows:

- A firm that employs architects, engineers, and construction professionals

- An architect who forms a joint venture with a contractor to create a design-build team

- A contractor who forms a joint venture with an architectural/engineering firm to create a design-build team

- Either an architect or a contractor who employs the other as a "subcontractor" for a specific project

Design-build has gained considerable favor in the past decade or so because of its cost-effective approach and more rapid completion of the construction cycle. Whichever form the design and construction team takes, its objective is the same: to provide a collaborative effort to transform an owner's building program into an aesthetic and cost-effective reality.

PROGRAM MANAGEMENT

As owners reduce staff and tighten their financial belts, they will rely on more and varied professional assistance in planning for the most cost-effective project.

Nowhere is this professional assistance more valuable than in the planning and conceptual stages of the project. The program manager is a function whereby an owner can hire a team of professionals to deal with the concept-to-completion process, beginning with the coordination of public officials and agencies in the project's planning stage through selection of a project-delivery system, architect/contractor selection, construction, and closeout. A program manager is important when an owner has embarked on multiple construction projects and does not have the staff to control and monitor them.

Construction contracts pros and cons

4

The three prime sources for standard construction contracts are the American Institute of Architects (*www.AIA.org*), the Associated General Contractors of America (*www.AGC.org*), and the Construction Management Association of America (*www.CMAAnet.org*). Each of these organizations has listings of the various types of standard construction contracts they offer for construction projects. The Design Build Institute of America (*www.dbia.org*) can provide design-build contracts.

Most of these "standard form" contracts are modified to meet the specific needs of the owner and the individual project, and the advice of an attorney specializing in construction law will provide assistance in the inclusion of those specific needs. Another form of contract, one with a very limited scope of work, is often employed as a precursor to the more formal contract for construction: the letter of intent.

THE LETTER OF INTENT

A letter of intent (LOI) is a contract of restricted scope that can authorize the start of construction to perform a limited amount of work; it is generally meant to be followed by a formal contract for construction. Limits are not only placed on the scope of work to be performed but also the dollar value of the work, often expressed as a cost "not to exceed" or a specific sum, as well as the time frame in which this limited work is to be completed. This is a legal document and should be prepared in consultation with an attorney. A letter of intent has six components:

1. A stated scope of work, either in the form of an architect/engineer-prepared sketch, plan, and/or specifications, or a complete narrative description of the work

2. A lump-sum cost, a cost "not to exceed," or a cost plus a fee with no set limit as the method of payment

© 2010 by Elsevier, Inc. All rights reserved.
Doi_No = 10.1016/B978-1-85617-548-7.00004-5

3. Payment terms, either one payment at completion of the job or according to a payment schedule

4. A date when the work is to commence and when the work is to be completed

5. A statement that the scope of work and related costs will be credited if its scope is included in a formal construction contract for the larger project that usually follows

6. The event that triggers termination, such as the issuance of a contract for an expanded project, completion of the described work, or expiration of the time frame in which the work was to be performed

Like all other contracts, the letter of intent is to be executed and signed by all parties. There are a number of reasons why an owner may elect to issue a letter of intent. An owner of a commercial building may wish to begin demolition of the space recently vacated by a tenant while negotiations are proceeding with the new tenant. If the new lease calls for either the landlord or the new tenant to reconfigure the new space, the existing workspace may have to be partially or totally demolished. The letter of intent would allow the building owner to engage a contractor to proceed with demolition and, when the new space is designed, to have the option of continuing to work with the owner or the tenant for the fit-up work.

Prior to completion of financing but after a verbal commitment has been issued by a lending institution, the owner may wish to proceed with a limited amount of construction work pending receipt of final loan approval before committing to the full construction contract. In that case the owner could issue a letter of intent to the contractor to, for example, cut down the trees on the site where the building is to be built.

When a project is "fast tracked," and certain materials or equipment must be pre-ordered before the final plans and specifications can be incorporated into a fully executed contract, a letter of intent will allow the builder to order the various pieces of long-lead-time materials or equipment such as structural steel, precast concrete components, electrical switch gear, or boilers to speed up the construction process once the final contract for construction has been executed.

A clear understanding of costs in a letter of intent is necessary to avoid any disagreements as invoices are submitted by the contractor. An owner should have the contractor present a list of the costs anticipated for the work so they can be reviewed and approved and attached to the LOI as an exhibit.

A list of hourly wage rates for all trades anticipated to work on the project can be submitted, reviewed, and modified if necessary and accepted by all parties. As an example of the details necessary to allow for review of an hourly wage rate, Figure 4-1 reflects a "straight time" and "premium time" (time and a half and double time) hourly rate for a laborer. This rate includes not only the worker's hourly pay but the addition of all fringe benefits and the agreed-upon markup for the general contractor—in this case, 10 percent overhead and 5 percent profit.

The letter of intent should include a provision that either sets a time limit or expires when triggered by some event. The termination may be as simple as

BREAKDOWN OF HOURLY RATES		MA	
Worker's Title:	LABORER	DRILLING JOBS	
	Straight Time	1½-Time Premium	Double-Time Premium
Base Wage Rate	23.35	11.68	23.35
FICA 7.65%	1.79	0.89	1.79
FUTA .80%	0.19	0.09	0.19
SUTA 7.42%	1.73	0.87	1.73
Gen. Liability	1.03		
Workers' Comp.	2.63		
Welfare Fund	3.50		
Pension Fund	6.90		
Apprentice Fund	0.35		
Vacation Fund			
Ed. & Cult. Fund	0.95		
Deferred Income Fund			
Paid Holidays			
Bond Premium			
Incidentals			
Other: Umbrella	0.57		
Subtotal	42.99	13.53	27.06
Overhead and Profit (10%) + (5%)	6.66	2.10	4.19
Total	49.65	15.62	31.25

FIGURE 4-1

A breakdown of labor rates for straight time, time and a half, and double time.

"for convenience," which allows either party the right to terminate upon written notification and a formula for establishing the cost of the work up to the date of termination. Termination can also occur upon the issuance of the formal contract for construction. Termination can also occur when the scope of work has been achieved—for example, using our previous example, when all of the trees in the building area have been cut down.

A typical letter of intent with a termination clause would look something like this:

Pursuant to the issuance of a formal contract for construction, the ABC Corporation (owner) hereby authorizes DEF Construction (contractor) to proceed with the tree removal in the areas designated on Drawing C-1, C-2, dated June 15, 2009, prepared by D&B Architects and including Specification Section 2 of the Project Specifications manual. DEF Construction to remove all debris, including tree stumps from the site. Prior to commencement of work, all erosion controls as shown on Drawing C-3, June 15, 2009, prepared by D&B Architects is to be installed, maintained, and left in good condition after this work has been completed. All of the above work to be performed on the basis of cost plus a contractor's fee for overhead and profit of 15%. All costs to be documented by daily work tickets and a copy of all subcontractor contracts and/or purchase orders (or in accordance with the attached labor rates, material costs, and equipment rates).

This letter of intent will terminate on September 15, 2009, or when the work has been completed prior to that date.

Signed: _____   Signed: _____

Owner   Contractor

TYPES OF CONSTRUCTION CONTRACTS

Most construction contracts in use today follow one of these formats:

- Stipulated or lump-sum contract
- Cost of work plus fee contract
- Cost of work plus fee with guaranteed maximum price (GMP) contract
- Construction management (CM) contract
- Design-build contract

Stipulated or lump-sum contract

A lump-sum contract is a rather straightforward contract approach that simply states, "You, the contractor, will build this building per the plans and specifications prepared by the architect/engineer for the sum of $_____." This type of contract, however, is a little more complicated than that.

A complete set of documents, plans, and specifications prepared and reviewed by the design consultants—the architect and consulting engineers—before being issued for bidding purposes is a critical component of the lump-sum contract approach. If there are missing details, inconsistencies, or lack of clarity in a portion or portions of these plans and specifications, the resultant prices would, most likely, reflect those missing details.

So when using the lump-sum contract approach, it is important to ensure that those plans and specifications are as complete and error-free as possible. And the owner shares in this responsibility if he does not allow the design team enough time (or money) for a thorough review.

The lump-sum or stipulated-sum format requires contractors who are submitting bids to include costs for all work accurately presented in the plans and specifications—no more, no less. It is basically a case of what you see is what you get.

Lump-sum contracts predominate in competitive bidding situations, particularly in the public sector, where all but blacklisted contractors can submit bids advertised by the public agency. Many government officials have sad tales about the low bidder who, after receiving a contract for construction, exploits some of the bid document shortcomings, having failed to advise the owner of any perceived errors and omissions beforehand, and begins to issue change order after change order.

And this may not necessarily reflect on the integrity of the contractor. Faced with a competitive bid situation, the bidder may not be willing to include costs to cover some obvious shortcomings in the plans and specifications because she is concerned that competitors may not include added costs. A high-quality set of bid documents is important when the lump-sum or stipulated-sum contract approach is contemplated. In the private sector, no such restraints on limiting bidding contractors are present, and a bidder's list composed of experienced, qualified builders with excellent reputations can reduce an owner's exposure to added costs due to vague or missing minor details.

For projects that are rather straightforward, the lump-sum contract format is an acceptable approach—again, with the caveat that the plans and specifications have been reviewed for quality purposes before issuance. When only experienced, prequalified contractors are invited to submit proposals, the nature of the competitive bid process makes it likely that the "forces of the marketplace" will produce a fair price, especially when there is a slowdown in construction activity.

The problem for the owner is how to determine whether the quality of the plans and specifications is sufficient to successfully apply the lump-sum approach. Perhaps hiring a consultant to review the bid documents prior to release may be one answer. Peer review on very complex projects is also an option.

As we mentioned previously, owners are often at fault when they prod an architect and engineer to produce a set of bid documents before they have had an opportunity to thoroughly review and vet them. I remember working on a project where the owner was pressing the mechanical engineer to complete the HVAC drawings too quickly. This engineer said simply, "Do you want accuracy or expediency? You can't have both at the same time." The owner, of course, chose accuracy.

Allowing sufficient time for plan and specification review is one of the most productive steps an owner and design team can take to avoid or lessen problems associated with what the professionals call E&Os—"errors and omissions." Any additional costs of this final review may be compensated by a reduction in change orders as the construction progresses.

When a lump-sum contract approach is being considered and competitive bids are to be pursued, it is a good idea to have a prebid conference conducted by the design team where all bidders are invited to attend a meeting and questions can be raised after receiving and reviewing the bid documents, including the plans and specifications. Contractors will often raise questions regarding interpretation of the work required, ask the attending architect and engineer questions about specific items of work, and allow the A/E to respond to these questions. The architect may also inject the "intent" with respect to certain areas or items of work.

At the conclusion of the prebid conference, minutes that include all discussions, questions, and answers will be prepared and distributed to all bidders. This creates a level playing field and allows all contractors attending the prebid conference to receive the same responses and adjust their bids accordingly.

Upon selection of the contractor, this prebid conference document can be appended to the construction contract as an exhibit, further amplifying and defining the scope of work of the project. Prior to awarding a lump-sum contract, a meeting with the architect and the contractor to review every drawing is another good idea. At that time the architect and engineers can ask the contractor specific questions, such as "Did you include all of this work?" "Did you understand what is required here?" "Are there any questions about the extent of this work?" Ask the contractor if any of the subcontractors or vendors raised any questions about the scope of work, and record the builder's response. When this meeting is concluded, minutes of the meeting should be prepared and transmitted to the contractor, stating that the memorandum of the meeting will also be included as an exhibit to the contract for construction. Work that can be defined and not left to interpretation will result in fewer misunderstandings as the contract is being administered.

There is another aspect to the lump-sum contract approach. An owner may, upon selection of the "best value" or lowest qualified bidder or bidders, engage them in negotiations to arrive at a contract sum and contract scope. Quite often the competitive bid produces a lowest responsible bid that exceeds the owner's budget. What should you do? A meeting with the selected contractor or short-listed contractors and the design consultants can result in a process whereby a contractor offers cost-saving measures that are fully accepted or modified by the design team, and the owner's budget requirements can be achieved. These changes in most cases can be documented by the architect/engineer with a few sketches (referred as to SKs) or a change in the specifications, either of which may result in little or no additional design change costs but will ultimately save the owner money.

Value engineering

Value engineering can apply to all types of contracts, and perhaps here is a good place to discuss the process. By the process referred to as "value engineering," the contractor can offer products, equipment, or construction techniques at some variance with the strict interpretation of the plans and specifications but acceptable to the architect and engineer. There are, however, some caveats to be aware of when value engineering suggestions are submitted by the contractor. What if the acceptance of what appears to be a savings in one component of construction actually results in increased costs to the owner in another related component? These possibilities should be explored when complex value engineering proposals are submitted. For example, replacing one large rooftop air-conditioning unit with two smaller ones may result in equipment cost savings and better interior climate control, but two smaller electrical circuits may be more expensive than one larger one, and two roof openings and roof frames may cost more than one large one. The cascading effect of a reduction in one area must be reviewed to determine how that change impacts other costs, possibly increasing operating costs while presenting a reduction in capital costs.

Cost of the work plus a fee contract

This is a rather open-ended type contract that simply states that the contractor will be directed to perform the work as specified by the owner and will be paid on the basis of the costs plus a preset fee, usually a percentage of the costs. This form of contract is often used for emergency work, such as flood repairs, fire, or weather-related damage where immediacy of work precludes assembling firm costs and when the actual cost to complete the work is difficult to determine.

The comfort level of a cost plus a fee contract depends on the quality and integrity of the contractor. A reputable one will perform the required work and accumulate costs representing the work accurately, while a not-so-reputable contractor may inflate costs or perform work that was not necessary and bill accordingly.

While contractor selection is a key ingredient in a successful cost-plus project, equally as important is defining the nature of the work as closely as possible. The other main ingredient is a mutual understanding of what constitutes reimbursable "costs." The nature of the work can be broadly described in a cost plus a fee contract, since rarely are there specific, concise plans or specifications to define that work. If we are dealing with demolition of an existing office space to make it ready for a new tenant's improvements, the contract requirements can be similar to the following example:

Demolition of all 2nd-floor existing drywall partitions in an area of approximately 15,000 square feet, including removal of all power electrical wiring back to the respective distribution panel circuitry. All low-voltage wiring in those partitions to be terminated at the location of the demolished partition. All debris to be removed from site by the contractor.

All suspended ceiling grids to be removed along with all ceiling tiles. All ceiling diffusers, lighting fixtures are to be removed and stored in a location designed by the owner. After removal of lighting fixtures, exposed wiring is to be taped with electrical tape, wire nuts applied, rolled up, and left in place. All flexible HVAC distribution ductwork, upon removal of distribution devices, to be rolled up and left in place. All HVAC rigid distribution ductwork to be removed back to the main duct. All flooring materials in this area are to be removed and discarded. All debris to be removed from the demolition area and taken offsite for disposal. All work areas to be left broom clean.

This kind of description, while not exact in nature, provides both the contractor and the owner with some rather concise guidelines and can be used to expand and explain other "cost-plus" work elected by the owner while that contractor is still on site.

At this point, the contractor should be able to assemble an order of magnitude of costs, but he or she may not agree to be held to those costs. If pressed by an owner for a guaranteed maximum price when the scope of the work is not fully defined and remains somewhat vague, the contractor, as a protective measure, will add a contingency amount to more than cover all unforeseen costs, defeating the purpose of a cost plus fee contract. We deal more fully with this subject in the following section in this chapter.

Once a more-or-less scope is defined, any additions or subtractions to that scope should be confirmed by the owner to avoid any misunderstandings of what constitutes extra work over and above the initially defined work. A letter to the contractor could simply state, "In addition to the work outlined in our agreement dated (date of agreement), you are directed to remove all flooring in the elevator lobby on the second floor and dispose of all materials offsite."

Instructions in the cost-plus contract should include a requirement for the contractor to submit daily work tickets (Figure 4-2), outlining the general nature of work performed each day. In addition to a description of the work performed, the names of the workers, the number of hours each worker expended on this task, and the signature of the contractor's foreman should be included on the ticket. In the case of this example, the actual labor costs are included along with some materials and equipment required for the work. These tickets will be presented to the owner's representative each day for review and acknowledgment or, at the owner's option, submitted to the office on a weekly basis.

Some owners request the contractor to submit weekly costs of work performed if the project is expected to continue for several weeks. This is to the advantage of both contractor and owner so there are few surprises when the final bill is submitted. Having an owner's representative stop at the work site each day to review that day's activities and accept the daily ticket is a sure way to avoid any misunderstanding of work performed and cost incurred upon completion of the work.

Particularly important in this type of contract is to define which "costs" are to be reimbursed and which are not reimbursable. One way to get a better fix on costs is to have the contractor submit the hourly billable rates for all trades that are anticipated to work on the project. The hourly rates shown in Figure 4-3 include the actual hourly rate paid to the worker and also the "fringe benefit" package and, in this case of an ironworker, an hourly rate for expendable supplies to be used during the welding operation. Some labor rates will include parking if the project is in an urban area where limited on-street parking is available and, in some instances, particularly when a foreman's rate is listed, an hourly rate for the truck the company supplies for travel to and from the site and picking up supplies during the workday. If there are questions about any element of the hourly rate breakdown, now is the time to resolve those questions.

EXTRA WORK FORM

Date: _10-9-08_   Job Location/Job #: _____

Job Description: _UNDER MINE FOOTER, FORM & POUR_
WITH CONCRETE. ALSO PUT REBAR IN NEW
FOOTER

Name	Time In	Lunch	Time Out	Total Hours	AJ	PPS	Invoice Amount
WOODY				8			440.00
SEAMUS				8			440.00
JIMMY				8			440.00

Equipment Used	Moving Charge	Arrival On Job	Lunch	Depart	Total Hours	AJ	PPS	Invoice Amount
VIBRATOR								
MIXER								

Materials	From	Invoice #	Price	M/U	Total	AJ	PPS	Invoice Amount
CONCRETE	LOWE'S				$426.00			$26.00
PORTLAND								
REBAR								
TIE'S								10.00
GAS					$10.00			

Grand Totals $ $ $1.756,00

Authorized Signature(s): _Joseph Sm_ _10/9/08_

FIGURE 4-2

A daily work ticket for extra work, including the worker's name, hours worked, tasks performed, materials consumed, and pricing for all work.

Cost plus a fee with a guaranteed maximum price contract

Although similar to the cost-plus contract, where the contractor will perform the work and be paid for the costs plus an agreed-upon fee, usually calculated as a

BOSTON, MA 02118   September 16, 200
                                                                                         TO

BREAKDOWN OF HOURLY RATE FOR FIELD LABOR:		FOREMAN	JOURNEYMAN
IRONWORKERS WAGE:		33.96	31.96

FUNDS:

PENSION	5.25	
WELFARE	7.50	
ANNUITY	6.25	
TRAINING	0.78	
INDUSTRY	0.03	
PENSION SUPPLEMENT	0.70	
DC LABOR MANAGEMENT	0.20	
IMPACT	0.21	

                                                                            20.92   20.92

INSURANCE:

WORKMENS COMPENSATION	42.370
MASS UNEMPLOYMENT	11.020
UNEMPLOYMENT HEALTH	0.120

GENERAL LIABILITY:

BODILY INJURY & PROPERTY DAMAGE	6.114		
F.I.C.A.	7.650		
F.U.T.A.	0.800		
UMBRELLA & PRODUCTS LIABILITY	4.167		
	72.241%	24.53	23.09
		79.41	75.97

EXPENDABLE SUPPLIES: 6.50   4.00
(welding rod, grinding wheels, drill bits)   85.91   79.97
(welding shields, fuel for welders, etc.)

10% OVERHEAD   8.69   8.00
                                                                         94.60   87.97

5% PROFIT   4.73   4.40
                                                                         99.23   92.37

FIGURE 4-3

Labor wage breakdown for iron worker foreman and journeyman, including list of expendable
supplies.

percentage of the costs, the GMP contract places a cap, or guarantee, on the maximum cost of the work. The GMP contract is often employed when an owner is desir-
ous of executing a contract with a general contractor or possibly a subcontractor
and getting started to order materials or equipment prior to the final completion of
the plans and specifications.

In this procedure, the architect will develop the plans and specifications to
a point that a contractor experienced in this type of construction can prepare an

estimate for not only the work shown in the plans and specifications but the cost of work that has yet to be designed and produced in the 100 percent complete documents. An experienced contractor can produce a reliable estimate when the drawings reach the 70 or 80 percent completion stage. The emphasis, however, is on the word *experienced*.

A contractor who is knowledgeable about mid- or high-rise commercial buildings has the background and database of costs to reasonably estimate the remaining 20 to 30 percent design work, which will largely consist of building finishes. But a contractor not well versed in this type of construction may not have enough experience to define and therefore include costs for, the work in those yet-to-be completed drawings. Although GMP contracts may be competitively bid, due to the nature of the unstated but assumed scope of work, the owner has an opportunity, after selecting the "best value" contractor, to negotiate the final terms, conditions, scope of work, and the cost with the contractor.

Two important components are also required for a successful GMP-type project, a qualification or exclusion/inclusion list and a contractor's contingency account. The contractor will have made certain assumptions as he or she prepared the total GMP estimate based on the incomplete drawings. The contractor may have assumed a certain quality of finish levels in specific areas of the building or added some items required by local building codes that had not yet appeared on the drawings. To quantify these assumptions, the contractor will prepare a list of pricing qualifications, sometimes called an exclusion/inclusion list. Figure 4-4 shows an example of such a list. It includes general conditions along with site work and demolition qualifications. The list will continue with qualifications for each division of work. Upon completion, the list should be reviewed by the owner and consultants, who can either agree with these assumptions or modify and negotiate other terms with the contractor. After this list is approved, it will become an exhibit to the construction contract, and all parties will have a clearer understanding of their mutual obligations.

The contractor and design consultants should be given an opportunity to meet and discuss the final drawing development as it progresses to ensure that the items on the qualifications list are being incorporated into the drawings. Upon final review of the fully completed drawings, any significant deviations from the qualification list must be resolved either by revising the drawings or increasing or decreasing the scope of work and hence the guaranteed maximum price.

The contractor will add a contingency, either stated, or unstated, to the estimate to cover items unforeseen but possibly to be included when the final drawings are completed. This contingency will generally be for the contractor's exclusive use, to be tapped in any limited fashion he or she may decide—for example, in cases when changing market conditions or inflation requires the contractor to pay a slightly higher price for some materials. If the contractor inadvertently failed to include a minor item of work that should have been anticipated, he or she can tap the contingency. As the contingency account is tapped, the contractor should be required to present the owner with the reasons for doing so and to create a running account of the balance remaining.

The Grey Falcon Stadium project
Fort Cartwell
Maryland

Pricing Qualifications

In an effort to ensure our team has reflected the design intent in our pricing, the following is intended to provide a division by division description.

Division 1 - General Conditions/General Requirements
Inclusions/Exclusions

1. This proposal is based on mutually agreeable contract terms.
2. Underpinning is said to be not required by the structural engineer is therefore not included
3. Asbestos, lead, or hazardous materials testing, removal and remediation are not included.
4. Utility and service company fees and charges for connections or meters are not included.
5. Temporary electric consumption cost during construction is not included.
6. A payment and performance bond is not included
7. Builders risk insurance is assumed provided by Owner and is not included
8. Any costs associated with the   .Health Department inspections are not included.
9. Construction site security, other than separation of work zones by construction fences is not included.
10. With timing going to be critical, we assume that the Owner will assist in obtaining permits

Allowances included within the General Conditions

1. *Police details and street permits: $15, 000 is included*
2. *Winter weather protection will most likely be required for the waterproofing, and concrete sidewalks as the current schedule does not allow for a start of this scope prior to Mid Oct 04. In addition, protection will likely be required for the façade restoration after the completion of the sidewalks. $25,000 is included.*

Division 2 - Site work / Demo
Inclusions/Exclusions

1. Below grade obstructions, rock, and/ or removals of these are not included
2. The furnishing and installing of the new gas line including the meter is by the gas company has included the excavation and backfill of this line as shown.
3. Relocation, repairs, or replacement of uncharted utilities.
4. Interior building demolition along with the removal of the existing floor slab in basement is currently underway and as such is not included.
5. Removal of any other topping slab at 1st floor level other than the area shown.
6. The work associated with the connection of existing roof drains to The Muddy River Conduit. was said to not be required and as such, we have not included.
7. The City of   will require flowable fill in the utility trenches and as such this is included
8. Based on the grade of the new floor slab, we believe that this job will require a small amount of imported fill, approx 500 cy, therefore, we assume that the existing soils will remain in place.
9. The current demo operation will included the removal of the existing walls into the former street vault areas
10. As discussed, the demolition will allow access to the new grease trap location for installation of the unit and of the piping

Allowances included within the Site work / Demo

1. As the full scope of the demo required for the remove of the existing concrete topping slab and unknown supports down to the assumed structural slab is included $10,000.
2. We believe that there will be a need for saw cutting of the existing column for new beam pocket: $2,500.
3. We have included an allowance for the installation of the new grease trap as do not have an elevation for this, $10,000. We assume that dewatering will not be required.
4. As we also do not have an elevation, we have included an allowance of $5,000 to support the

FIGURE 4-4

A pricing qualification statement, also called an exclusion/inclusion list.

Contract language can be prepared to define those events and occasions when the contingency account can be tapped. Here is an example.

1. Changes resulting from refinement of details of design within the scope and quality standards indicated in the contract documents, or changes required because of concealed or unknown conditions encountered during the performance of the work that do not constitute differing conditions, as defined in the contract.

2. Additional costs due to unanticipated delays in the delivery of materials and/or equipment, unless those delays are attributable to the contractor.

3. Cost overruns by contractor for subcontracted work or vendor purchases provided these cost overruns were not caused by gross negligence on the part of the contractor or the contractor's willful misconduct or breach of contract by the contractor, their vendors, or their subcontractors.

4. Expediting or acceleration costs necessary to meet the construction schedule provided that any delays requiring expediting or acceleration were caused by the contractor's gross negligence, willful misconduct or breach of contract by either the contractor, their vendors or their subcontractors.

5. Costs to correct nonconforming work, but only to the extent that the owner determines, in writing, that the nonconforming work was caused by conditions so unclear that the responsibility for that work could not be reasonably placed with the owner, their design consultants, or the contractor, or any of their subcontractors.

Another aspect of the contingency account that must be addressed is if a balance remains in the account at the conclusion of the job. Does the contractor receive the full amount of the remaining funds, or are they counted as a cost savings contributing to the overall project actual costs versus GMP, any savings of which are to be shared with the owner? Standard practice is that such savings are lumped into total savings.

The GMP contract usually includes a savings clause so if actual final costs are less than the GMP, some form of savings distribution between owner and contractor is included. There are arguments on both sides as to how any savings should be distributed; some argue that savings should be split 50-50; others are of the opinion that a greater percentage should accrue to the contractor to better motivate him or her to seek out savings. A typical savings clause would be stated as follows:

If the actual total price of the work, as established by the contractor's final accounting and owner's review and audit thereof, is less than the GMP, then (insert the savings percentage: 40%, 50%, or all) savings shall accrue to the owner. The owner will then be responsible for paying only the actual total price.

As we discussed in the case of a cost plus a fee contract, differing interpretations of what constitutes costs to be reimbursed and those not to be reimbursed must be included in the contract. The following can act as a guide, and other items can be added, subtracted, or modified to fit a particular project.

Costs that are reimbursed

Such costs include the following:

1. Labor costs: basic hourly rates; premium or overtime rates; labor "burden," which in the case of union workers will be union benefits plus employer contributions to Social Security; unemployment taxes (federal and state); worker-compensation costs; and collective-bargaining-agreement benefits

2. Wages for the project superintendent, onsite project engineer, project manager, and other personnel as submitted and approved by the owner; generally these costs include only field-based personnel, not office-based employees (with the exception of an office-based project manager).

3. Taxes, employee contributions, sales or use taxes, assessments

4. Costs for all subcontracted work

5. Costs of materials and equipment, temporary facilities (field office, storage trailers), and other items fully consumed in the performance of the work

6. Rental costs for machinery, equipment, and tools *not customarily owned by construction workers,* whether rented from the contractor or from a rental company (does not tools such as hammers, screwdrivers, and so forth that would normally be in a worker's toolbox; power tools, saws, heavy-duty drills, welding machines, and so forth are acceptable reimbursable costs).

7. Cost of debris removal from the site, including disposal of same, not including hazardous materials unless so stated elsewhere in the contract

8. Costs of documentation reproduction, faxes, postage, expedited delivery service, other shipping costs

9. Reasonable petty-cash disbursements such as parking meter fees, minor purchases from a lumberyard, bottled water, travel expenses by the contractor in the course of performing and discharging duties connected with the work

10. Cost of materials and equipment suitably stored offsite, *only if approved in advance by the owner*

11. A portion of insurance and bond premiums, if applicable

12. Fees, assessments for building permits, and other project-related permits required by the governing authorities

13. Royalties and license fees for use of a particular design, process, or product

14. Data-processing costs related to the work, including any wireless network installation, operation, and maintenance costs at the construction site

15. Deposits lost for causes other than by the contractor's negligence

16. Expenses incurred by the contractor for temporary living expenses or allowances

17. Costs to repair or correct damaged work, provided that such work was not damaged due to negligence of the contractor or was deemed nonconforming work by the architect

On projects requiring large pieces of excavating or hoisting equipment, hourly rental costs can range from $150 and up per hour, exclusive of the equipment operator's wages, and some restrictive language can be inserted in the "costs to be reimbursed" category to control these costs.

Two conditions must be considered: when the contractor will be furnishing his or her own equipment and renting it to the owner on an hourly basis and when the equipment will be employed for extended periods of time and warrant an adjustment in hourly rental costs. Contractor-furnished and owned equipment should have hourly rates lower than the rates posted by an equipment-rental company; the following contract language will help in that regard:

The rental rates charged for any piece of equipment from contractor's own stock or from an affiliated contractor entity should not exceed 75% of the published Associated Equipment Distributors (AED) published rate, or the rental tares for similar equipment, whichever is lower, for the geographic area in which the project is located.

(*Note:* This AED publication is also known as the *Green Book*.)

Rental rates can vary depending on the length of the rental period. There are hourly, daily, weekly, and monthly rental rates for most pieces of equipment, and each extended period of time yields lower rates. By inserting a time frame in which each rate will be in effect, an owner will be able to take advantage of those lower rates. The following will be helpful in that regard:

The rates charged for equipment rental shall conform to the following periods of rental:

1. If on any date a piece of equipment is used for more than three (3) hours, the daily rate of rental shall apply instead of the hourly rate.

2. If in any week any piece of equipment is used for three (3) days or more, the weekly rental rate shall apply instead of the daily rate.

3. If in any month, a piece of equipment is used for three (3) weeks or more, the monthly rental rate shall apply instead of the weekly rate.

Costs that are not reimbursed

The following are nonreimbursable costs:

1. Salaries and other compensation of the contractor's personnel stationed at the contractor's principal office, *except as specifically agreed upon and provided for in the contract;* an exception would be a project manager assigned to the project who is generally based in the home office and may be responsible for managing multiple projects. In such a case, the project manager should maintain a daily log showing time spent on each of the projects he or she is managing.

2. Expenses of the contractor's principal office or satellite offices

3. Overhead and general corporate expenses *except as specifically provided for in the contract*

4. The contractor's capital expenses

5. Rental costs for machinery or equipment *except as specifically called for in the contract*

6. Costs due to the negligence of the contractor

7. Any costs *not specifically included in the costs to be reimbursed;* this transfers to the contractor the responsibility to be specific about costs to be included and is meant to defray any arguments later that the contractor "forgot" to include one cost or another

8. Premium pay for salaried employees of the contractor, sick pay, or vacation or holiday pay

9. Costs that would cause the guaranteed maximum price to be exceeded (not including any approved change orders, which by their very nature change the GMP)

The exclusion of reimbursable costs for contractor's negligence and nonconforming work can also be included in the body of the contract, as this typical statement indicates:

Costs due to negligence or breach of contract by the contractor, or any subcontractor, or anyone directly or indirectly employed by any of them, or for whose acts any of them may be liable, including costs to correct damaged or nonconforming work; costs for disposing of or replacing materials and equipment incorrectly ordered or supplied, costs or losses resulting from lost, damaged, or stolen tools and equipment; costs for repairing damage to property not forming a part of the work; penalties, fines, or similar costs imposed by government agencies in connection with violation or noncompliance with applicable laws, ordinances, or regulations by the contractor or their subcontractors or employees thereof.

These lists of costs are not meant to be all-inclusive, but they can serve as a guide to the most accepted costs—both reimbursed and nonreimbursed—in the GMP contract format.

A construction contract cannot include and anticipate all potential problems; only the goodwill and integrity of the architect, contractor, and owner working together to establish a harmonious relationship can minimize those unforeseen events, Here are some pitfalls to avoid when considering a GMP contract:

- The scope of the work must be as clearly defined as possible. The design-development drawings and the specification manual will provide one such definition; the qualifications and/or exclusion/inclusion list is another.

- A complete statement of reimbursable and nonreimbursable costs is important to avoid misinterpretation of what is or is not reimbursable.

- The contract should include a provision allowing the owner to audit the contractor's book to ascertain final project costs.

- The contractor should present his or her procedures for documenting project costs to ensure that all costs are appropriate and apply to the specific project. The project superintendent and project manager should explain how they intend to segregate costs associated with the project to avoid comingling costs of another project.

- A periodic accounting method to track "costs to date" versus budget and projected "cost to complete" should be established by the contractor for review with the owner. This can be done on a monthly basis or some other agreeable time frame, but these periodic cost reviews will allow all parties to discern how actual costs compare with projected costs and whether the potential for any cost savings is realistic.

The procedure for the documentation of costs to be included in the contractor's monthly request for payment should be established. Receipted bills, delivery receipts, copies of subcontractor requisitions, the contractor's payroll register reflecting the individual, function, and corresponding labor costs, purchase orders, and the like are to be attached. A simple statement in the contract such as "Sufficient detailed documentation of costs to allow for a reasonable approval of such costs shall be provided with the contractor's monthly (or whatever period) request for payment" will suffice.

Construction-management contract

A construction-management contract is often referred to as an "agency" contract, since the construction manager acts as the owner's agent, performing work on the owner's behalf on construction-related matters. This differs from the other contracts we have discussed, where the relationship between contractor and owner is at arm's length and the contractor acts as an independent operator hired to build a project for an owner. In that scenario, the general contractor will negotiate and award subcontract agreements in the name of the *general contractor*, whereas the construction manager will negotiate subcontract agreements and, after consultation with the owner, issue agreements in the *owner's* name. CMs essentially provide a professional service; a contractor delivers a product—the construction project.

There are two basic types of construction managers (CM): a CM for a fee, who provides all services included in the contract and receives a fee based on a percentage of the final project costs (which are not, however, guaranteed by the CM), and a CM at risk, who provides all services included in the contract and receives a fee, generally based on a percentage of costs, with the final cost of the project guaranteed.

CM fees are lower than those of general contractors because their risk is greatly reduced, particularly in the case of the CM for fee contract; in both cases, more of the CM's costs are reimbursed by the owner instead of being absorbed into the fixed estimate and included by the general contractor in the general conditions category of costs.

The CM offers a wide range of services for the preconstruction phase, the construction phase, and even the postconstruction phase of a project. In the preconstruction phase, which some consider the most important phase, the CM will work with the architect and engineers to monitor costs as design progresses, offer information about local costs and materials availability, draw on their experience with certain building systems and components, and assist in developing progress schedules for both design documents and construction activities.

During construction, the CM will negotiate purchase orders for materials and equipment and will subcontract agreements, all of which will be issued in the owner's name after consultation with the owner. The CM will provide supervisory staff such as project managers, project superintendents, and project engineers to manage the project in the field and from the office. He or she will scrutinize change orders for validity of scope and costs and advise the owner accordingly, assist in dealing with disputes and claims, and supervise all project closeout procedures. He or she can also coordinate and supervise separate contracts issued by the owner, such as those for furniture, voice and data equipment installation, interior decoration, and the like. The full range of construction-management services can be found in CMAA Document A-1, Standard Form of Agreement Between Owner and Construction Manager, which is included in Appendix A.

An expansion of construction-management services is often referred to as program management (PM), which is basically the management of multiple ongoing, interdependent projects for the owner. The PM will coordinate and prioritize all of the owner's interrelated requirements relating to the construction project, such as those just mentioned, but also including installation of equipment, demountable office partitions and furniture, artwork, relocation activities, permitting, environmental issues, and other regulatory agency matters.

The CM approach has grown considerably in the past several decades for two basic reasons: It allows the owner to hire a construction professional to work with the architect during the crucial design phase when the CM's up-to-date pricing and construction component information and means and methods of construction can assist the architect/engineers in producing a more inclusive set of plans and specifications that more closely meet the owner's budget. The CM can be hired solely for this purpose with an option to extend the contract for services during the construction phase and beyond into postconstruction activities such as establishing and monitoring equipment maintenance schedules.

The services of a CM during construction allow the owner to maintain a professional management team whose goal and allegiance are to look after the owner's interests in those instances when an owner lacks such professional staff. The general contractor, by contrast, actually serves two masters, the owner and his or her own construction company.

The Construction Management Association of America (CMAA) defines CM work as follows:

- A project-delivery system comprised of a program of management services
- A project-delivery system that defines the specific needs of the project and the owner
- A project-delivery system to control time, cost, and quality, applied to a project from conception to completion
- Professional services, under contract, provided to an owner
- A professional selected on the basis of experience and qualifications in the type of project under consideration
- A professional compensated on the basis of a negotiated fee for the scope of services provided

As just mentioned, the services of a CM can be obtained for each phase of the project's development. In predesign, the CM will work with the owner to develop a project-management plan, assist in the selection of an architect, develop a master plan for the entire project, develop a budget and estimate based upon the owner's program, and develop a management-information system if required.

In the design phase the CM will monitor the design consultant's compliance with the owner's program, review the design documents, revise the master schedule if required, prepare estimates as needed, perform any value engineering required, prequalify bidders, advise on contract awards, conduct any pre- or postbid conferences, and develop management-information services (MIS) programs for schedules, cost reporting, and cash flow.

In the construction phase, the CM will provide project management, conduct cost-administration procedures, monitor safety on site, review and comment on change orders, supervise the project's closeout and occupancy procedures, and stay involved through the entire warranty period. In the postconstruction phase, the CM can provide life-cycle analysis of major pieces of HVAC equipment along with a recommended maintenance schedule and assist in the preparation of those maintenance agreements.

In the beginning of our discussion on CMs, we mentioned that fees are lower than those for general contractors because of their limited exposure to risk in some instances and the reimbursement of many expenses that the general contractor includes as part of the lump-sum or GMP contract.

The CM fee does not represent the total cost to an owner. Other costs, called "reimbursable expenses," are billed at a cost-plus add-on known as a "multiplier." Multipliers can ranged from 1 to 2 or higher and relate to the increase of a basic cost to a billable one that includes the CM's overhead and profit. A multiplier of 1.5 means that the CM will bill the actual cost multiplied by 1.5, so a $100 cost will be billed at $150, which ostensibly includes the cost of processing those costs plus a small profit. So it is necessary to ask the CM what multiplier she uses for reimbursable costs and which reimbursable costs will be incurred on this project.

These added costs can be considerable and should be identified during the contract negotiation stage. These costs are related to the field office, site expenses, and maintenance services and can include the following:

- The field office complex, trailers, delivery to the site, removal from the site, security fencing
- Field office equipment: duplicating machines, computers, printers, miscellaneous items such as hole punchers, staplers, telephones, scanners
- Utility connections for telephone, data communication, water, sewer, electricity
- Signage at the site entrance and for interior directions
- Office furniture: desks, chairs, file cabinets, conference tables, whiteboards
- All IT-related costs, Internet connections, monthly charges, wireless costs, miscellaneous like batteries, memory cards
- Supplies for all office equipment, periodic maintenance costs
- Copier, printer, and fax supplies—paper, cartridges
- Monthly utility costs: power, telephone, sanitary, water, gas (if applicable)
- Reproductions: copies of plans, specifications, sketches
- Automobile and trucking expenses and repairs attributable to the project
- Travel expenses
- Security required for office trailers: installation and monthly costs
- Office maintenance and cleaning services, debris removal
- Engineering, site surveys if requested/required
- Initial site-layout survey, interim surveys, final site survey (as-built)
- Testing and inspection services as required
- Shop drawings: receipt, review, processing to architect/engineer, return to appropriate subcontractor/vendor, and follow-up action as required
- As-built drawings (either preparation or review and coordination with architect/engineer and appropriate subcontractors)
- Safety and first aid
- Photographs, videos as required
- Site security, fencing and security patrols, security lighting installation, maintenance, and monthly electrical costs
- Erosion control: installation, maintenance, and removal
- Access roads: installation and maintenance, snow removal
- Fire extinguishers: purchase, replacement, and maintenance

- Personal safety equipment for CM staff: hard hats, ear and eye protection, shoes, boots, raingear
- Portable toilets in number sufficient to accommodate number of workers, on site: delivery, maintenance, removal
- Site cleanup, debris removal, dust control
- Dumpster services if not provided by appropriate subcontractors
- Trash chutes to allow removal of trash from multistoried buildings to central point
- Pest control

These general conditions costs are included as a lump sum in a general contractor's stipulated-sum contract and in the cost-plus and GMP contracts, but unless these costs are capped in the CM contract, they will be accumulated and billed as they occur at the multiplier rate. It is important to discuss and possibly negotiate the terms of the multiplier and even ask for a lump sum for some other costs.

Design-build contract

Owners, contractors, architects, and engineers have always explored project delivery systems that may prove to be more equitable to all parties. The lump-sum, competitively bid project was seen as fraught with dangers when unqualified contractors were allowed to participate in the bidding process. Risk aversion, sought by each party to the construction process, in many cases shifted too much risk from one party to another, resulting in claims, disputes, and litigation. The GMP contract approach provided a more equitable solution, since it is based on documented actual costs, and it caps the final cost of the project (exclusive of any change orders). The CM approach gained favor, in part, because it was more owner-oriented.

Design-build carries this search for the perfect project delivery system a little further, allowing owner, architect, and contractor to work in a team environment rather than as adversaries. As the name implies, this form of contract includes both the design and the construction components, and owners will find themselves dealing with one entity, a design-build firm, instead of contracting separately with an architect and engineer for design and a builder for construction. The design-build team can be created in a number of ways:

- A contractor who has architect(s) on staff or, conversely, an architect who has a staff of construction professionals can form the design-build team. In either case, both design and construction will be performed "in-house."

- A contractor who has a relationship with a design firm may decide to team up with an architectural firm and offer an owner services as a design-build team, generally as a joint-venture or limited-liability corporation (LLC) with the design consultants hired as a subcontractor.

- An architect may form the joint venture or LLC and hire a contractor as the participant.

The design-build process has many advantages that have been reinforced by results in both the private and public sectors. Design-build was virtually unknown in public works projects several decades ago, but it has now become an accepted option.

A survey conducted by Pennsylvania State University's College of Engineering reached the following conclusions:

1. Design-build project unit costs were 4.5 percent lower that those using a CM at risk and 6 percent lower than conventional design-bid-build projects. This may be due to the provision of real-time cost estimates as the design-build project unfolds.

2. Design-build projects increased construction progress 7 percent faster than a CM at risk and 12 percent faster than design-bid-build projects. As the design progresses, the design-builder can order long lead items; equipment such as elevators, heating, and air-conditioning equipment; and even the building's structural system instead of having to wait until the design is completed and a contractor selection and contract award have been made. This results in a just-in-time delivery of key construction components, thereby substantially decreasing the length of construction.

3. When the speed of design is factored into the more rapid construction cycle, design-build was 24 percent faster than a CM at risk system and 33 percent faster than the design-bid-build process. Under a design-build process, changes in design due to budget restraints are mainly held in check, and because design and construction planning and preordering of key materials track along with design, the entire cycle is foreshortened.

Other professed claims of design-builders include additional control over costs impacted by inflation because of the shortened time of the design and construction cycle. The higher cost of construction funding is reduced, since construction time is shortened, and, just as important, the facility is available for occupancy more quickly. And if the project is a rentable commercial project, this shortened time frame allows more rapid occupancy, representing real dollar value.

But before everyone rushes to the door to call a design-builder, you must be aware that not all projects are appropriate for design-build. The following conditions need to be reviewed by an owner before deciding to make that phone call:

- Does the owner have the staff available to adequately extract the building program, define those needs, and present them clearly and precisely to the design-build team?

- Will this team be able to dedicate all of its time to a lengthy, intense process beginning with program development and carrying through to design and construction?

- The design-build process will bypass any type of competitive bidding, and the owner may not be realizing the lowest price in the marketplace.

- Unless the owner has staff experienced in both the design and construction process, he or she may have to hire consultants to perform these functions at added costs, which could be more than inconsequential.

- Not all states have legislation or licensing laws that permit combining architectural services and construction in one firm.

Probably the most important component of the design-build process is the ability to succinctly convey the owner's program to the design-build team. An owner embarking on a new project similar to a previous one is a good candidate for design-build. The selection of the design-build team is another key component.

In most design-build teams, the contractor is the lead team member, but many architect-led design-build teams exist as well. The design-build team must have considerable experience in similar projects; for example, a firm that has successfully completed a half-dozen midrise commercial buildings would be a candidate for your midrise office building, whereas a design-build firm experienced in distribution warehouses would not be.

Bridging contract

Another form of what could be called a "contract" that approaches a construction project obliquely is bridging. Bridging is an option open to owners primarily involved in the design-build process, but it is equally adapted to other forms of project delivery systems. Bridging is a process whereby an owner hires a design firm to produce a set of partially complete design documents (schematics) or design development drawings that can be used to test the market for pricing or when a less than 100 percent commitment has been made to proceed with a new project. With these schematic documents, an owner can control the financial exposure to the project under consideration if he or she is unsure of the economic viability or is uncertain as to the extent of financing available.

By inviting contractors to submit proposals based on these bridging documents, an owner can also elicit suggestions from the contractors to possibly improve or change the design or effect cost savings. Once again, only those contractors who exhibit experience in the type of project under consideration should be invited to submit bids.

The bridging bidding process can be divided into two phases:

- The request for proposal, which includes the bridging documents; bidders submit their qualifications in similar projects and other design-build projects.

- A short list of bidders will be requested to submit proposals to expand the project concept and to provide approximate costs and other pertinent information to show that they understand the owner's program.

Upon receipt of the bids, an owner has several choices: abandon the project, direct the architect to complete the design in anticipation of proceeding to construction via design-build, continue to complete the design with a design-build firm, or hire another architect to develop a modified or entirely new design.

Bridging includes some legal issues. If another architect is engaged, the question of who owns the design will be raised: Who is the architect of record and who assumes liability for the design? These issues should be addressed with the architect employed for the preliminary design and, if the project goes forward, should be reviewed with the design-builder.

The Design-Build Institute of America (DBIA), the Associated General Contractors of America (AGC), and the American Institute of Architects (AIA) are all good sources for a list of design-builders in your area. Owners will generally look for the following qualifications when interviewing design-build firms:

- The firm's financial and bonding capabilities
- The firm's experience in the type of construction being considered
- A track record of successful projects and technical engineering competence
- The experience of key design and construction personnel and support staff
- The firm's overall experience in the design-build process
- The firm's quality control program and administration
- The firm's record of on-budget performance and on-time delivery performance

A design-build selection process can be done in several ways:

- Direct selection: An owner can select a design-build firm after interviewing and vetting the qualifications.
- A negotiated contract can follow one of the standard design-build contract formats available from AIA, AGC, or DBIA.
- Competitive negotiation: A short list of design-build firms will be requested to comply with the owner's program and submit preliminary design criteria, cost, time frame for design and construction, and accomplishments regarding similar projects.
- Weighted criteria: An owner will establish a point system for evaluating proposals, assigning points for various criteria. How well does the design meet my program? How well do the project systems (electrical, HVAC, etc.) fit into my program? How does the cost enter into my selection decision? How is the bidder's conformance to schedule? An owner may establish that design is worth 25 points, systems 15, price 40, schedule 15, and other considerations 5. Evaluating proposals with this point system can provide a rationale for design-builder selection.

THE GENERAL CONDITIONS TO THE CONTRACT FOR CONSTRUCTION

Another standard contract usually attached to the construction contract is known as the general conditions. This document is actually like a "rules and regulations," game plan statement. It defines the roles of the owner, the architect, and the contractor through construction, listing each one's obligations and responsibilities and elaborating on certain events that take place during the process. The AIA Document is known as A201 General Conditions of the Contract for Construction; the AGC includes the general conditions in each type of contract form.

The AIA general conditions document

The AIA document is the most prevalent form of general conditions contract. AIA A201 is divided into 14 parts, called articles, and we will examine each one, pointing out those sections that contain information of importance to an owner. However, it is suggested that an owner read this general conditions document from cover to cover.

Article 1.1, general provisions, discusses the roles of the owner, the contractor, and the architect in administering the contract. This article should be read thoroughly because it contains some important safeguards for owners.

Article 1.2, the "intent" of the plans and specifications, lists all of the items necessary for the completion of the work and what can be "reasonably inferable from them as being necessary to produce the indicated results." An owner may argue that if something is missing from the plans and specs but was "reasonable inferable," the contractor may have to provide that work at no cost.

Article 1.5.2 requires the contractor to visit the site and become familiar with local conditions under which he or she will be operating. For example, there may not be enough room to store all materials on the site, so the contractor must make provisions accordingly. Tight access to the site may preclude deliveries by tractor-trailer, or there may be noise-abatement provisions to follow in an adjoining site(s), precluding work late at night or on weekends.

Article 2.3 notes that the owner has the right to stop the work if the contractor fails to correct faulty work; *Article 2.4* notes that if the contractor fails to correct that work after receiving proper notification, the owner can proceed with corrective action and deduct those costs from any payments due the contractor.

Article 3.2.2 states that the contractor is required to review the plans and specifications and to take field measurements of any existing conditions relating to new work. If he or she discovers any inconsistencies with the plans and specs, he or she must notify the architect promptly. This is particularly important when the project at hand is an addition to an existing building or involves renovation or remodeling of an existing structure.

Article 3.8, which covers allowances is important because it deals with those costs that must be included in an allowance item (See Glossary). Other parts of this article deal with construction schedules, shop drawing submissions, and permits and fees, all important issues.

Article 4.2.1 states that the architect will visit the site at various times and for various purposes, including the right to reject work and conduct routine inspections.

Article 4.3, which covers claims and disputes, should be read in its entirety by an owner to become familiar with steps to take if a contractor raises an issue that may result in a dispute or claim. *Article 4.4* deals with the resolution of a dispute or claim. The first step is mediation, the second is arbitration, and the third is litigation, which should be avoided if at all possible.

Articles 5.1 through 5.4 relate to subcontractor relations, in particular Article 5.4—contingent assignment of subcontractors. If a general contractor defaults and the subcontract agreements are not assigned to an owner, the subcontractors may have no obligation to continue to fulfill their contract with the general contractor on that construction site for work required by their trade. If the contract does not have a subcontractor assignment clause, and if a general contractor default occurs, the owner or the replacement general contractor may need to acquire a new crew of subcontractors if the current group decides to quit work. If this were to occur, the cost of completing the remaining work will most likely increase greatly. But if assignment has been affected, these original subcontractors have cause to fulfill their original contract obligations, with the likelihood that some costs will be controlled. This is not to say that when a contractor defaults, even with existing subcontractors who are willing to assume their contractual relationship with a new contractor, additional costs will not be incurred but will only be a matter of degree.

Article 6.1.1 allows an owner to engage other contractors for other portions of work not under contract with the general contractor. However, a note of caution: If the general contractor is a union contractor, he or she may object to the hiring of a nonunion contractor to work alongside the union crews. It is best to check it out *before* bringing the other contractor on board.

Article 7, which covers changes in the work, should be read all the way through. The method by which a change order is prepared, processed, and resolved is something that most owners will encounter more than once. I have devoted an entire chapter to change orders because of their nearly universal occurrence and complexity.

Article 8 includes progress and completion time definitions, as well as delays and extensions, another important section to read and understand in case the contractor has made the intention known to claim a delay in the project.

Article 9 outlines the procedures for submission and payment of the contractor's requests for payment. It also defines "substantial completion" in Article 9.8, a term often misunderstood by both owner and contractor. Basically, when a project has reached the completion stage where it can be utilized for the purpose for which it was intended, it has reached substantial completion, and this will trigger a payment request by the contractor. Not all of the work must be completed in order to claim substantial completion; not all painting must be completed or some flooring may be missing, but if these items do not impact the occupancy of the building, the project has reached substantial completion.

Article 9.9 also includes the conditions under which an owner may occupy a portion of the building even though it may not have achieved substantial completion. For example, an owner may wish to install kitchen equipment in the employees' lounge while that area is under construction or begin to install low-voltage data communications wiring before areas are painted, ceilings installed, and so forth.

This section establishes the rules for such partial occupancy. Final completion and final payment are also included in this section.

Article 10 contains information regarding protection of persons and property and safety concerns. Hazardous material discovery, handling, responsibility (usually excluded from a general contractor's scope of work), and procedures are included in this article.

Article 11 covers insurance and bond matters, which are generally spelled out more specifically in the contract for construction or included in the specification manual.

Article 12 discusses correction of work deemed unsatisfactory by the architect and should be required reading for the owner.

Article 13 includes a miscellaneous provisions section relating to legal issues, rights and remedies, successors and assigns, tests and inspections, and statutory limitations.

Article 14 describes the ways in which the contract can be suspended or terminated; if an owner gets to that point, however, surely an attorney will take over and review this section.

STANDARD CONTRACT FORMS

The American Institute of Architects, the Associated General Contractors of America, the Design-Build Institute of America, and the Construction Management Association of America offer many different forms of standard contracts. Although an owner's architect or engineer will provide them to the client, an owner may also wish to purchase a copy beforehand to become more familiar with its contents. You can find all of the standard forms on the organization's website. The next sections discuss the most frequently used forms.

American Institute of Architects contracts

By far the most widely employed contracts are those issued by the American Institute of Architects. The A series contains owner-contractor contracts, and the B series consists of owner-architect contracts. A listing and synopsis of these two types of AIA contracts are included at the end of this chapter.

Associated General Contractors of America contracts

In September 2007, the AGC issued a series of construction documents that included comments from 23 industry organizations representing, owners, contractors, subcontractors, sureties, and designers. These documents were identified by AGC as ConsensusDOCs. Here are some of them:

No. 200, Owner/Contractor Agreement and General Conditions (Lump Sum), a stipulated or lump-sum contract that includes general conditions

No. 205, Owner/Contractor Agreement, Short Form (Lump Sum), similar to the preceding but for small, limited-scope projects

No. 22, Contractor's State of Qualifications for Specific Project, similar to AIA Contractor Qualification Statement

No. 222, Architect/Engineers Statement of Qualifications, similar to AIA's architect qualifications statement but includes the engineers qualification statement

No. 230, Owner/CM Agreement and General Conditions (Cost of Work), a contract between an owner and a CM where work is to be performed on a cost plus a CM fee basis

No. 250, Owner/Contractor Agreement and General Conditions for a GMP, a contract at cost plus a fee with a guaranteed maximum price, with general conditions included

No. 260, Performance Bond

No. 261, Payment Bond

No. 262, Bid Bond

No. 270, Instructions to Bidders on Private Work, lists items to be addressed when contractors are to submit bid proposals

No. 291, Application for Payment for GMP project

No. 292, Application for Payment for Lump Sum

No. 293, Schedule of Values, similar to AIA G703 Continuation Sheet

Construction Management Association of America contracts

These are the relevant forms:

A-1, Standard Form of Agreement Between Owner and CM with CM as Owner's Agent

A-2, Standard Form of Agreement Between Owner and Contractor

A-3, General Conditions Between Owner and Contractor

A-4, Standard Form of Agreement Between Owner and Designer

Design-Build Institute of America contracts

These are the most commonly used DBIA contracts:

Document 501, Contract for Design-Build Consultant Services

Document 520, Standard Form of Preliminary Agreement Between Owner and Design-Builder

Document 525, Standard Form of Agreement Between Owner and Design-Builder, payment is lump-sum

Document 530, Standard Form of Agreement Between Owner and Design-Builder, payment is cost plus a fee with option for GMP

Document 535, Standard Form of General Conditions of Contract Between Owner and Design-Builder.

Documents 540, 550, 555, 560, 565, and 570 pertain to contracts between a Design-Builder and either a general contractor or a subcontractor

DOCUMENTS SYNOPSES BY SERIES

The documents listed in these synopses are organized according to their letter series, a system of classification that refers to the specific purpose of each document.

Reproduced with permission of The American Institute of Architects, 1735 New York Avenue, Washington, D.C. 20006

A series

A101–2007 (formerly A101–1997) Standard form of agreement between owner and contractor where the basis of payment is a stipulated sum

AIA Document A101-2007 is a standard form of agreement between owner and contractor for use where the basis of payment is a stipulated sum (fixed price). A101 adopts by reference, and is designed for use with, AIA Document A201-2007, General Conditions of the Contract for Construction. A101 is suitable for large or complex projects. For projects of a more limited scope, use of AIA Document A107-2007, Agreement Between Owner and Contractor for a Project of Limited Scope, should be considered. For even smaller projects, consider AIA Document A105-2007, Agreement Between Owner and Contractor for a Residential or Small Commercial Project.

A101CMa–1992 Standard form of agreement between owner and contractor where the basis of payment is a stipulated sum, construction manager-adviser edition

AIA Document A101CMa-1992 is a standard form of agreement between owner and contractor for use on projects where the basis of payment is a stipulated sum (fixed price), and where, in addition to the contractor and the architect, a construction manager assists the owner in an advisory capacity during design and construction. The document has been prepared for use with AIA Document A201CMa-1992, General Conditions of the Contract for Construction, Construction Manager-Adviser Edition. This integrated set of documents is appropriate for use on projects where the construction manager only serves in the capacity of an adviser to the owner, rather than as constructor (the latter relationship being represented in AIA Documents A121CMc-1991 and A131CMc-1991). A101CMa-1992 is suitable for projects where the cost of construction has been predetermined, either by bidding or by negotiation.

A102–2007 (formerly A111™–1997) Standard form of agreement between owner and contractor where the basis of payment is the cost of the work plus a fee with a guaranteed maximum price

This standard form of agreement between owner and contractor is appropriate for use on large projects requiring a guaranteed maximum price, when the basis of payment to the contractor is the cost of the work plus a fee. AIA Document A102-2007 is not intended for use in competitive bidding. A102-2007 adopts by reference and is intended for use with AIA Document A201-2007, General Conditions of the Contract for Construction.

A103–2007 (formerly A114™–2001) Standard form of agreement between owner and contractor where the basis of payment is the cost of the work plus a fee without a guaranteed maximum price

AIA Document A103-2007 is appropriate for use on large projects when the basis of payment to the contractor is the cost of the work plus a fee, and the cost is not fully known at the commencement of construction. A103-2007 is not intended for use in competitive bidding. A103-2007 adopts by reference, and is intended for use with, AIA Document A201-2007, General Conditions of the Contract for Construction.

A105–2007 (formerly A105–1993 and A205–1993) Standard form of agreement between owner and contractor for a residential or small commercial project

AIA Document A105-2007 is a stand-alone agreement with its own general conditions; it replaces A105-1993 and A205-1993. A105-2007 is for use on a project that is modest in size and brief in duration, and where payment to the contractor is based on a stipulated sum (fixed price). For larger and more complex projects, other AIA agreements are more suitable, such as AIA Document A107-2007, Agreement Between Owner and Contractor for a Project of Limited Scope. A105-2007 and B105-2007, Standard Form of Agreement Between Owner and Architect for a Residential or Small Commercial Project, comprise the Small Projects family of documents. Although A105-2007 and B105-2007 share some similarities with other agreements, the Small Projects family should *not* be used in tandem with agreements in other document families without careful side-by-side comparison of contents.

A107–2007 (formerly A107–1997) Standard form of agreement between owner and contractor for a project of limited scope

AIA Document A107-2007 is a stand-alone agreement with its own internal general conditions and is intended for use on construction projects of limited scope. It is intended for use on medium-to-large sized projects where payment is based on either a stipulated sum or the cost of the work plus a fee, with or without a guaranteed maximum price. Parties using A107-2007 will also use A107 Exhibit A, if using a cost-plus payment method. AIA Document B104-2007, Standard Form of Agreement Between Owner and Architect for a Project of Limited Scope, coordinates with A107-2007 and incorporates it by reference.

For more complex projects, parties should consider using one of the following other owner/contractor agreements: AIA Document A101-2007, A102-2007, or A103™-2007. These agreements are written for a stipulated sum, cost of the work with a guaranteed maximum price, and cost of the work without a guaranteed maximum price, respectively. Each of them incorporates by reference AIA Document A201-2007, General Conditions of the Contract for Construction. For single-family residential projects or smaller and less complex commercial projects, parties may wish to consider AIA Document A105-2007, Agreement Between Owner and Contractor for a Residential or Small Commercial Project.

A121CMc–2003 Standard form of agreement between owner and construction manager where the construction manager is also the constructor (agc document 565)

AIA Document A121CMc-2003 is intended for use on projects where a construction manager, in addition to serving as adviser to the owner, assumes financial responsibility for construction of the project. The construction manager provides the owner with a guaranteed maximum price proposal, which the owner may accept, reject, or negotiate. Upon the owner's acceptance of the proposal by execution of an amendment, the construction manager becomes contractually bound to provide labor and materials for the project. The document divides the construction manager's services into two phases: the preconstruction phase and the construction phase, portions of which may proceed concurrently in order to fast track the process. A121CMc-2003 is coordinated for use with AIA Documents A201-1997, General Conditions of the Contract for Construction, and B151-1997, Standard Form of Agreement Between Owner and Architect.

Caution: To avoid confusion and ambiguity, do not use this construction management document with any other AIA or AGC construction management document.

A131CMc–2003 Standard form of agreement between owner and construction manager, where the construction manager is also the constructor and where the basis of payment is the cost of the work plus a fee and there is no guarantee of cost (agc document 566)

Similar to AIA Document A121CMc-1991, this construction manager as constructor agreement is intended for use when the owner seeks a construction manager who will take on responsibility for providing the means and methods of construction. However, in AIA Document A131CMc-2003, the construction manager does not provide a Guaranteed Maximum Price (GMP). A131CMc-2003 employs the cost-plus-a-fee method, wherein the owner can monitor costs through periodic review of a control estimate that is revised as the project proceeds. The agreement divides the construction manager's services into two phases: the preconstruction phase and the construction phase, portions of which may proceed concurrently in order to fast track the process. A131CMc-2003 is coordinated for use with AIA Documents A201-1997, General Conditions of the Contract for Construction, and B151-1997, Standard Form of Agreement Between Owner and Architect.

Caution: To avoid confusion and ambiguity, do not use this construction management document with any other AIA or AGC construction management document.

A141–2004 Agreement between owner and design-builder

AIA Document A141-2004 replaces A191-1996 and consists of the agreement and three exhibit: Exhibit A, Terms and Conditions; Exhibit B, Determination of the Cost of the Work; and Exhibit C, Insurance and Bonds. Exhibit B is not applicable if the parties select to use a stipulated sum. A141-2004 obligates the design-builder to execute fully the work required by the design-build documents, which include

A141-2004 with its attached exhibits, the project criteria, and the design-builder's proposal, including any revisions to those documents accepted by the owner, supplementary and other conditions, addenda and modifications. The Agreement requires the parties to select the payment type from three choices: (1) Stipulated Sum, (2) Cost of the Work Plus Design-Builder's Fee, and (3) Cost of the Work Plus Design-Builder's Fee with a Guaranteed Maximum Price. A141-2004 with its attached exhibits forms the nucleus of the design-build contract. Because A141-2004 includes its own terms and conditions, it does not use A201-1997.

A142–2004 Agreement between design-builder and contractor

AIA Document A142-2004 replaces A491-1996 and consists of the agreement and five exhibits: Exhibit A, Terms and Conditions; Exhibit B, Preconstruction Services; Exhibit C, Contractor's Scope of Work; Exhibit D, Determination of the Cost of the Work; and Exhibit E, Insurance and Bonds. Unlike B491-1996, A142-2004 does not rely on AIA Document A201 for its general conditions of the contract. A142-2004 contains its own terms and conditions. A142-2004 obligates the contractor to perform the work in accordance with the contract documents, which include A142-2004 with its attached exhibits, supplementary and other conditions, drawings, specifications, addenda, and modifications. Like AIA Document A141-2004, A142-2004 requires the parties to select the payment type from three choices: (1) Stipulated Sum, (2) Cost of the Work Plus Design-Builder's Fee, and (3) Cost of the Work Plus Design-Builder's Fee with a Guaranteed Maximum Price.

A151–2007 (formerly A175ID–2003) Standard form of agreement between owner and vendor for furniture, furnishings, and equipment, where the basis of payment is a stipulated sum

AIA Document A151-2007 is intended for use as the contract between Owner and Vendor for Furniture, Furnishings, and Equipment (FF&E), where the basis of payment is a stipulated sum (fixed price) agreed to at the time of contracting. A151-2007 adopts by reference and is intended for use with AIA Document A251-2007, General Conditions of the Contract for Furniture, Furnishings, and Equipment. It may be used in any arrangement between the owner and the contractor where the cost of FF&E has been determined in advance, either through bidding or negotiation.

A195–2008 Standard form of agreement between owner and contractor for integrated project delivery

AIA Document A195-2008 is a standard form of agreement between owner and contractor for a project that utilizes Integrated Project Delivery. A195-2008 primarily provides only the business terms and conditions unique to the agreement between the owner and contractor, such as compensation details and licensing of instruments of service. A195-2008 does not include the specific scope of the contractor's work; rather, it incorporates by reference AIA Document A295-2008, General Conditions of the Contract for Integrated Project Delivery, which sets forth the contractor's duties and obligations for each of the six phases of the project, along with the duties and obligations of the owner and architect. Under A195-2008 the contractor provides a guaranteed maximum price. For that purpose, the agreement includes a guaranteed maximum price amendment at Exhibit A.

A201–2007 (formerly A201–1997) General conditions of the contract for construction

The General Conditions are an integral part of the contract for construction for a large project and they are incorporated by reference into the owner/contractor agreement. They set forth the rights, responsibilities, and relationships of the owner, contractor, and architect. Though not a party to the contract for construction between owner and contractor, the architect participates in the preparation of the contract documents and performs construction phase duties and responsibilities described in detail in the general conditions. AIA Document A201-2007 is adopted by reference in owner/architect, owner/contractor, and contractor/subcontractor agreements in the A201 family of documents; thus, it is often called the "keystone" document.

A201CMa–1992 General conditions of the contract for construction, construction manager-adviser edition

AIA Document A201CMa-1992 sets forth the rights, responsibilities, and relationships of the owner, contractor, construction manager, and architect. A201CMa-1992 is adopted by reference in owner/architect, owner/contractor, and owner/construction manager agreements in the CMa family of documents. Under A201CMa-1992, the construction manager serves as an independent adviser to the owner, who enters into multiple contracts with prime trade contractors.

Caution: Do not use A201CMa-1992 in combination with agreements where the construction manager takes on the role of constructor, gives the owner a guaranteed maximum price, or contracts directly with those who supply labor and materials for the project, such as AIA Document A121CMc-2003 or A131CMc-2003.

A201SC–1999 Federal supplementary conditions of the contract for construction

AIA Document A201SC-1999 is intended for use on certain federally assisted construction projects. For such projects, A201SC-1999 adapts AIA Document A201-1997 by providing (1) necessary modifications of the General Conditions, (2) additional conditions, and (3) insurance requirements for federally assisted construction projects.

A251–2007 (formerly A275ID–2003) General conditions of the contract for furniture, furnishings, and equipment

AIA Document A251-2007 provides general conditions for the AIA Document A151-2007, Standard Form Agreement between Owner and Vendor for Furniture, Furnishings, and Equipment, where the basis of payment is a stipulated sum. A251-2007 sets forth the duties of the owner, architect, and vendor, just as AIA Document A201-2007, General Conditions of the Contract for Construction, does for building construction projects. Because the Uniform Commercial Code (UCC) governs the sale of goods and has been adopted in nearly every jurisdiction, A251–2007 recognizes the commercial standards set forth in Article 2 of the UCC, and uses certain standard UCC terms and definitions. A251–2007 was renumbered in 2007 and modified, as applicable, to coordinate with A201–2007.

A295–2008 General conditions of the contract for integrated project delivery

AIA Document A295–2008, provides the terms and conditions for AIA Documents B195–2008, Standard Form of Agreement Between Owner and Architect for Integrated Project Delivery, and A195–2008 Standard Form of Agreement Between Owner and Contractor for Integrated Project Delivery, both of which incorporate A295–2008 by reference. Those agreements provide primarily only business terms and rely upon A295–2008 for the architect's services, the contractor's preconstruction services, and the conditions of construction. A295–2008 not only establishes the duties of the owner, architect, and contractor, but also sets forth in detail how they will work together through each phase of the project: Conceptualization, Criteria Design, Detailed Design, Implementation Documents, Construction, and Closeout. A295–2008 requires that the parties utilize building information modeling.

A305–1986 Contractor's qualification statement

An owner preparing to request bids or to award a contract for a construction project often requires a means of verifying the background, references, and financial stability of any contractor being considered. These factors, along with the time frame for construction, are important for an owner to investigate. Using AIA Document A305–1986, the contractor may provide a sworn, notarized statement and appropriate attachments to elaborate on important aspects of the contractor's qualifications.

A310–1970 Bid bond

This simple, one-page form establishes the maximum penal amount that may be due to the owner if the selected bidder fails to execute the contract and/or fails to provide any required performance and payment bonds.

A312–1984 Performance bond and payment bond

This form incorporates two bonds: one covering the contractor's performance and one covering the contractor's obligations to pay subcontractors and others for material and labor. In addition, AIA Document A312–1984 obligates the surety to act responsively to the owner's requests for discussions aimed at anticipating or preventing a contractor's default.

A401–2007 (formerly A401–1997) Standard form of agreement between contractor and subcontractor

This agreement establishes the contractual relationship between the contractor and subcontractor. It sets forth the responsibilities of both parties and lists their respective obligations, which are written to parallel AIA Document A201–2007,

General Conditions of the Contract for Construction, which A401–2007 incorporates by reference. A401–2007 may be modified for use as an agreement between the subcontractor and a sub-subcontractor, and must be modified if used where AIA Document A107–2007 or A105–2007 serves as the owner/contractor agreement.

A441–2008 Standard form of agreement between contractor and subcontractor for a design-build project

AIA Document A441–2008 is a fixed-price agreement that establishes the contractual relationship between the contractor and subcontractor in a design-build project. A441–2008 incorporates by reference the terms and conditions of AIA Document A142–2004, Standard Form of Agreement Between Design-Builder and Contractor, and was written to ensure consistency with the AIA 2004 Design-Build family of documents. Because subcontractors are often required to provide professional services on a design-build project, A441–2008 provides for that possibility.

A503–2007 (formerly A511™–1999) Guide for supplementary conditions

AIA Document A503–2007 is not an agreement, but is a guide containing model provisions for modifying and supplementing AIA Document A201-2007, General Conditions of the Contract for Construction. It provides model language with explanatory notes to assist users in adapting A201–2007 to specific circumstances. A201–2007, as a standard form document, cannot cover all the particulars of a project. Thus, A503–2007 is provided to assist A201–2007 users either in modifying it, or developing a separate supplementary conditions document to attach to it.

A511CMa–1993 Guide for supplementary conditions, construction manager-adviser edition

Similar to A503–2007, AIA Document A511CMa–1993 is a guide for amending or supplementing the general conditions document A201CMa–1992. A511CMa–1993 should only be employed, as should A201CMa–1992, on projects where the construction manager is serving in the capacity of adviser to the owner (as represented by the CMa document designation), and not in situations where the construction manager is also the constructor (CMc document-based relationships). Like A503–2007, this document contains suggested language for supplementary conditions, along with notes on appropriate usage.

A701–1997 Instructions to bidders

This document is used when competitive bids are to be solicited for construction of the project. Coordinated with AIA Document A201, General Conditions of the Contract for Construction, and its related documents, A701–1997 provides instructions on procedures, including bonding requirements, for bidders to follow in preparing and submitting their bids. Specific instructions or special requirements, such as the amount and type of bonding, are to be attached to, or inserted into, A701–1997.

A751–2007 (formerly A7751D–2003) Invitation and instructions for quotation for furniture, furnishings, and equipment

AIA Document A751-2007 provides (1) the Invitation for Quotation for Furniture, Furnishings, and Equipment (FF&E) and (2) Instructions for Quotation for Furniture, Furnishings, and Equipment. These two documents define the owner's requirements for a vendor to provide a complete quotation for the Work. The purchase of FF&E is governed by the Uniform Commercial Code (UCC), and A751-2007 has been developed to coordinate with the provisions of the UCC.

B Series

B101–2007 (formerly B151–1997) Standard form of agreement between owner and architect

AIA Document B101-2007 is a one-part standard form of agreement between owner and architect for building design and construction contract administration. B101-2007 was developed to replace AIA Documents B141-1997, Parts 1 and 2, and B151-1997, but it more closely follows the format of B151-1997. Services are divided traditionally into Basic and Additional Services. Basic Services are performed in five phases: Schematic Design, Design Development, Construction Documents, Bidding and Negotiation, and Construction. This agreement may be used with a variety of compensation methods, including percentage of construction cost and stipulated sum. B101-2007 is intended to be used in conjunction with A201-2007, General Conditions of the Contract for Construction, which it incorporates by reference.

B102–2007 (formerly B141–1997 Part 1) Standard form of agreement between owner and architect without a predefined scope of architect's services

AIA Document B102-2007 replaces and serves the same purpose as B141-1997 Part 1. B102-2007 is a standard form of agreement between owner and architect that contains terms and conditions and compensation details. B102-2007 does not include a scope of Architect's services, which must be inserted in Article 1 or attached as an exhibit. The separation of the scope of services from the owner/architect agreement allows users the freedom to append alternative scopes of services. AIA standard form scopes of services documents that may be paired with B102-2007 include AIA Documents B203-2007, Site Evaluation and Planning; B204-2007, Value Analysis; B205-2007, Historic Preservation; B206-2007, Security Evaluation and Planning; B209-2007, Construction Contract Administration; B210-2007, Facility Support Services; B211-2007, Commissioning; B214-2007, LEED® Certification; B252-2007, Architectural Interior Design; and B253-2007, Furniture, Furnishings, and Equipment Design.

B103–2007 Standard form of agreement between owner and architect for a large or complex project

AIA Document B103-2007 is a standard form of agreement between owner and architect intended for use on large or complex projects. B103-2007 was developed to replace AIA Documents B141-1997, Parts 1 and 2, and B151-2007 specifically with respect to large or complex projects. B103-2007 assumes that the owner will retain third parties to provide cost estimates and project schedules, and may implement fast-track, phased, or accelerated scheduling. Services are divided along the traditional lines of Basic and Additional Services. Basic Services are based on five phases: Schematic Design, Design Development, Construction Documents, Bidding and Negotiation, and Construction. The architect does not prepare cost estimates, but designs the project to meet the owner's budget for the cost of the work at the conclusion of the Design Development Phase Services. This document may be used with a variety of compensation methods. B103-2007 is intended to be used in conjunction with AIA Document A201-2007, General Conditions of the Contract for Construction, which it incorporates by reference.

B104–2007 Standard form of agreement between owner and architect for a project of limited scope

AIA Document B104-2007 is a standard form of agreement between owner and architect intended for use on medium-sized projects. B104-2007 is an abbreviated version of AIA Document B101-2007. B104-2007 contains a compressed form of Basic Services with three phases: Design, Construction Documents, and Construction. This document may be used with a variety of compensation methods. B104-2007 is intended to be used in conjunction with AIA Document A107-2007, Standard Form of Agreement Between Owner and Contractor for a Project of Limited Scope, which it incorporates by reference.

B105–2007 (formerly B155–1993) Standard form of agreement between owner and architect for a residential or small commercial project

AIA Document B105-2007 is a standard form of agreement between owner and architect intended for use on a residential or small commercial project that is modest in size and brief in duration. B105-2007 and AIA Document A105-2007, Standard Form of Agreement Between Owner and Contractor for a Residential or Small Commercial Project, comprise the Small Projects family of documents. B105-2007 is intended for use with A105-2007, which it incorporates by reference. B105-2007 is extremely abbreviated and is formatted more informally than other AIA agreements. Although A105-2007 and B105-2007 share some similarities with other AIA agreements, the Small Projects family should NOT be used with other AIA document families without careful side-by-side comparison of contents.

B141CMa–1992 Standard form of agreement between owner and architect, construction manager-adviser edition

AIA Document B141CMa-1992 is a standard form of agreement between owner and architect for use on building projects where construction management services are to be provided under a separate contract with the owner. It is coordinated with AIA Document B801CMa-1992, an owner/construction manager-adviser agreement where the construction manager is an independent, professional adviser to the owner throughout the course of the project. Both B141CMa–1992 and B801CMa–1992 are based on the premise that one or more separate construction contractors will also contract with the owner. The owner/contractor agreement is jointly administered by the architect and the construction manager under AIA Document A201CMa-1992, General Conditions of the Contract for Construction, Construction Manager-Adviser Edition.

B142–2004 Agreement between owner and consultant where the owner contemplates using the design-build method of project delivery

AIA Document B142-2004 provides a standard form for the upfront services an owner may require when considering design-build delivery. The consultant, who may or may not be an architect or other design professional, may perform a wide ranging array of services for the owner, including programming and planning, budgeting and cost estimating, project criteria development services, and many others, commencing with initial data gathering and continuing through to post occupancy. B142-2004 consists of the agreement portion and two exhibits, Exhibit A, Initial Information, and Exhibit B, Standard Form of Consultant's Services. Exhibit B provides a menu of briefly described services that the parties can select and augment to suit the needs of the project.

B143–2004 Agreement between design-builder and architect

AIA Document B143-2004 replaces AIA Document B901-1996 and establishes the contractual relationship between the design-builder and its architect. B143-2004 consists of the Agreement, Exhibit A, Initial Information and Exhibit B, Standard Form of Architect's Services. Exhibit B provides a menu of briefly described services that the parties can select and augment to suit the needs of the project.

B144ARCH-CM–1993 Standard form of amendment to the agreement between owner and architect where the architect provides construction management services as an adviser to the owner

AIA Document B144ARCH-CM-1993 is an amendment to AIA Document B141-1997 for use in circumstances where the architect, already under contract to perform architectural services for the owner, agrees to provide the owner with a package of construction management services to expand upon, blend with, and supplement the architect's design and construction contract administration services described in B141-1997.

B152–2007 (formerly B171ID–2003) Standard form of agreement between owner and architect for architectural interior design services

AIA Document B152-2007 is a standard form of agreement between the owner and architect for design services related to Furniture, Furnishings, and Equipment (FF&E) as well as to architectural interior design. B152-2007 divides the architect's services into eight phases: Programming, Pre-lease Analysis and Feasibility, Schematic Design, Design Development, Contract Documents, Bidding and Quotation, Construction

Phase Services, and FF&E Contract Administration. B152-2007 was re-numbered in 2007 and modified to align, as applicable, with AIA Documents B101-2007 and A201-2007. B152-2007 is intended for use in conjunction with AIA Documents A251-2007, General Conditions of the Contract for Furniture, Furnishings, and Equipment, and A201-2007, General Conditions of the Contract for Construction, both of which it incorporates by reference.

B153–2007 (formerly B175ID–2003) Standard form of agreement between owner and architect for furniture, furnishings, and equipment design services

AIA Document B153-2007 is a standard form of agreement between the owner and architect for design services related solely to Furniture, Furnishings, and Equipment (FF&E). B153-2007 divides the architect's services into six phases: Programming, Schematic Design, Design Development, Contract Documents, Quotation, and FF&E Contract Administration. B153-2007 was renumbered in 2007 and modified to align, as applicable, with AIA Document B101-2007. B153-2007 is intended for use in conjunction with AIA Document A251-2007, General Conditions of the Contract for Furniture, Furnishings, and Equipment, which it incorporates by reference.

B161–2002 (formerly B611INT–2002) Standard form of agreement between client and consultant for use where the project is located outside the united states

AIA Document B161-2002 is designed to assist U.S. architects involved in projects based in foreign countries, where the U.S. architect is hired on a consulting basis for design services and the owner will retain a local architect in the foreign country. The document is intended to clarify the assumptions, roles, responsibilities, and obligations of the parties; to provide a clear, narrative description of services; and to facilitate, strengthen, and maintain the working and contractual relationship between the parties. Because of foreign practices, the term Owner has been replaced with Client throughout the document. Also, since it is assumed that the U.S. architect is not licensed to practice architecture in the foreign country where the project is located, the term Consultant is used throughout the document to refer to the U.S. architect. B161-2002 was re-numbered only in 2007; its content remains the same as in AIA Document B611INT-2002.

B162–2002 (formerly B621INT–2002) Abbreviated standard form of agreement between client and consultant for use where the project is located outside the united states

AIA Document B162-2002 is an abbreviated version of AIA Document B161-2002, Standard Form of Agreement between Client and Consultant. The document is designed to assist U.S. architects involved in projects based in foreign countries where the U.S. architect is hired on a consulting basis for design services and a local architect will be retained. The document is intended to clarify the assumptions, roles, responsibilities, and obligations of the parties; to provide a clear, narrative description of services; and to facilitate, strengthen, and maintain the working and contractual relationship between the parties. Because of foreign practices, the term Owner has been replaced with Client throughout the document. Also, since it is assumed that the U.S. architect is not licensed to practice architecture in the foreign country where the project is located, the term Consultant is used throughout the document to refer to the U.S. architect. B162-2002 was renumbered only in 2007; its content remains the same as in AIA Document B621INT-2002.

B163–1993 Standard form of agreement between owner and architect with descriptions for designated services and terms and conditions

AIA Document B163-1993 is discontinued and will be available only until May 31, 2009. This three-part document contains a thorough list of 83 possible services divided among nine phases, covering predesign through supplemental services. This detailed classification allows the architect to estimate more accurately the time and personnel costs required for a particular project. Owner and architect benefit from the ability to establish clearly the scope of services required for the project as responsibilities and compensation issues are negotiated and defined. The architect's compensation may be calculated on a time/cost basis through use of the worksheet provided in the instructions to B163-1993.

B181–1994 Standard form of agreement between owner and architect for housing services

This document, developed with the assistance of the U.S. Department of Housing and Urban Development and other federal housing agencies, is primarily intended for use in multiunit housing design. AIA Document B181-1994 requires that the owner (and not the architect) furnish cost-estimating services. B181-1994 is coordinated with and adopts by reference AIA Document A201-1997, General Conditions of the Contract for Construction.

B188–1996 Standard form of agreement between owner and architect for limited architectural services for housing projects

AIA Document B188-1996 is intended for use in situations where the architect will provide limited architectural services for a development housing project. It anticipates that the owner will have extensive control over the management of the project, acting in the capacity of a developer or speculative builder of a housing project. As a result, the owner or consultants retained by the owner will likely provide the engineering services, specify the brand names of materials and equipment, and administer payments to contractors, among other project responsibilities. B188-1996 is not coordinated for use with any other AIA standard form documents.

B195–2008 Standard form of agreement between owner and architect for integrated project delivery

AIA Document B195-2008 is a standard form of agreement between owner and architect for a project that utilizes Integrated Project Delivery. B195-2008 primarily provides only the business terms unique to the agreement between the owner and architect, such as compensation details and licensing of instruments of service. B195-2008 does not include the specific scope of the architect's services, but rather it incorporates by reference AIA Document A295-2008, General Conditions of the Contract for Integrated Project Delivery, which sets forth the architect's duties and scope of services for each of the six phases of the project, along with the duties and obligations of the owner and contractor.

B201–2007 (formerly B141–1997 Part 2) Standard form of architect's services: design and construction contract administration

AIA Document B201-2007 replaces AIA Document B141-1997 Part 2. B201-2007 defines the architect's traditional scope of services for design and construction contract administration in a standard form that the owner and architect can modify to suit the needs of the project. The services set forth in B201-2007 parallel those set forth in AIA Document B101-2007: The traditional division of services into Basic and Additional Services, with five phases of Basic Services. B201-2007 may be used in two ways: (1) incorporated into the owner/architect agreement as the architect's sole scope of services or in conjunction with other scope of services documents, or (2) attached to AIA Document G802-2007, Amendment to the Professional Services Agreement, to create a modification to an existing owner/architect agreement. B201-2007 is a scope of services document only and may not be used as a stand-alone owner/architect agreement.

B203–2007 (formerly B203–2005) Standard form of architect's services: site evaluation and planning

AIA Document B203-2007 is intended for use where the architect provides the owner with services to assist in site selection for a project. Under this scope, the architect's services may include analysis of the owner's program and alternative sites, site utilization studies, and other analysis, such as planning and zoning requirements, site context, historic resources, utilities, environmental impact, and parking and circulation. B203-2007 may be used in two ways: (1) incorporated into the owner/architect agreement as the architect's sole scope of services or in conjunction with other scope of services documents, or (2) attached to AIA Document G802-2007, Amendment to the Professional Services Agreement, to create a modification to an existing owner/architect agreement. B203-2007 is a scope of services document only and may not be used as a stand-alone owner/architect agreement. B203-2007 was revised in 2007 to align, as applicable, with B101-2007.

B204–2007 (formerly B204–2004) Standard form of architect's services: value analysis

AIA Document B204-2007 establishes duties and responsibilities when the owner has employed a Value Analysis Consultant. This document provides the architect's services in three categories: Pre-Workshop Services, Workshop Services and Post-Workshop Services. The services include presenting the project's goals and design rationale at the Value Analysis Workshop, reviewing and evaluating each Value

Analysis Proposal, and preparing a Value Analysis Report for the owner that, among other things, advises the owner of the estimate of the cost of the work resulting from the implementation of the accepted Value Analysis Proposals. B204-2007 may be used in two ways: (1) incorporated into the owner/architect agreement as the architect's sole scope of services or in conjunction with other scope of services documents, or (2) attached to AIA Document G802-2007, Amendment to the Professional Services Agreement, to create a modification to an existing owner/architect agreement. B204-2007 is a scope of services document only and may not be used as a stand-alone owner/architect agreement. B204-2007 was revised in 2007 to align, as applicable, with AIA Document B101-2007.

B205–2007 (formerly B205–2004) Standard form of architect's services: historic preservation

AIA Document B205-2007 establishes duties and responsibilities where the architect provides services for projects that are historically sensitive. The range of services the architect provides under this scope spans the life of the project and may require the architect to be responsible for preliminary surveys, applications for tax incentives, nominations for landmark status, analysis of historic finishes, and other services specific to historic preservation projects. B205-2007 may be used in two ways: (1) incorporated into the owner/architect agreement as the architect's sole scope of services or in conjunction with other scope of services documents, or (2) attached to AIA Document G802-2007, Amendment to the Professional Services Agreement, to create a modification to an existing owner/architect agreement. B205-2007 is a scope of services document only and may not be used as a stand-alone owner/architect agreement. B205-2007 was revised in 2007 to align, as applicable, with B101-2007.

B206–2007 (formerly B206–2004) Standard form of architect's services: security evaluation and planning

AIA Document B206-2007 establishes duties and responsibilities where the architect provides services for projects that require greater security features and protection than would normally be incorporated into a building design. This scope requires the architect to identify and analyze the threats to a facility, survey the facility with respect to those threats, and prepare a Risk Assessment Report. Following the owner's approval of the Report, the architect prepares design documents and a Security Report. B206-2007 may be used in two ways: (1) incorporated into the owner/architect agreement as the architect's sole scope of services or in conjunction with other scope of services documents, or (2) attached to AIA Document G802-2007, Amendment to the Professional Services Agreement, to create a modification to an existing owner/architect agreement. B206-2007 is a scope of services document only and may not be used as a stand-alone owner/architect agreement. B206-2007 was revised in 2007 to align, as applicable, with AIA Document B101-2007.

B207–2008 (formerly B352–2000) Standard form of architect's services: on-site project representation

AIA Document B207-2008 establishes the architect's scope of services when the architect provides an onsite project representative during the construction phase. B207-2008 provides for agreement on the number of architect's representatives to be stationed at the project site, a schedule for the onsite representation, and the services that the onsite representative will perform. The onsite representative's services include attending job-site meetings, monitoring the contractor's construction schedule, observing systems and equipment testing, preparing a log of activities at the site, and maintaining onsite records. The owner will provide an onsite office for the architect's onsite representative. B207-2008 is a scope of services document only and may not be used as a stand-alone owner/architect agreement. B207-2008 replaces AIA Document B352-2000.

B209–2007 (formerly B209–2005) Standard form of architect's services: construction phase administration

AIA Document B209-2007 establishes duties and responsibilities when an architect provides only Construction Phase services and the owner has retained another architect for design services. This scope requires the architect to perform the traditional contract administration services while design services are provided by another architect. B209-2007 may be used in two ways: (1) incorporated into the owner/architect agreement as the architect's sole scope of services or in conjunction with other scope of services documents, or (2) attached to AIA Document G802-2007, Amendment to the Professional Services Agreement, to create a modification to an existing owner/architect agreement. B209-2007 is a scope of services document only and may not be used as a stand-alone owner/architect agreement. B209-2007 was revised in 2007 to align, as applicable, with AIA Document B101-2007.

B210–2007 (formerly B210–2004) Standard form of architect's services: facility support service

AIA Document B210-2007 focuses attention on providing the owner with means and measures to ensure the proper function and maintenance of the building and site after final completion. This scope provides a menu of choices of services, including initial existing condition surveys of the building and its systems, evaluation of operating costs, and code compliance reviews. B210-2007 may be used in two ways: (1) incorporated into the owner/architect agreement as the architect's sole scope of services or in conjunction with other scopes of services documents, or (2) attached to AIA Document G802-2007, Amendment to the Professional Services Agreement, to create a modification to an existing owner/architect agreement. B210-2007 is a scope of services document only and may not be used as a stand-alone owner/architect agreement. B210-2007 was revised in 2007 to align, as applicable, with AIA Document B101-2007.

B211–2007 (formerly B211–2004) Standard form of architect's services: commissioning

AIA Document B211-2007 requires that the architect, based on the owner's identification of systems to be commissioned, develop a Commissioning Plan, a Design Intent Document, and Commissioning Specifications. It also requires that the architect review the contractor's submittals and other documentation related to the systems to be commissioned, observe and document performance tests, train operators, and prepare a Final Commissioning Report. B211-2007 may be used in two ways: (1) incorporated into the owner/architect agreement as the architect's sole scope of services or in conjunction with other scope of services documents, or (2) attached to AIA Document G802-2007, Amendment to the Professional Services Agreement, to create a modification to an existing owner/architect agreement. B211-2007 is a scope of services document only and may not be used as a stand-alone owner/architect agreement. B211-2007 was revised in 2007 to align, as applicable, with B101-2007.

B214–2007 (formerly B214–2004) Standard form of architect's services: leed® certification

AIA Document B214-2007 establishes duties and responsibilities when the owner seeks certification from the U.S. Green Building Council's Leadership in Energy and Environmental Design (LEED®). Among other things, the architect's services include conducting a predesign workshop where the LEED rating system will be reviewed and LEED points will be targeted, preparing a LEED Certification Plan, monitoring the LEED Certification process, providing LEED specifications for inclusion in the contract documents and preparing a LEED Certification Report detailing the LEED rating the project achieved. B214-2007 may be used in two ways: (1) incorporated into the owner/architect agreement as the architect's sole scope of services or in conjunction with other scope of services documents, or (2) attached to AIA Document G802-2007, Amendment to the Professional Services Agreement, to create a modification to an existing owner/architect agreement. B214-2007 is a scope of services document only and may not be used as a stand-alone owner/architect agreement. B214-2007 was revised in 2007 to align, as applicable, with AIA Document B101-2007.

B252–2007 (formerly B252–2005) Standard form of architect's services: architectural interior design

AIA Document B252-2007 establishes duties and responsibilities where the architect provides both architectural interior design services and design services for Furniture, Furnishings, and Equipment (FF&E). The scope of services in B252-2007 is substantially similar to the services described in AIA Document B152-2007. Unlike B152-2007, B252-2007 is a scope of services document only and may not be used as a stand-alone owner/architect agreement. B252-2007 may be used in two ways: (1) incorporated into the owner/architect agreement as the architect's sole scope of services or in conjunction with other scope of services documents, or (2) attached to AIA Document G802-2007, Amendment to the Professional Services Agreement, to create a modification to an existing owner/architect agreement. B252-2007 was revised in 2007 to align, as applicable, with AIA Document B101-2007.

B253–2007 (formerly B253–2005) Standard form of architect's services: furniture, furnishings, and equipment design

AIA Document B253-2007 establishes duties and responsibilities where the architect provides design services for Furniture, Furnishings, and Equipment (FF&E). The scope of services in B253-2007 is substantially similar to the services described in AIA Document B153-2007. Unlike B153-2007, B253-2007 is a scope of services document only and may not be used as a stand-alone owner/architect agreement. B253-2007 may be used in two ways: (1) incorporated into the owner/architect agreement as the architect's sole scope of services or in conjunction with other scope of services documents, or (2) attached to AIA Document G802-2007, Amendment to the Professional Services Agreement, to create a modification to an existing owner/ architect agreement. B253-2007 was revised in 2007 to align, as applicable, with AIA Document B101-2007.

B305–1993 (formerly B431–1993) Architect's qualification statement

AIA Document B305-1993 is a standardized outline form on which the architect may enter information that a client may wish to review before selecting the architect. The owner may use B305-1993 as part of a Request for Proposal or as a final check on the architect's credentials. Under some circumstances, B305-1993 may be attached to the owner-architect agreement to show—for example, the team of professionals and consultants expected to be employed on the project. B305-1993 was renumbered only in 2007; its content remains the same as in B431-1993.

B352–2000 Duties, responsibilities and limitations of authority of the architect's project representative

When and if the owner wants additional project representation at the construction site on a full- or part-time basis, AIA Document B352-2000 establishes the project representative's duties, responsibilities, and limitations of authority. The project representative is employed and supervised by the architect.

B503–2007 (formerly B511–2001) Guide for amendments to aia owner-architect agreements

AIA Document B503-2007 is not an agreement but is a guide containing model provisions for amending owner/architect agreements. Some provisions, such as a limitation of liability clause, further define or limit the scope of services and responsibilities. Other provisions introduce a different approach to a project, such as fast-track construction. In all cases, these provisions are provided because they deal with circumstances that are not typically included in other AIA standard form owner/ architect agreements.

B727–1988 Standard form of agreement between owner and architect for special services

AIA Document B727-1988 provides only the terms and conditions of the agreement between the owner and architect—the description of services is left entirely to the parties, and must be inserted in the agreement or attached in an exhibit. Otherwise, the terms and conditions are similar to those found in AIA Document B151-1997. B727-1988 is often used for planning, feasibility studies, and other services that do not follow the phasing sequence of services set forth in B151-1997 and other AIA documents. If construction administration services are to be provided using B727-1988, which is not recommended, care must be taken to coordinate it with the appropriate general conditions of the contract for construction.

B801CMa–1992 Standard form of agreement between owner and construction manager

AIA Document B801CMa-1992 provides the agreement between the owner and the construction manager, a single entity who is separate and independent from the architect and the contractor, and who acts solely as an adviser (CMa) to the owner throughout the course of the project. B801CMa-1992 is coordinated for use with AIA Document B141CMa-1992, Standard Form of Agreement Between Owner and Architect, Construction Manager-Adviser Edition. Both B801CMa-1992 and B141CMa-1992 are based on the premise that there will be a separate, and possibly multiple, construction contractor(s) whose contracts with the owner will be jointly administered by the architect and the construction manager under AIA Document A201CMa-1992. B801CMa-1992 is not coordinated with, and should not be used with, documents where the construction manager acts as the constructor for the project, such as AIA Document A121CMc-1991 or A131CMc-1991.

Preparing the bid documents

5

Contractors can obtain bids by several methods, including by advertising in a local publication, which will elicit responses from a wide variety of contractors, or by selecting a small group of reliable contractors from business associates or local architects and engineers. Some companies prepare weekly bidding information to alert contractors that a particular project is in the bidding stages. The *Dodge Report* is probably the best known. This McGraw-Hill subsidiary distributes a weekly report in print as well as electronically to its subscribers, in which the project type, location, architect, and owner are listed along with the bid date, availability of plans and specifications, and a brief description of the project. Both public agencies and private owners use this source.

Most private-sector owners want a more restricted list of potential bidders, with desirable and qualified contractors sought out to submit bids. One way to achieve that goal is to prepare a contractor's qualification document and develop a concise and structured set of documents to establish the ground rules for the bidding process. This requires preparing bid documents.

Depending on the type of contract to be awarded, a slightly different set of bid documents will be required: one for a lump-sum contract award, one for a cost-plus contract, one for a construction-manager contract, and one for design-build. In any case, the first step in the bidding process involves selecting a group of qualified contractors, and that is accomplished by creating and issuing a request for qualification (RFQ) form.

Some owners side-step the bidding process by selecting a qualified contractor whom they have worked with before, providing them with a complete set of plans and specifications and negotiating a contract sum after a series of give-and-take sessions to resolve both the scope of work and the contractor's fee.

THE CONTRACTOR QUALIFICATION PROCESS

When preparing an open invitation to bid to the local construction community at large or to a selected group of contractors, a form called a Contractor's Request for

© 2010 by Elsevier, Inc. All rights reserved.
Doi_No = 10.1016/B978-1-85617-548-7.00005-7

Qualification (RFQ) can be obtained from the American Institute of Architects or the Associated General Contractors of America. This form lists the information that the contractor submitting the bid on the project must provide. Whether the standard form or a modified version for a specific type of project is used, the RFQ should contain requests for at least the following information:

1. The contractor's company name and mailing address, if it is a branch office; the address of the main office; and the phone numbers and e-mail addresses for both offices

2. The legal structure of the organization: sole proprietor, partnership, corporation, limited-liability corporation

3. Length of time in business; if it is a branch office, the length of time the branch has been operational

4. If the company has conducted business under any other name, the former name and the reason for the name change

5. The name of the senior executive at the main office or branch office, whichever is applicable

6. Current backlog: type of project, approximate dollar value, and dates of potential start

7. Annual revenue for the past three years

8. Three largest projects completed within the last five years

9. Names of corporate officers and experience

10. Number of employees, both office- and site-based

11. Type of work, if any, performed by employees

12. Names of the company's primary bank and the executive in charge of that account

13. Ability to provide a payment and performance bond and the limits of the company's bonding capacity

14. Limits of the company's general, personal, and property insurance coverage

15. List of any projects not completed and description of circumstances

16. Notification of whether the company has been blacklisted from any government bidders lists

17. Existence of a quality control program, and if so, attach to document

18. Existence of a safety program, and if so, attach to document

19. List of reportable accidents within the last year, and a copy of any OSHA violations within that period

20. Indication of whether the company operates primarily as a union or merit shop or both

Safety and quality issues

More owners are expressing concern about quality control and safety issues. The two seem to be intertwined: A company with an outstanding safety record quite often devotes more attention to the quality of the work.

Safety becomes an issue for several reasons. Serious accidents can impact the contractor's performance, such as when a key member of a work crew is injured and must be replaced by an assistant. Quality can suffer, and schedule delays can occur during this transition period. No owner wants to see his or her name and new construction project on the front page of the local newspaper announcing an accident or fatality at that construction site.

The bid documents should include a provision requiring the contractor to furnish his or her past five-year safety record and workers' compensation experience modification rate (EMR), which is the "multiplier" used in establishing the workers' compensation insurance rates based on previous accident rates. A modification rate that is lower than 1.0 would be deemed acceptable.

Contractors with a poor safety record will incur high EMRs, which in turn will increase their overhead; it takes three years of a good safety record to lower the EMR. The company's safety program, to be submitted along with its safety record, should include the following:

- A statement of company policy
- The objectives of the accident-prevention program
- The duties and responsibilities of the safety director or safety coordinator
- The relationship of the field supervisors administering the plan to the safety director or safety coordinator
- Procedures for reporting job-related injuries and illnesses
- Working rules of the safety program
- A hazard communications program (Hazmat), as required by OSHA
- Procedures for dealing with safety violations and safety violators

The contactor must also report any state or federal OSHA violations for the past five-year period.

The ten most frequently cited OSHA violations are as follows:

1. Scaffolding and construction falls
2. Nonprotection of employees during construction
3. Hazard communication
4. Lockout/tagout (tags on defective equipment have been ignored)
5. Machine guards (missing or removed at the site)
6. Respiratory equipment (not being used when required)
7. Electrical wiring
8. Powered industrial trucks (hitting workers)
9. Mechanical power transmission (failures)
10. Ladders (inappropriately secured and poorly made)

A company with a formal quality control/quality assurance program has recognized the need to maintain control over the quality levels of its work and the work of its subcontractors.

High-quality work results in the elimination or substantial reduction of rejected work and less intense scrutiny by the architect and engineer, which may have the effect of reducing the owner's inspection costs. A high-quality structure will result in reduced maintenance costs over the life of the building.

More owners are recognizing the need for contractors to submit a quality control/quality assurance program with their bid to inform the owner of the programs they have in place to monitor the quality of construction. The terms *quality control* (QC) and *quality assurance* are not synonymous. Quality control (QC) is the standard to which the construction material, equipment, or component has been designed to be incorporated into the project. Quality assurance (QA) is the process that verifies that these quality standards have been met.

A typical quality control and quality assurance program that appears in several project specification manuals is shown in Figure 5-1. Each section in the specifications manual will usually include a section on quality control for the work included in that section. Figure 5-2 is a part of the Division 4 Masonry specifications for this project and includes quality control standards for construction tolerances.

The bid documents may simply state that each bidder is to submit a QC/QA program that will be implemented if he or she is the successful bidder. The architect/engineer can rule on the value of each bidder's program and use this as a measure in the evaluation of each bid, along with the contractor's other qualifications, safety record, and, of course, cost to determine "best value."

References

Have you ever known anyone who provided a reference, either personal or business, that did not give the candidate a good report? A better approach is to call those listed references and ask them if they know of any other owners for whom the contractor worked, and then contact them. After their response, ask if they are aware of any other owners who hired the contractor. If any subcontractors who have previously worked with the general contractors can be located, they can also be a valuable source of information. Your architect and engineer can also make telephone calls to obtain information about the bidders.

It is a good idea to prepare a matrix to be used when contacting contractor references so the same questions are asked of all respondents. This list can include the following:

1. How would you rate the contractor's overall performance on your project?
2. Did the contractor impact the project schedule negatively or fail to complete it on time?
3. Did the contractor have a negative impact on your construction budget?
4. Did contractor's management and supervisory teams have continuity?

Section 01450
QUALITY CONTROL

PART 1 - GENERAL

1.1 SECTION INCLUDES

A. General quality assurance.

B. Field samples and mockups.

C. Source quality control.

D. Manufacturer's field services and quality control.

E. Testing laboratory and inspection services.

1.2 QUALITY ASSURANCE AND CONTROL OF INSTALLATION

A. Monitor quality control over suppliers, manufacturers, products, services, site conditions, and workmanship, to produce Work of specified quality.

B. Comply fully with manufacturers' instructions, including performance of each step in sequence. Notify Architect when manufacturers' instructions conflict with the provisions and requirements of the Contract Documents; obtain clarification before proceeding with the work affected by the conflict.

C. Comply with specified standards as a minimum quality for the Work except when more stringent tolerances, codes, or specified requirements indicate high standards or more precise workmanship.

D. Perform work by persons qualified to produce workmanship of specified quality.

E. Secure products in place with positive anchorage devices designed and sized to withstand stresses, vibration, physical distortion or disfigurement.

1.3 FIELD SAMPLES

A. Install field samples demonstrating quality level for the Work, at the site as required by individual specifications Sections for review and acceptance by Architect. Remove field samples prior to date of Final Inspection, or as directed.

1.4 MOCK-UPS

A. Where requested by Architect, or as specified in individual specification sections, assemble and erect specified items, with specified attachment and anchorage devices, flashings, seals, and finishes. Remove mock-up assemblies prior to date of Final Inspection, or as directed.

B. Mock-ups, when approved by the Architect/Engineer, will be used as datum for comparison with the remainder of the Work for the purposes of acceptance or rejection.

Quality Control
01450-1

8/15/07

FIGURE 5-1

A quality control general specification.

C. Demolish and remove from site prior to requesting inspection for certification of Substantial Completion, all Mock-ups which are not permitted to remain as part of the finished work.

1.5 MANUFACTURER'S FIELD SERVICES AND REPORTS

A. When called for by individual Specification Sections, provide at no additional cost to the Owner, manufacturers' or product suppliers' qualified staff personnel, to observe site conditions, start-up of equipment, adjusting and balancing of equipment, conditions of surfaces and installation, quality of workmanship, and as specified under the various Sections.

    1. Individuals shall report all observations, site decisions, and instructions given to applicators or installers. Immediately notify Architect of any circumstances which are supplemental, or contrary to, manufacturer's written instructions.

    2. Submit full report within 30 calendar days from observed site conditions to Architect for review.

1.6 TESTING LABORATORY AND INSPECTION SERVICES

A. Owner will appoint, employ, and pay services of an independent firm to perform inspection and testing and other services specified in individual specification Sections and as required by the Architect.

B. Cooperate with independent firm; furnish samples of materials, design mix, equipment, tools, storage and assistance as requested.

    1. Notify Architect and independent firm 48 hours prior to expected time for operations requiring services.

    2. Make arrangements with independent firm and pay for additional samples and tests required for Contractor's use.

C. Retesting required because of non-conformance to specified requirements shall be performed by the same independent firm on instructions by the Architect. Payment for retesting will be charged to the Contractor by deducting inspection or testing charges from the Contract Sum.

PART 2 - PRODUCTS (Not Used)

PART 3 - EXECUTION (Not Used)

End of Section

FIGURE 5-1

Continued.

5. Would you hire the contractor for your next project?
6. Did the contractor work collaboratively with your architect and engineer?

When more than three contractors have submitted bids, the selection process becomes a little more difficult. After reviewing and analyzing all requests for qualifications, an owner, in consultation with the architect, should consider creating a shortlist of bidders, generally confined to three qualified builders. These shortlisted contractors will then be invited to submit their bids.

3. Verify that the foundation elevation is such that the bed joint thickness shall not vary from specified thickness, and that the foundation edge is true to line with masonry not projecting over more than 1/4".

C. Provide temporary bracing during installation of masonry work. Maintain in place until building structure provides permanent bracing.

D. Protect surfaces of windows, door frames, louvers and vents as well as similar finish products with painted and integral finishes from mortar droppings and stains.

3.3 INSTALLATION - GENERAL

A. Build chases and recesses as shown or required to accommodate items specified in this and other Sections of the Specifications. Provide not less than 8 inches of masonry between chase recess and jamb of openings and between adjacent chases and recesses.

B. Leave openings for equipment to be installed before completion of masonry. After installation of equipment, complete masonry to match construction immediately adjacent to the opening.

C. Establish lines, levels and coursing indicated. Protect from displacement.

D. Maintain masonry courses to uniform dimension. Form vertical and horizontal joints of uniform thickness.

E. Isolate masonry partitions from vertical structural framing and where indicated on the Drawings. Maintain joints free from mortar, ready to receive sealant and joint bead back-up.

F. Provide compressible filler at tops of interior masonry partitions abutting structural steel trusses.

3.4 CONSTRUCTION TOLERANCES

A. Maximum variation from true surface level for exposed to view walls and partitions:

1. Unit-to-unit tolerance: 1/16 inch.

2. Surface, overall tolerance: 1/4 inch in 10 feet in any direction and 1/2 inch in 20 feet or more.

    a. Where both faces of single wythe wall or partition will be exposed to view, request and obtain decision from the Architect as to which face will be required to conform to the specified surface level tolerance.

B. Maximum variation from plumb: For lines and surfaces of walls do not exceed 1/4 inch in 10 feet, 3/8 inch in any story up to 20 feet maximum. At expansion joints and other conspicuous lines, do not exceed 1/4 inch in 20 feet.

C. Maximum variation from level: For lines of sills, tops of walls and other conspicuous lines, do not exceed 1/8 inch in 3 feet, or 1/4 inch in 10 feet and 1/2 inch in 30 feet.

D. Maximum variation of linear building line: For position shown in plan relating to columns, walls and partitions, do not exceed 1/2 inch in 20 feet or 3/4 inch in 40 feet.

E. Maximum variation in specified height: 1/2 inch per story.

Unit Masonry Assemblies
04810-7

8/15/07

FIGURE 5-2

A quality control section in each specification division. This one is in Division 4: masonry.

procedures proposed for cleaning the masonry including, but not limited to: method of application, dilution of application, temperature of application, length of time of surface contact, method of rinsing surface (temperature, pressure, and duration), repetition of procedure, etc.

3. Methods of Protection: Prior to commencing the cleaning operations, the Contractor shall submit a written description of proposed materials and methods of protection for preventing damage to any material not being cleaned, for review.

4. Methods of Effluent Control: Prior to commencing the cleaning operations, the Contractor shall submit a written description of proposed materials and methods for the containment, neutralization and disposal of all effluent.

5. Manufacturer's use instructions for filtration equipment and cleaning equipment.

6. Warranty: Provide sample copies of manufacturers' actual warranties for all materials to be furnished under this Section, clearly defining all terms, conditions, and time periods for the coverage thereof.

1.5 QUALITY ASSURANCE

A. Field Supervised Work: Contractor shall notify Architect before beginning masonry cleaning. Obtain the Architect's approval of Contractor's cleaning procedures, before proceeding with the work.

B. The masonry cleaning work shall be performed by a firm which, for the past 10 years has been continuously and primarily engaged in masonry cleaning work of the type required for this project. Each bidder shall submit evidence that, during the above period, the proposed firm has satisfactorily performed on at least each of three projects, work equivalent to 50 percent or more of the dollar value of his bid for the work for this project.

1. All subcontractors shall meet the same requirements as the masonry cleaning Contractor.

2. The Contractor shall maintain a steady work crew consisting of skilled workers who are experienced with the materials and methods specified and familiar with the requirements, and a full time foreman, on site daily, with a minimum of five (5) years successful experience. The Contractor shall confirm that all workmen under his direction fully understand the requirements of the job.

C. Source of Materials: Obtain materials for masonry cleaning from a single source for each type of material required to ensure a match in quality, color and texture.

1.6 DELIVERY, STORAGE AND HANDLING

A. Deliver and store cleaning materials in original, sealed, containers clearly showing manufacturer's identification and brand name, date and location of production.

B. Store cleaning materials in accordance with the manufacturer's written directions, above ground, under cover, with sufficient ventilation to prevent the buildup of flammable vapors and at temperatures between 35 and 95 degrees F.

FIGURE 5-2

Continued.

The invitation to bid

The invitation to bid process is also known as a request for proposal (RFP) and is accompanied by a number of instructions, which the AIA incorporates in the form A701 Instructions to Bidders. Specific instructions or requirements, such as bond requirements, specific levels of insurance to be met, and any unique requirements of the project at hand, will be included.

An urban construction site with residential buildings close by may require the contractor to start and stop work within prescribed time limits. The construction site may have limited parking areas available, so contractors are alerted to the fact that offsite parking many be required and that parking may be restricted to certain designated areas offsite (to appease neighbors).

The invitation to bid will include the location and costs to obtain copies of the bid documents and the plans and specifications, which are available at the architect's office or at the printing company designated by the architect. Reproduction of a set of plans and specifications can be expensive: for large, complex projects, a cost of $500 is not unusual. Most RFPs state that the contractor must pay a deposit for the bid documents; if he or she is not the successful bidder, the documents may be returned to the architect for a full refund. This statement must include a notice that the plans and specifications must be returned in good condition with no notations or comments on them.

While owners usually provide the successful contractor with a certain number of copies of the plans and specifications, documents returned in good condition by the unsuccessful bidders can be given to the successful contractor. The architect will make the decision as to whether any returned drawings fail to meet acceptable conditions and whether a deposit should be returned or denied.

Elements of an invitation to bid include the following:

1. The name and location of the project

2. A brief description of the project

3. The time and place where bids are to be returned with the bid form

4. A statement as to whether the bid opening will be private or public (public works projects are always public bid openings, where each bidder's price is announced along with any bid qualifications; most bids for private work are opened in private)

5. Where a copy of the plans and specifications can be obtained, as well as the cost, indicated as either nonrefundable or refundable if in good condition

6. A statement as to whether a bid bond and/or payment and performance bonds are required (generally, public works projects require a bid bond and payment and performance bonds; private-sector work does not require bid bonds)

7. A statement that the owner may reject any or all bids at his or her discretion

8. The specific requirements that the project demands

Other elements in a bid document include the following:

- Notification of a prebid meeting (if applicable), where it will be held, and who will attend from the owner's team (architect/engineer/owner's representative)

- The bid format to be distributed at the prebid meeting or included in the package if no meeting is held, usually in the form of a schedule of values breakdown

- Bid-submission information: number of copies required, whether it is to be in an Excel or PDF format, and the time and place of submission of the bid

- The form of contract anticipated—for example, lump-sum or GMP

- Schedule of interviews, if applicable, with either shortlisted contractors or presumed "best value" contractor

- Date of proposed contract award to the successful contractor

- Requirement for the contractors to submit a staffing plan with names, positions, and job experience

- A preliminary milestone schedule to be provided by the owner, setting the anticipated date of the start of construction

- A request for bidders to submit a proposed list of subcontractors, which may alert the owner or the designer to acceptable subcontractors

- A list of preferred subcontractors and vendors with whom they have worked successfully in the past and wish to have the bidders consider if awarded the job; in some cases, the owner will actually direct bidders to select a subcontractor or subcontractors as a condition of contract

- Whether MBE (minority business enterprises) or WBE (women-owned business enterprises) are intended to participate or are required (if government funding is involved); this can be included, as well as the percentage the owner would like to see participate in the project

- A statement as to whether value-engineering suggestions will be accepted and evaluated as part of the selection process

- If any items are going to be prepurchased by the owner, a list of those items and anticipated purchase and delivery dates, along with a statement that the contractor will be required to accept and incorporate these items into the project

One important provision in the invitation to bid is a "site visit" requirement to ensure that all existing site conditions have been observed by each bidder, whether or not they have been accurately shown in the bid documents (the plans). This is important for a number of reasons. Let's say there was an abandoned well in the location of the proposed building and it was not shown on the plans (for whatever reason) but was clearly visible if the contractor toured the site. The contractor, by virtue of the site visit, will have the responsibility to query the architect about this missing detail or be responsible for all work associated with dealing with this condition if awarded the job.

A typical site-visit statement in an invitation to bid would look something like this:

By executing and submitting this proposal, the contractor represents that he has visited the site, familiarized himself with the local conditions under which the work is to be performed and correlated his observations with the requirements of the bid documents.

Without this statement, the contractor could have a strong argument for an "extra" claim to remove and fill that abandoned well, because it was not indicated on the drawings and it represented additional work. Also, using this well situation as an example, if the well was clearly visible on a site visit and was not included in the drawings, the contractor would be obligated to request clarification from the architect. The architect in turn would advise all bidders of this change and request them to include all work to remove the well in their bid or to note all such work as an exclusion.

When the project requires extensive renovation and/or rehabilitation of an existing building, this site-visit statement or disclaimer takes on added significance, and the requirement can be expanded to add the following:

Prior to bidding on the scope of the work, the contractor is to become familiar with the existing facility in order to fully understand the project. Special attention should be paid to (name and point out the areas of concern). Bidders are to notify (name of project architect) of any discrepancies, omissions, or variations noted on the drawings or in the specifications that are discovered during the bidding process.

When rehab or renovation projects are being considered, a mandatory prebid conference, preferably conducted at the site, will give all bidders the opportunity to ask questions of the architect and engineer to further define the scope of work. There are usually significant unknowns in rehab and renovation work, and most of these "unknowns" can be addressed at a walk-through at the site:

- What is the extent of removal of existing finishes back to a "sound substrate"? In other words, at what point does the contractor stop when he or she uncovers more unstable finishes than indicated on the bid documents?

- Areas on the plans or in the specifications where a particular portion stipulates that "any loose or unstable wall finishes are to be removed, whether indicated in the plans and/or specifications or not" should require close attention by the inspecting bidder.

- Can the existing utilities be used by the contractor, and if so, how are the charges for power and water to be paid? If the contractor determines that new electrical or water service is required during demolition and prior to installation of permanent power, whose responsibility is it to supply that "temporary power"?

- Do the bid documents correctly indicate the items of work that will remain and the items of work that will be removed during the demolition process? Should the architect/engineer be more specific, possibly by spray-painting with a dab of fluorescent paint the questionable areas?

- Is it possible to apply a new surface over an existing one in some areas where the existing surface appears to be unsound? If not, what is the contractor's responsibility?

- Do any of the items have scrap value, and if so, who will retains ownership of the scrapped items?

- Have any bidders observed hazardous materials in the building other than those indicated on the drawings? Is the contractor's responsibility after encountering hazardous materials clear? Generally, the architect requires the contractor to cease work immediately in areas where hazardous materials are suspected, notify the owner, and await instruction before proceeding.

As questions are raised at the prebid meeting, the architect will take careful note of those questions, after which he or she will respond, in writing, to all attendees. The results of this walk-through, with any cost or time implications, will be added to the project's scope and become an attachment to the contract for construction.

INSURANCE AND BONDS

The architect, the lender, and the owner's insurance agent all have a hand in determining the amount of insurance needed for a project. The types of insurance required for a construction project are as follows:

- Commercial general liability (CGL) provides the contractor with bodily injury and property damage liability coverage while contractors are working on the owner's construction project. Generally an owner will require a minimum of $1 million in CGL insurance from a contractor.

- Contractor's professional liability (CPL) protects contractors from claims of negligent acts and errors and omissions incurred in the course of their professional business. It includes loss of client data, software and related system failures, and claims of fraud. Owners generally require a minimum $1 million CPL policy from a contractor.

- Umbrella coverage provides coverage in excess of that included in the basic liability insurance policy.

- Builder's risk, also known as course of construction insurance, provides coverage for loss or damage to any part of the project under construction that has already been paid for via the normal monthly requisitions submitted by the contractor. In effect, as the owner pays for a portion of construction, he or she actually "owns" that portion. There are two basic types of builder's risk insurance: all-risk, which covers all risks except those specifically excluded, and name-peril, which covers only those risks identified in the policy. Generally, owners either exclude the builder's risk policy or have the bidding contractors include the cost as an option or alternate.

Traditionally an owner and a general contractor execute a construction contract with insurance limits established in the bid documents and the specifications. Included in the notification of insurance limits, the contractor is also given notice that he or she cannot commence work at the site until the insurance certificates have been received and reviewed to ensure compliance with the contract documents.

Contractors are to include coverage for workers' compensation, employer's liability, and general liability. In addition, the contractor must name the owner as an "additional insured" on the contractor's liability policies, which ensures that the contractor will defend the owner in the event of a third-party action.

The construction contract includes an indemnification clause that requires the contractor to hold the owner blameless for any loss arising out of the execution of the contract for which the contractor may be held liable.

Owner- and contractor-controlled insurance programs

For large construction projects in the tens or hundreds of millions, an owner-controlled insurance program (OCIP) may be a cost saver. In most construction contracts, the general contractor is required to furnish all insurance coverage as required by that contract, and in turn he or she requires the subcontractors to comply with specific insurance requirements, so there are several layers of often repetitive coverage and associated costs.

The OCIP process basically states, "General contractor and your subcontractors provide me (the owner) with the cost of insurance premiums for all of the required coverage on my project, and I will provide a blanket policy to cover everyone." The owner can then expect a credit to the construction contract sum, since both the general contractor and the subcontractors will have reduced their costs to exclude insurance coverage. A variation on this scheme is one in which the general contractor provides coverage for all of the subcontractors, which is called a contractor-controlled insurance program (CCIP). In both cases, these insurance products are meant to provide adequate coverage at a reduced cost.

Here are some of the advantages of OCIP:

- Potentially lower insurance costs because of volume discounts

- Insurance requirements filled by the owner, allowing some contractors who couldn't furnish policies with high limits to fulfill the job requirements

- Improved management of claims and loss control due to centralization

- Broader coverage with higher limits

- Faster provision of insurance certificates and more precise monitoring of expiration dates

And here some of the disadvantages:

- Additional administrative burden placed upon the owner

- Additional time and effort required to extract, analyze, and approve the cost of insurance from the general contractor and all subcontractors to ensure that the proper credit has been provided

- Significant owner's responsibility to monitor on-site safety because a contractor's inadequate safety program will cause more accidents and therefore increase premiums

For certain projects, however, an OCIP or CCIP program may be beneficial and, if considered, should be made a part of the bid documents.

Construction surety bonds

If the contractor can provide a payment and performance bond, even when it may not be a specific requirement, this is an indication that the contractor has financial strength, since the bonding company requires certain minimum financial, management, and cost-control standards. Bonds are not the same as insurance. Insurance is a loss-sharing mechanism that protects the policy holder from damages that may occur in the future, whereas bonds provide guarantees of contract performance, payment, and project completion. High bonding limits or capacities are prized by contractors, since they exhibit the three C's of bonding: character, capital, and capacity.

"Character" denotes the reputation of the contractor in the community where he or she operates. This requires a track record of successful projects with satisfied owners and participation in community activities. "Capital" refers to strong financial statements and evidence of acceptable accounting practices. "Capacity" indicates strong management, a history of completing profitable projects, a strong estimating department, and the presence of strong cost-control procedures. Capacity reflects the ability of the contractor to successfully complete new projects.

The Surety Association of America has developed a series of qualities that they attribute to a solid contractor organization:

- Contractors who have formal and on-the-job training for all levels of employees
- Logical, incentive-based compensation programs
- Tenure for proven field supervisors and internal promotion wherever possible
- Depth at all levels of the organization
- Succession planning (since many small to midsize construction firms are family businesses)
- An up-to-date distributed organization chart
- A culture of loyalty, ownership, and urgency
- Visionary, inspirational leadership

- Low turnover
- Solid management of cash flow and overhead
- Profit-focused company
- Timely payment of bills
- Reasonable growth without overextending resources
- Superior estimating skills and systems to manage costs
- Satisfied customers
- A well-defined market niche and a 12- to 36-month growth plan
- Closely managed projects with early warning systems to catch potential problems
- Litigation avoidance
- Only qualified and interested family members in management
- A continuity plan with adequate life insurance

How does *your* contractor fit in this profile?

A bond involves three parties: the contractor, the owner of the construction project, and the issuer of the bond. When two interested parties are involved, such as the bank or lending institution that is financing the project, these two interested parties are known as "dual obliges." The following are the main types of construction bonds:

- A *bid bond* is a bond that guarantees that the principal (the contractor) will honor the bid if selected by the owner. If the contractor declines a contract when offered, the bid bond is forfeited and will be used to reimburse the owner for any additional costs required to select another contractor. This bond type is frequently used in public project bidding but rarely in bidding in the private sector. Bid bonds are usually required to cover 10 percent of the contract price. A proposed $10 million project would in effect require a bid bond of $1 million.

- A *payment bond* guarantees that the contractor will pay all monies received from the owner to all subcontractors and for all material and equipment used to construct the project. Subcontractors and vendors are referred to as "beneficiaries" of the bond.

- A *performance bond* guarantees that the contractor will complete the project in accordance with the contract documents, including completion on the date included in the contract, adjusted by mutual agreement—for example, by the addition of "extra work" change orders or deletion of work.

Bonds contain certain terms:

- *Calling the bond* occurs when the owner notifies the bonding company (the guarantor) that the contractor (the principal) has failed to live up to the terms of the contract by either failing to pay all subcontractors and vendors or failing to "perform" as required by the performance bond.

- *Consent of surety* occurs upon successful completion of the project, with all debts paid and all contract obligations fulfilled; the contractor will request the bonding company (surety) to acknowledge same, sign off, and cancel the bond.

- A *penal sum* is the amount of the bond, which is generally the amount of the contract. It is the amount of money the bonding company will pay in the event that the contractor fails to complete the terms and conditions of the contract.

- *Dual obligee* denotes parties other than the contractor and owner that have an interest in the project, such as the bank or financing institution. They will be added to the bond as another party having an interest in whether the contractor succeeds or fails.

- The *guarantor* is the underwriter or surety company

- The *obligee* is the project owner and/or others if there is a dual obligee

- The *premium* is cost of the bond.

- The *principal* is the entity requesting the bond: the contractor, subcontractor, architect, or engineer.

Subguard

Subguard is an alternative to a bond—actually an insurance policy provided by general contractors to guard against default by subcontractors. This form of insurance protects contractors and owners against a defaulting subcontractor. While subguard does not cover the subcontractor's losses, it does cover the losses incurred by an owner or general contractor in case of subcontractor failure. If failure occurs, the owner and general contractor will receive funds to pay for a successor subcontractor.

Developed by Zurich U.S. Construction in 1995, subguard is less expensive thant a bond but generally carries a large deductible—as big as six figures—making it economically feasible only on very large projects.

Letter of credit

A further instrument for owner protection is the letter of credit. A bank-issued letter of credit is a cash guarantee that would freeze a portion of all of the contractor's assets to pay an owner in case of contractor default. The letter of credit (LOC) is sometimes used as a substitute for a bond when a subcontractor or general contractor, for some reason, may not be able to obtain a bond or has reached the limits of bonding capacity and cannot obtain another one until he or she successfully completes a project, thereby freeing up the bond capacity. A new contractor who doesn't have a lot of experience may not meet certain bond qualifications and need a letter of credit.

GENERAL CONDITIONS

A contractor's estimate will be broken down into several basic components:

- General conditions: costs associated with the administration of the construction project

- Costs of work to be performed by the contractor's own forces, including labor, materials, and equipment

- Materials and other components purchased by the general contractor to be installed by a subcontractor, including items such as doors, hardware, and windows

- Subcontracted work

- Bonds, if required, insurance, and building permits (which may be listed as separate line items or included in the general conditions)

- The contractor's overhead and profit, generally indicated as separate items

The general conditions categories are often displayed in different ways by different contractors. But first let's look at a typical listing that most contractors use in the formulation of that portion of their estimate. Project administration costs include the following:

1. Project management salaries
2. Project superintendent salaries
3. Field engineer(s), basically a project superintendent assistant who is employed for large, complex projects
4. Field office clerical personnel

Field office equipment and supplies include the following:

1. Temporary office trailer rentals, including costs to transport to and from site and to set in place

2. Temporary office utilities, including light, heat, power, water, and portable toilets, and the cost of installing and monthly charges for connecting field office utilities to existing sources such as water mains and offsite electric lines

3. Telephone services, including service connections, monthly charges, and data-communication services

4. Office cleaning and debris removal

5. Office furniture: desks, chairs, tables, filing cabinets

6. Office equipment, owned or rented: copy machines and supplies, plotters, fax machines, computers, PDAs, shredders

7. Office supplies: envelopes, stationery, pads, pens and pencils, expedited delivery services, postage

Security services include the following:

1. Alarm systems for the field office(s)
2. Site alarm and floodlights
3. Security fencing: installation, maintenance, and removal
4. Security guard services

Safety and first-aid equipment and supplies include the following:

1. Safety manager, either full- or part-time
2. First-aid supplies
3. Drug-testing supplies if required
4. Fire extinguishers in the building under construction and in the field office(s)
5. Barricades in the building or on the site to block off hazardous areas; labor and materials to install and remove
6. Safety materials such as personal protection equipment, hard hats, goggles, hearing protection, safety vests, raincoats, and foot gear
7. Pest control

Other items include the following:

1. Progress photographs or videos
2. Preparation of record (as-built) drawings
3. Schedule updates
4. Electronic data-processing costs not associated with the contractor's general office accounting work

A listing of general conditions is important when analyzing and evaluating the contractor's bid proposal. Some contractors include such general conditions items as project management costs in their overhead percentage, along with some related project administration costs. This will tend to increase overhead percentages but decrease the general conditions costs. Project management costs can range from $85,000 per annum to $150,000 and even higher for some very experienced managers, plus fringe benefits and possibly year-end bonuses as well.

When evaluating bid proposals, you must look, of course, at not only the total project cost but the percentage of overhead and profit that applies to both the contract sum and future change orders, unless contract language dictates otherwise. How does the value of the general conditions compare with the competition?

Different contractors may apply their overhead and profit figures separately, while others will merely apply them together as one percentage. The two methods produce the same bottom line; if we assume the total is 15 percent, a single application is slightly lower:

	$4,000,000		$4,000,000	$4,000,000
10% OH&P	400,000	5% OH&P	200,000	(applied as one)
	$4,400,000		$4,200,000	$4,000,000
5% profit	220,000	10% profit	$20,000	$4,000,000
Total	$4,620,000		$4,620,000	$4,600,000

We discussed the need for a prebid conference when renovation and/or rehabilitation projects are concerned, but this important event is not limited to only those types of projects. Many other topics may come up at this meeting of all potential bidders, the design consultants, and the owner's representative.

The process for bid assembly will be discussed at the prebid conference. Bidders may need clarification of some details or procedures that appear vague. This questioning period will be followed up with responses from the architect and engineer and possibly sketches to clarify the unclear details. This document will not only become a part of the official bid documents but will also be included as an attachment to the contract for construction.

One important aspect of the prebid conference is to provide an opportunity for the builders to meet the architect and owner's representatives for what may be the first time and for the A/E team and the owner to meet the group of contractors.

From experience, I can tell you that the tone of the architect and/or owner's presentation is important because if it appears to be one of cooperation and reasonableness, contractors will be more interested in working with the design and owner team. But if they sense that the design team or owners will be too demanding or arrogant, some contractors will walk away. A tone of fairness and reasonableness exhibited by the architect and owner can definitely impact the way in which these contractors prepare their estimate. And the same would be true of the architect/engineer and owner if a contractor appears to be argumentative or strongly opinionated. Exhibiting an air of professionalism is required by both contractor and owner team at the prebid conference.

EVALUATION OF THE BID

The bid documents are the place where instructions can be included that will allow the owner to better evaluate each contractor's bid. It is not enough to merely require the contractors to include a bottom-line price. They should be directed to provide sufficient details to allow for proper comparison with their competition.

Bids do not normally fall within a very narrow price range—several may be 2 to 5 percent apart—but one bid may be exceptionally low and another exceedingly high. When a bid is very high, the contractor for one reason or another may have made an estimating mistake simply by transposing numbers. Instead of punching in $30,000 for a particular component, the estimator may have inserted $300,000 by mistake. When a builder is very busy and does not need too much new work, he or she may submit a high bid with the thought that if they get the job, great, but if not, it is no big deal.

In either case, when a very low or a very high bid is submitted, a call to the contractor drawing attention to the seemingly excessive variation between the bid and other bidders may be necessary to review the estimate, uncover and correct any errors, resubmit, and thank you for the call.

The architect may include a preprinted form on which the cost of each component of construction is listed separately or simply ask each bidder to break down all components and price separately. A precise list of component breakdowns required is preferable because this will allow comparison of the total bid price and comparison of each element within the schedule of values from all bidders.

A schedule of values may include these elements:

- General requirements
- Site work
- Concrete
- Masonry
- Structural steel and miscellaneous metals
- Carpentry
- Thermal and moisture protection
- Doors and windows
- Finishes (painting, flooring, ceilings), either listed together or by each type of finish
- Specialties (toilet partitions, louvers, lockers)
- Elevators
- Mechanical (HVAC)
- Plumbing
- Fire protection
- Electrical

Each of these components can be expanded to further allow for comparison of bids, and this is the opportunity to do so. Concrete work, for example, if a major component in the project, can be further broken down into these categories:

- Footings and foundations
- Concrete slab on grade
- Suspended concrete slabs (those above the first floor)
- Equipment pads (for transformers and other equipment)
- Concrete walks and curbs (if not included in the site work scope)

When bids are returned, comparison can be more effectively analyzed when an extensive schedule of values is required for each major and even some minor components of construction. This analysis would be similar to Table 5-1.

Two bidders have general requirements in the same general range, but Bidder C's general conditions are 10 percent higher than Bidder B and about 19 percent higher than Bidder A. But set this aside for the moment. Bidder C has significantly higher plumbing and HVAC prices—$1.09 million versus $890 for Bidder A and $974 for Bidder B. The architect and engineer can query all three bidders to ensure that they understood the mechanical plans and specifications and, depending on their answers, a "best value" contractor can be selected.

In the case where the apparent low bidder may have had some components priced significantly lower than the other competitors, the reliability of the bid is in question. This scenario, with an abbreviated schedule of values, might look like Table 5-2.

It would appear that Bidder D's steel price and plumbing price are too high, so the $2.75 million bid is questionable. The bid price of $2.75 million is low, but why

Table 5-1 Comparing Bids

Schedule of Value	Bidder A	Bidder B	Bidder C
General Requirements	$150,000	$165,000	$185,000
Site Work	$85,000	$92,000	$87,000
Concrete	$125,000	$115,000	$110,000
HVAC	$650,000	$685,000	$750,000
Plumbing	$240,000	$289,000	$340,000
Electrical	$260,000	$265,000	$272,000
Total Bid	$2.5 million	$2.7 million	$2.8 million

Table 5-2 Comparing Bids

Schedule of Value	Bidder D	Bidder E	Bidder F
Structural Steel	$275,000	$240,000	$238,000
Plumbing	$295,000	$245,000	$252,000
Total	$2.75 million	$2.95 million	$3.05 million

are the plumbing and steel prices so high? Is the contractor unable to attract competitive subcontractors? In spite of the high prices in two important components, has the contractor underestimated some other items? If awarded the contract for construction, will the bidder suddenly find himself in financial trouble, which means that the owner will be in trouble?

In this case, it would be advisable to have the architect meet with Bidder A and review in detail each element of the proposal to determine if it is proper, complete, and priced accordingly. Based on that review, the contractor will either remain in consideration or be eliminated. There may well be valid reasons for the seeming disparity in the value of some of the line items, but without some exploration, the bid should be suspect.

Depending on which form of contract is being considered, including a line item for the contractor's overhead and profit is advisable. If a lump-sum contract format is the one selected, the need for the contractor to list the overhead and profit separately may not be so important except that an owner should have the contractor state his or her overhead and profit percentages for any change order work. When a lump-sum or stipulated-sum contract is being administered, the contractor's actual overhead and profit will be determined by the final tally of actual costs versus the estimate. If final costs are higher than the estimated costs, obviously the profit percentage will be reduced; on the other hand, if through efficiencies or market conditions the contractor is able to better the costs and therefore the profit, he or she will keep that higher sum. However, in a cost-plus or GMP contract, stipulating the percentage of overhead and profit is essential and will be incorporated into the winning bidder's contract.

The overhead costs are the costs to run the contractor's business and are not particularly associated with the cost to perform and administer the contract work; the billable general conditions will reimburse the contract for project operational costs in the field and some office expenses such as accounting and legal. It is really the total of both that is important.

A higher overhead percentage and a lower profit percentage will generally vary slightly. There is a slight difference for contractors applying one percentage for both instead of separately. Let's look at the following, assuming the cost of work is $2.5 million:

	$2,500,000		$2,500,000
7% OH	175,000	8%	200,000
Subtotal	$2,675,000		$2,700,000
5% profit	133,750	4%	108,000
Total	$2,808,750		$2,808,000

Take the same project "cost" of $2,500,000; by applying a combined 12 percent overhead and profit, the bottom line becomes $2,800,000.

If the project is a cost plus a fee or cost plus a fee with a GMP and any high-value items were purchased specifically for the project, they may have some residual value at the end of the project. The owner should be given the option to either take possession of those items or have the contractor establish a residual value and offer a credit to the owner. This is nothing more or less than the concept of the cost-plus contract, in which the contractor is to invoice costs. If a desk, chairs, or computer has been purchased, the cost is the initial cost less residual value at project closeout.

Certain items of equipment purchased by the general contractor in the course of construction will also have some residual value. For example, a $450 radial saw purchased new for a 12-month project will certainly have some residual value at the end of the project, and the contractor will use that saw on other future projects. The residual value should accrue to the owner when a cost-plus or GMP project is in effect.

CONTRACTOR SELECTION

Assuming that a contractor qualification process has taken place, the bids have been evaluated, and the "low, responsible" bidder has been selected, a few more steps must be taken before issuing the contract. The architect and owner should meet with the builder's team even if they met during a prebid conference.

From experience as a consultant to an owner, I can state that evaluating the team assigned to the project is as important as the initial contractor evaluation process. The construction company, in the eyes of the owner and architect, is only as good as the project manager and project superintendent assigned to the job. Using an inexperienced builder's team on a complex project will create a great deal of problems, and the managers may fail to recognize and alert the architect or owner to any impending delays or mounting problems on the site. They may lose control over subcontractor operations and, in general, just be in over their heads.

The architect/owner should request that the company's project executive bring the proposed team members to a meeting, and they should bring their resumes. The architect and the owner should lead the discussion and question the project executive, the project manager, and the project superintendent:

- How do they plan to organize the project? Are they planning to have an assistant super on site? Does the scope of work require one?

- Who will staff the field office?

- How can the architect contact the project superintendent and project manager: land line, cell phone, e-mail?

- What additional supervisory staff will be on site?

- Can the superintendent provide a sample of the daily log? (The log documents job activity on a daily basis and should be fairly detailed.)

- How do they inspect the project for proper manpower requirements?

- How do they inspect the work to ensure conformance with the contract documents?

- What is the project superintendent's daily schedule of work? When does he or she start and when does he or she leave the job? (This is not an eight-hour job!)

- How do they work with subcontractors? (The answer to that question should be "fairly and offering assistance whenever needed.")

- Do they prepare a one- or two-week look-ahead schedule that projects short-term work-in-place goals?

These are the people whom the owner and the architect will be working with over the next year or two, and if a comfort level is not established during these initial meetings, it is not out of the question to request another contractor team.

The construction contract

6

Although one of the standard AIA, AGC, or CMAA contract forms may be used, the owner should consider a number of clarifying statements, attachments, and additions when discussing the preparation of the contract for construction with an attorney. Few construction contracts are limited to "boilerplate" provisions, and they can be customized to fit the project at hand.

A standard contract format contains the following broad categories:

- The Work (capitalized) of the contract is the description of the work as set forth in the plans, specifications, and other documents that define or expand them, as well as addenda, revisions, and so on.

- The date of commencement of Work and the date of completion

- The contract sum

- Contractor payment schedules

- Final payment requirements

- Procedures for contract modifications, increasing or decreasing the scope of work, and associated cost- and time-related implications

Each one of these broad segments can be expanded with the insertion of specific contract language regarding the owner, architect, and contractor obligations and responsibilities, contract administration processes, safety and insurance issues, and specific project-management requirements.

An attorney who is well versed in construction contracts will guide the owner through the complicated process of developing a contract that is not only fair to all parties but includes provisions to protect the owner from unforeseen events, contractor and subcontractor defaults, and assorted claims presented during the course of construction.

Many of the provisions in the contract between owner and general contractor are passed on to the subcontractor via the general contractor's subcontract agreement, so

© 2010 by Elsevier, Inc. All rights reserved.
Doi_No = 10.1016/B978-1-85617-548-7.00006-9

many of the articles in that owner-contractor agreement may also apply to subcontractors and vendors engaged by the general contractor. As previously discussed, the terms and conditions in the owner-contractor agreement are made available by the general contractor for inspection by any subcontractor or supplier wishing to do so.

All of the special conditions customizing the project at hand can be inserted within the contract by the use of attachments referred to as "exhibits." One of the first exhibits to be presented by the contractor for review and approval by the owner and subsequent inclusion in the contract will be the schedule of values.

SCHEDULE-OF-VALUES EXHIBIT

After execution of the contract and prior to commencement of work, the architect will request the contractor to submit what is called a schedule of values for the project. This schedule will list the assigned value the contractor places on each component of construction. Figure 6-1 is a sample schedule of values where each item is identified by its CSI specification number: Section 1—General conditions; Section 2—Sitework; Section 3—Concrete; and so forth, each listed with a brief description.

When discussing proposed contract requirements with the general contractor, he or she should be asked to submit a schedule of values for review and approval by the architect and owner; upon approval, it will be included as a contract exhibit. If a value is in question, the general contractor should provide sufficient supporting documents to justify it.

The designation of component costs is important because, once approved by the architect and owner, it becomes the basis for establishing values used to calculate progress payments. If sufficient detail is not provided or more detailed breakdowns are required, the contractor should be advised to produce them.

Quite often when a few construction components are of significant cost, further detail may be required in order to analyze their value properly. And since the approved schedule of values will be used as the basis of the contractor's application for payment, more information may be required so the architect and owner can review the status of work-in-place and compare against the total value of a particular line item.

For example, giving a value of $125,000 for doors and windows may be insufficient to discern that 35 percent of that item has been completed as of the date of the application for payment, but if the schedule of values was broken down further into $75,000 for windows and $50,000 for doors, the analysis is easier. An even further breakdown of labor and materials would make it simpler still:

Windows: Labor $30,000   Materials $45,000   Total $75,000

Doors and Hardware: $50,000

Labor $12,000   Materials $30,000   Total $50,000

Hardware $8,000

A	B	C
ITEM NO.	DESCRIPTION OF WORK	ORIGINAL SCHEDULED VALUE
01000	General Conditions	900,000.00
01601	Winter Conditions	75,000.00
01801	Plant & Equipment	314,630.00
02061	Demolition	588,115.00
02081	Hazardous Material Removal	279,643.00
02121	Alterations	500,831.00
02171	Shoring/Scaffolding	614,934.00
02201	Sitework/Earthwork	104,500.00
02351	Piles & Caissons	148,557.00
03001	Concrete/Formwork/Rebar	407,151.00
03401	Precast Concrete	566,900.00
05121	Structural Steel	1,746,694.00
05501	Misc. Metals	826,082.00
06101	Rough Carpentry	124,445.00
06201	Finish Carp/Millwork	122,126.00
07111	Waterproofing/JT Sealants	321,002.00
07411	Metal Panels	161,250.00
07501	Roofing System	157,681.00
08121	Doors/Frames/Hardware	17,607.00
08311	OH Doors/Frames/Shutters	58,140.00
08501	Windows/Glass/Glazing	245,917.00
09251	Gypsum Drywall	413,193.00
09511	Acoustical Ceilings	8,645.00
09601	Interior Stonework	72,628.00
09686	Wood Flooring	17,566.00
09901	Painting & Wallcovering	120,589.00
10161	Toilet Partitions	9,804.00
10521	Fire Extinguishers	2,137.00
	Cumulative Subtotals	8,925,767.00

FIGURE 6-1

A Schedule of Values with each Item No. listed according to the Construction Specifications Institute (CSI) numbering classification.

Since the owner will be submitting payment to the contractor based on the percentage of work completed as represented by the schedule of values, paying more than the actual value of the work may create a problem if the general contractor defaults on the project; the owner will have paid more than the work is worth, leaving less funds to complete the project.

The cutoff date for the value of work in place for requisition purposes is another topic to be addressed. A standard contract provision regarding progress payments is as follows:

> *The Contractor shall submit the Application for Payment to the Architect on the 7th day (or whatever early date is determined) of each month for the cost of work performed during the preceding month.*

Quite often a contractor will "project" the amount of work he or she expects to complete by that day or whatever date is indicated, and in many instances, he or she will be close to that projected value—but not always. If the project is in the site work stage and the contractor has projected the amount to be completed for payment purposes but it rains for five days, he or she will not have put in place the value of the work initially committed. In countering a claim like this, the contractor will state that the 10 percent retainage withheld from each payment request will compensate for any shortfalls in the estimate of completed work. In some ways that explanation offers some security to the owner (but not really).

Retainage and completed work are two different issues: Retainage is money withheld to ensure satisfactory completion of the project. The amount approved for payment is the value of work in place at the time of the payment request, verified by the architect.

There is another way to address the percentage. The contract can stipulate that the contractor is to request payment for all work completed as of the date of the application for payment. If that is the case, the architect and owner have the responsibility to promptly process the contractor's request for payment, because, by continuing to work while the payment request is being processed, the builder will actually be placing much more value in place. The practice of "projecting" costs should be discouraged for the reasons just mentioned.

The other concern when reviewing the contractor's proposed schedule of values is determining whether there is any evidence of "front-end loading." Owners may hear this term when a discussion of the schedule of values takes place. Certain work tasks take place early in the construction process: Site work, underground site utility installation, and concrete foundations are all early activities.

When a contractor "front-end loads" her schedule of values, she assigns more value (cost) to those early activities than their true value. If the progress payment is approved on that front-end-loaded basis, the contractor will have received more money than the value of the work put in place at that time. This affects an owner in several ways. The owner will have borrowed more construction financing funds for a longer period of time than necessary, thereby incurring more interest costs.

But just as important, if the contractor had front-end-loaded the project and subsequently defaulted on the contract, the owner may have paid out more money than the actual value of work in place at the time of the default. The remaining funds may not be sufficient to pick up the pieces and negotiate a contract with another contractor. Reviewing the contractor's proposed schedule of values with the front-end-loading concept in mind is another task for the architect and owner to consider.

EXHIBITS

Standard contracts are frequently expanded by adding specific terms and conditions to standard provisions by the use of exhibits to elaborate on the unique requirements of the current project. Exhibits list many of the modifications, amplifications, and additions to the basic contract form. One or more exhibits, labeled either Exhibit 1, 2, 3, or A, B, C, become integral parts of the standard contract and elaborate and expand on qualifications, inclusions and exclusions, approved management hourly rates, contract alternates and allowance items, milestone completion schedules, the detailed schedule of values, unit prices, special insurance requirements, and lien waiver requirements.

From the general contractor's viewpoint, the need to explain and clarify various portions of a contract and lessen or avoid disputes and disagreements is welcomed. Figure 6-2 shows Exhibit J, an exclusion and clarification list. Figure 6-3 shows a more detailed clarification and exclusion list developed according to each specification section—in this case, for Division 1: General Conditions.

An alternate exhibit can be created to serve many purposes. In a cost-plus-fee GMP contract, an owner may wish to have the option to add work if it looks as if there may be some savings accruing from the contractor or if added funding becomes available as the project progresses. This can also apply to lump-sum and construction-management (CM) contracts when there are no savings provisions but an owner would like the option to add work at predetermined costs at certain times during the project.

Alternate exhibits

Alternates can be included in the bid documents where contractors' proposals are requested to include costs for potential added work, or these items can be negotiated with the selected contractor as a value assigned to each alternate and incorporated in the contract for construction. In fact, if the alternates have been included in the bid documents, the owner will receive several prices on each alternate, which might help determine the true value for that work when negotiating the contract with the successful bidder. A slightly higher contract price but a lower price for one or more alternates that the owner expects to accept might actually result in a better deal.

Qualification or Exclusion and Clarification Exhibit

(This happens to be for a renovation project in an urban area)

Exhibit J- Exclusion and Clarifications

The following items are excluded from our proposal:

1.Abatement of any hazardous materials is excluded with the exception of lead based paint on the structural steel as required for the new installation.

2. We have excluded handling or shipment of any contaminated soils. Removal of soil and urban fill to an unlined site is included.

3.We have excluded any underground obstructions

4. We have assumed that the owner will furnish all required temporary power

5.We have included an Allowance for Winter Protection in the amount of $50,000 and will notify owner prior to reaching this amount.

6. Police protection, if required for any street closings, is excluded from this proposal

7. We have included Temporary Protection from all areas other that 2nd floor from Cols A-1 to A-7 to B1 to B-7 on the , 3rd floor- same as 2nd floor, 4th Floor Col C-6 to C-10 to D-6 to D-10.

8. Masonry restoration was not included except as need for the new structural steel installation.

9. No spray fireproofing is included other than as required for the new structural steel

10.We have assumed that all furniture, art work, wall hanging, equipment will be removed by the owner and stored out of the construction area.

11.We have included removal of finishes with the "Area of Work" only

12. We have not included Specification Section 07199 since limits of area where work is to be included is unclear.

13. We have not included an overhead coiling door as specified in Specification Section 08334 inasmuch as none is shown

14. We assume all signage and furniture is furnished and installed by others

15. Fire protection is included in Rooms 235, 248. This work is not shown on the drawings

16. Costs for builder's risk insurance premium are to be paid directly by the owner

17.All building permit costs are excluded and are to be paid directly by owner to appropriate city authority.

FIGURE 6-2

A contract Exhibit. This one an Exclusion and Clarifications listing for an urban renovation project.

The use of alternates allows the owner to consider adding work while not committing to the work in the contract. It gives an owner a little breathing room to consider adding more scope and more money to the project as work progresses or as funds become available. A simple alternate clause would be similar to this:

Alternate No.1—$28,000—Add a kitchenette to the Employee's Lounge (Room 244) to include cabinets, sink, dishwasher, disposal, laminate countertop,

Clarifications & Exclusions

Exclusions and Clarifications

General

1. Proposal is based on 95% documents dated 01-04-2008.
2. Proposal is based on a June 2008 construction start.
3. General conditions are based on a 12 month schedule.
4. Only individual trade permits are included. Building and grading permits and MDE Notice of Intent are excluded.
5. Alta survey (final property survey) is excluded.
6. Certified survey per specification section 01700-3.4 D. is excluded. _____ will have surveyor verify location of building foundations after they are poured, but will not issue a certified, as-built survey.
7. All utility tap applications and fees (including plumbing impact fees) are excluded.
8. Agreements and application/new service and connection fees with local utility companies (e.g. BGE, water and sewer connection fees) are excluded.
9. Relocation of overhead electric (as shown on the civil drawings) is excluded and is to be by others. Any fees and agreements associated with this work are also excluded. Relocation of any other existing utilities that may be required, whether they are shown on the drawings or not, is excluded.
10. All land development bonds, including but not limited to, public and private utility bonds, stormwater management bonds, landscaping bonds and grading bonds, are excluded.
11. Any requirements (such as investigation, testing, abatement, removal, etc.) for dealing with hazardous materials are excluded.
12. _____ has a standard mold prevention and management plan that will be incorporated as part of our quality control program. Implementation of this plan does not imply that _____ will be responsible for costs associated with mold remediation if cause of mold is due to circumstances beyond _____ s control (e.g. humidity in building due to delay in permanent power due to no fault of _____ to provide Owner with date permanent power required).
13. Changes required as a result of lack of coordination of the contract documents are excluded.
14. General Note No. 1 on drawing A0.1 is excluded; _____ is not responsible for discrepancies on the contract documents and is not responsible for any delays and/or cost impact associated with changes required as a result of discrepancies in the contract documents.
15. All allowances are listed as such in the 215, and will be listed in the contract exhibits.
16. Regarding temporary fire protection: Fire extinguishers will be provided in the building during construction. Standpipes and sprinkler system will be made active when the permanent water supply to the building is in place, when the standpipes and sprinkler system are installed, and when it is feasible to make the system active in the realm of the overall construction schedule.

FIGURE 6-3

Another Exclusions and Clarifications Exhibit, this one was more detailed and began with CSI Specification Section 1-General Conditions.

plumbing, exhaust fan and lighting, five 120/240-volt outlets, including all related mechanical, plumbing, and electrical rough-ins per Plainfield Architect drawing SK-25, dated June 15, 2009.

Note that some work can be added via a narrative and a small a sketch, but other alternates require an architect/engineering drawing to fully define the work.

The time frame in which an alternate can be accepted or rejected should be included somewhere in the contract or directly after the description of the work. The owner may be required to accept or decline the alternate work in a time frame where it can be incorporated in the normal flow of contract work. In the preceding kitchenette example, the contractor prepared the estimate for this alternate based on having the authority to proceed before the walls in the lounge were completed in order to install the necessary plumbing and electrical rough-in work required before those walls were sheetrocked.

Also in the case of this kitchenette work, the contractor might qualify the alternate by stating that the work must be accepted before the date indicated in the progress schedule when this area is to be sheetrocked, or the contractor can include a specific date when it must be accepted or rejected. If the alternate is accepted after this date or when the work in that area has been completed, the owner should be prepared to incur extra costs if this window of acceptability was not met and work in place must be removed to accommodate the alternate.

Allowance exhibits

Another method of creating a delay in deciding to add or change an item of work can be accomplished with a separate allowance exhibit or combined in an alternate-allowance exhibit. Let's say that an owner had not made a decision on the quality level and type of carpet to be installed in a conference room because she needed to coordinate this material with the interior designer or needs a decision delay for some other reason. The scope of work to carpet that room would have been established and represented on the contract drawings, and the contractor would have included a cost based on a quality level included in the specifications and the time frame for installation in the project schedule. The value of the "contract" carpet materials and labor to install, upon review and approval by the architect, will be the basis for inclusion in an exhibit as an "allowance" item. This simply means that the contractor will include that amount in the contract schedule of values for the material and installation of conference room carpet as shown on the plans and to a quality level included in the specifications.

When the owner's interior designer makes the final selection, the contractor will submit the specifications to the flooring contractor for pricing. If the final price for the material is less than the allowance, the owner will be due a credit. If the actual cost is higher than the allowance, the contractor is entitled to an "extra" if the owner decides to use the more expensive carpet. Unless this accepted carpet change requires more or less installation labor, demonstrated to the architect's satisfaction, the installation costs will not change.

Reconciling an allowance item is a little more complicated than this description. What did that allowance actually include? Did the "allowance" include labor, material, and related taxes and the contractor's overhead and profit percentage? This should have been spelled out when the architect specified the allowance item in the bid documents or when the contractor responded to the request.

If the full terms and conditions of the allowance item are unclear, all parties can refer to AIA 201 Document—General Conditions if that document was made a part of the construction contract or, if not, use those parameters to resolve the matter. Article 3 addresses allowances along with other issues. Article 3.8 states that unless otherwise provided in the contract, an allowance includes the following:

- Cost of material, equipment, and all associated taxes

- Cost to unload, distribute, and install the work

- The contractor's overhead and profit and other contemplated expenses for that allowance item

- Whenever costs are more or less than the allowance, the contract sum is adjusted by change order to reflect the difference between actual costs and the allowance

Selection of an allowance item is time-sensitive, much like the election to accept an alternate, and this article in the General Conditions document recognizes that fact. Article 3.8.3 requires the architect or owner to select the material or equipment *in time to avoid any delay in the work.*

Unit prices

Another exhibit regarding unit prices can be prepared and attached to the construction contract if the work presents an opportunity to use unit prices. A unit price is a price established to install a unit of some item of work. It can be a cubic yard of topsoil, a square foot of drywall, a light fixture, or some other readily definable item. When it is anticipated that some items of work may be added as construction proceeds but the number of units have not been firmly established at the time the construction contract was executed, the inclusion of unit prices, one for each item listed, can be included in an exhibit.

For example, the owner's data-communications consultant may suggest that more 120-volt receptacles should be added in the information technology area if new equipment is purchased, or the owner may decide to add another small office or two; these items of work can be defined, unit-priced, and included in a unit-price exhibit.

Unless stated otherwise, a unit price will include all labor, materials, equipment, and contractor's overhead and profit. The items for which unit prices are requested can either be included in the bid documents or negotiated with the successful general contractor as the terms and conditions of the construction contract are being prepared. But there is another consideration when unit prices are developed: the approximate quantity of the items that could be installed. Many general contractors provide just one unit price without including any limitations on the quantity. As quantities of unit prices increase, the unit price should change somewhat; more units should be less expensive, and conversely fewer items should be priced higher. For example, a unit price for one 120-volt electrical receptacle may be worth $275 per outlet, but if 20 are added, that unit price should be substantially reduced. By establishing a sliding scale for unit prices, both contractor and the owner will have been treated fairly. Using those electrical receptacles as an example, the sliding-scale unit price may look like this:

> 120 volt electrical receptacles:
>
> 1–5 $300
>
> 6–10 $265
>
> 11–20 $225
>
> 21 and over $200

The same can be applied to other work: excavation and/or backfill as cubic-yard unit prices; painting, drywall, and partitions as square-foot unit prices; and doors, frames, and hardware as "each" unit prices.

Once again, the architect's General Conditions A201 document weighs in on unit prices. Article 4.3.9 of that document states that although unit prices have been established in the contract, when they are incorporated in a change order or a contract change directive (CCD) and result in a substantial inequity to the contractor or owner, "The applicable unit prices shall be equitably adjusted." This problem can be averted when a sliding scale of unit prices is proposed and accepted by all parties.

GUARANTEED MAXIMUM PRICE (GMP) CONTRACT

The estimate in a GMP contract is generally prepared by the general contractor on incomplete drawings and is based on projecting or anticipating what those final plans and specifications will contain and what their associated costs will be. This estimating process often occurs when the plans reach the 70 to 80 percent completion stage. At that point, the specification manual will have been completely assembled, even though the plans are not.

The contractor has some financial exposure for that unknown but reasonably anticipated final scope of the project's plans, and it is not unreasonable for the contractor to include a contingency clause in the estimate. The amount of the contingency will be held separate in the estimate—often calculated at 5 to 10 percent of the estimate—and will appear as a separate line in the schedule of values and the application for payment.

Even though this is an accepted practice, an owner would be prudent to include contract provisions that place limits on the use of this contingency, whose purpose is to cover unanticipated circumstances encountered by the contractor. Some contractors, in the absence of restrictive language, will tap the contingency account to correct work deemed substandard or nonconforming by the architect or include costs for work inadvertently omitted from the estimate but required by the plans and specifications. An owner needs to construct contract language defining how and under what circumstances the contingency can be used by the contractor.

The following clause(s) inserted into the owner-contractor agreement may be helpful in defining the purpose of the contingency. The contingency can be tapped for any of the following reasons:

1. Changes resulting from refinement to the contract documents that could not have been reasonably anticipated

2. Changes required due to an encounter with unknown or concealed conditions

3. Added costs due to delays in receipt of materials caused by strikes or other unforeseen events, unless directly attributable to the contractor

4. Cost overruns for procurement or purchase of materials as a result of a defaulting trade contractor (or subcontractor), provided that the default was not caused by the gross negligence of the contractor or subcontractors or by willful misconduct or breach of contract on either the part of the contractor or the subcontractors

5. Deductible amounts a contractor is required to pay for claims under insurance policies obtained by either the owner or contractor, except in cases where these claims are a result of the contractor's or subcontractor's negligence, willful misconduct, or any breach of contract

6. Costs to expedite or accelerate the progress of work to meet the contract schedule, provided that these costs were not due to gross negligence, willful misconduct, or by breach of contract among owner, contractor, or subcontractors

7. Any unforeseen events and/or conditions stated elsewhere in the contract for construction as reimbursable costs

The contractor should notify the owner, in writing prior to incurring any cost, that he or she proposes to charge against the contingency account. Before such costs can be charged against this account, written permission from the owner is required, indicating that the nature of the costs are reasonable and proper and meet the requirements of whatever paragraph or section appears in the contract.

Under the terms and conditions of the GMP contract, contingency costs are not reimbursable as the cost of work once the contingency account has been exhausted. In the event that the contingency account has not been exhausted, any residual monies are to be treated as a credit to the GMP and cannot be applied to overages in other line items. The contingency is not to be used if overruns are experienced in a line item of work that is indicated in the approved schedule of values. It is standard procedure to allow the contractor to apply a cost overrun in one line item in the schedule of values to a line item that reflects a savings, as long as the contract sum is unaffected.

Limiting an owner's exposure to final plan development costs

A contract provision can limit to some extent one of the disagreements that often occurs in GMP contract administration. The contractor may argue that the final plans and specifications exceeded expectations or required additional work. The contractor will claim that the limits of work or quality of materials as represented in the completed plans and specification were in excess of what a standard project design would have included.

One way to avoid this kind of situation is to have the general contractor review the plans and specifications as they develop and either accept them or resolve any "scope" and cost issues with the architect as they occur. The contract language, which the contractor reviews and approves, can be used to limit any such claims. Here is an example of a clause:

> *The Guaranteed Maximum Price (GMP) for the complete performance of the Work set forth in the Contract Documents, including the Contractor's fee shall be _%_. The GMP shall be decreased or increased, as provided in accordance with this Agreement. To the extent the Cost of Work plus the Contractor's fee exceeds the GMP, the Contractor shall be responsible for all excess Cost of Work.*
>
> *To the extent that the Contractor has based the GMP on any qualifications, exclusion and assumptions, they have been set forth in Exhibit _X_. Notwithstanding anything herein, Contractor shall have no claim for an increase in the GMP arising out of the further refinement and detailing of the drawings and Specifications listed in this Agreement, if such drawing and specification is incomplete. Such further development does not include such items as changes in scope, systems, kinds and quality of materials, finishes or equipment, all of which, if required, will be incorporated by Change orders.*

A cost plus a fee with a guaranteed maximum price is an "open book" contract. A common provision in a GMP contract is one that gives the owner the option to audit the contractor's books to verify that all reimbursable costs are justified. Although rarely enforced, this provision does reinforce the concept of having the contractor reveal all costs to the owner. This requirement extends to the review of all subcontract proposals submitted to the general contractor and to purchases of major pieces of equipment to ensure that the most competitive, competent source has been tapped. This procedure can be memorialized in the contract for construction as follows:

> *The Contractor shall obtain bids from at least three (3) qualified subcontractors for each phase of subcontracted work. The Contractor shall furnish the Owner with a list of all such subcontractors to whom they plan to award a subcontract agreement prior to that award. The Owner will notify the Contractor of an objection to any Subcontractor proposed by the Contractor.*

The Owner may elect to participate in any subcontractor negotiation meetings, buyout meetings or other such meetings between Contractor and Subcontractors after being given at least three (3) days' prior notice.

Copies of all bids received by the Contractor from the subcontractors are to be delivered to the Owner, within 48 hours of receipt, with a proposed selected award and the reason for recommending that Subcontractor.

If the Owner directs the Contractor to accept a Subcontractor's bid proposal that is in excess of that bid recommended by the Contractor, any such excess will be treated as cause for a Change Order that will, in effect, increase the GMP.

Documenting general conditions costs

General conditions costs can amount to anywhere from 5 to 12 percent of the actual cost of construction or even higher, depending on the nature of the work. In a GMP contract, the extent of the reimbursable costs should be clearly enumerated in the contract. These costs vary from contractor to contractor; some include a few of these items in their overhead. Figure 6-4 is a sample of a typical listing of general conditions items.

Some of the equipment in the general conditions category may be purchased by the contractor instead of rented or leased. Laptop or desktop computers, scanners, printers, fax machines, and other equipment may be purchased because it is more cost effective than leasing for the extended period of time of construction. It is good to remember that the GMP is a "cost plus" contract, and some of these purchased items may have residual value at the end of the project. A laptop or desktop computer will probably have considerable remaining life after an 18-month or two-year project.

As these types of items are purchased, the general contractor should begin to develop a list of all such products so it is easy to effect a final settlement at the end of the project. Figure 6-5 is an example of a small tools list included in a GMP contract. At the conclusion of the project, a residual value for each item retaining value will be established.

The contractor can present the owner with two options: turn over that equipment to the owner at no cost or, when they mutually arrive at the equipment's residual value, provide the owner with the appropriate credit and keep the equipment for the contractor's use on another project.

CHANGE ORDERS

Change orders, which we discuss in more detail in Chapter 8, can occur for many reasons: a change in scope as directed by the owner, a request from the general contractor for added costs due to a recognized and an accepted omission in the contract documents or some unforeseen condition arising generally in the site

Exhibit L

General Conditions Costs

General Conditions Costs include the costs associated with the following items:

Item Description

Administration
    Project Management
    Field Supervision
    Punch List Supervision
    Clerical

Field Office & Equipment
    Temporary Office Trailers
    Telephone Service
        Service Connection
        Hardware & Software
        Service/Usage/Long Distance
    Office Cleaning
    Office Furniture
    Office Equipment (Hardware/Service)
        Copy Machines (Lease/Service)
        Drawing Copy Machine (Lease/Service)
        Plotter (Lease or Purchase)
        Facsimile Machines
        Computers (Lease)
    Office Supplies
        Postage/Overnight Service
        Radios

Security Services

Safety & First Aid
    Safety Manager
    First Aid Supplies
    Drug Testing Supplies
    Fire Extinguishers
    Barricades (Labor)
    Barricades (Materials)

Safety Materials

Temporary Job Construction

Exhibit L – Page 1

FIGURE 6-4

A Typical listing of each item in this contractor's General Conditions Schedule of Values, attached to the contract as an Exhibit.

Project Signs (with approval)

Progress Photos

Record drawings (Includes ½ Scale As-builts and CAD)

Schedules and Monthly Schedule Updates

Purchasing Costs

EDP Services

Monthly Accounting Fees

Pest Control

Drinking Water

Exhibit L – Page 2

FIGURE 6-4 *Continued*

No.	Quantity	Manufacturer	Description	Model No.	Serial No.
			Small Tools List		
1	1	Bosch	4.5 Paddle Switch Grinder - 8Amp	1810	791003399
2	6	CEP	50' extension cords	CEP-50-123	
3	1	Bosch	1" SDS-Plus Rotary Hammer D Handle	Bosch 1125	790002832
4	7	Bosch	1/4 15 Amp Circular Saw	Bosch - CS10	886002439
5	1	Milwaukee	Milwaukee Super Saw Zall	Mil-6509-22	A17EL08221445
6	1		8' Fiberglass Step Ladder	Ladder -S-8	
7	1		6' Fiberglass Step Ladder	Ladder -S-6	
8	1		4' Fiberglass Step Ladder	Ladder -S-4	
9	1	Bosch	14.4V Brut Tough Drill/Driver	Bosch 35614	981000236
10	1	Bosch	14.4V Fastener Driver	Bosch 23614	5601909422
11	1		Classic Geyser 3 1/8 HP Sump Pump Simmer	PUMP -2300	
12	1		HD Contractor Water Hose 3/4 x 50" Rated @ 250 PSI	Water Hose	
13	4	Dohla	Hardwood Dollys w/ Carpeted Ends 30"x10	Dolha	
14	1	Bosch 1590	Bosch Top Handle Jig Saw Kit 6.4 AMP	Bosch 1590	884000995
15	1	CEP	500 Watt Portable Halogen Work Light	CEP-51251	
16	1	Bosch	Bosch 120V Heat Gun	Bosch 1944	783000961
17	4	Razor Back	Razor Back Shovels	UNI43201	
18					
19					
20					
21					
22					
23					
24					
25					
26					
27					
28					
29					
30					
31					
32					
33					
34					
35					
36					
37					
38					

FIGURE 6-5

A List of Small Tools purchased under a Cost plus or GMP contract to be used to establish residual value at the conclusion of the project.

work phase of the project, or, in the case of a renovation or rehabilitation project, concealed conditions uncovered as work progresses. Although there is little the owner can do to handle some of these claims for extra work, specific language can be inserted into the contract that will make those requests somewhat easier to resolve.

To deal with some changes, an owner may request a contractor to perform some work on a time and material (T&M) basis, especially when it must be performed quickly to avoid creating a delay in the work progress or where it is difficult to fairly estimate the lump-sum cost of the work. The use of T&M work, if documented properly and supervised by the general contractor, is a fair way of proceeding with extra work, but some ground rules must be established to account for both the time expended and the materials used.

Procedures for documenting all T&M work and specific percentages for allowable overhead and profit will provide a degree of control over this work. There are occasions when an owner might authorize premium time work, and a clarification statement in the contract will be helpful. Premium time work is defined as any work beyond the normal eight-hour workday and beyond the normal five-day workweek. Premium time work is divided into two categories: Time and a half generally applies to all work extending beyond the normal eight-hour workday or the normal 40-hour workweek; double time usually applies to Sunday or holiday work.

Some union collective bargaining agreements have specific requirements that define both stages of premium time work, and it is best to inquire about those union rules beforehand. An owner can request a copy of the union agreement.

Another area where additional costs are likely to be expended is a General Conditions item that refers to winter or weather conditions. When working in a geographic area where cold temperatures prevail in the fall and winter, a contractor will incur additional costs to enclose and heat portions of weather-sensitive work. The contractor must also include costs for ice and snow removal and other cold-weather-related activities. Since the contractor has no control over the weather and cannot predict what costs to include in the estimate for these "winter conditions," he or she can approach them in a number of ways:

■ Include winter conditions as an allowance item in the contract

■ Exclude all winter conditions and request a change order to cover the actual costs incurred

■ Include sufficient money to cover all costs (and you can be sure that more than enough money will be added for that purpose)

Most general contractors and owners will elect an allowance option to deal with winter conditions, and in some geographic locations, provisions for hot weather may be required as well.

When change-order work is required in geographic areas where hot weather prevails either most of the year or in typical summer months, the building may be air-conditioned, and working premium time hours will require the general contractor to incur additional utility costs to operate the cooling equipment. Concrete placed in hot weather may require water misting during the curing cycle, and if this is extra work, these misting costs will be added to the cost of work.

If winter or other weather conditions are specified as an allowance item, the contractor must set forth measures to fully document not only all labor, material, and equipment costs required but an accurate recording of daily temperatures to justify the need for these extra costs

Controlling overhead and profit

General contractors include an overhead and profit percentage on all change-order work, and their subcontractors also include a percentage for overhead and profit in their cost proposals to the GC for the change-order work they perform. These are known as second- and third-tier subcontractors: specialty contractors who work for either the prime subcontractor or a second-order sub.

This frequently occurs in heating, ventilating, and air-conditioning work, where the prime subcontractor may purchase and install all of the HVAC equipment but contracts other portions of the work such as sheet-metal duct fabrication and testing and balancing of fans, air-conditioning, and heating equipment to other firms. By adding specialty contractors to the prime subcontractor's costs, which are subsequently folded into the general contractor's change order, the amount of overhead and profit can really add up.

Let's use an example of a change-order request from the owner for the addition of a kitchenette in an employee lounge. This may require installation of an exhaust fan, ductwork, ceiling modifications, and electrical connections. Subcontractors customarily add 15 to 20 percent to their cost of work. This change order will require work by the HVAC sub and the sheet metal duct subcontractor; the air-balancing contractor; and other subs, including the electrician—all of which, when included in the general contractor's proposal, will add multiple overhead and profit costs.

The HVAC contractor subcontracts ductwork to ABC sheet metal:

Cost: $200 + 15% OH&P ($30) = $230

The balancing subcontractor charges:

Cost: $50 + 20% OH&P ($10) = $60

The HVAC subcontractor's total price, including ABC and the balancing sub:

Cost: $3,300 + 15% OH&P ($495) = $3,795

The electrical contractor's total cost:

Cost: $750 + 15% OH&P ($112.50) = $862.50

The general contractor's total price, including all of the above, the drywall and ceiling work, and buying and installing the kitchen cabinets and appliances:

Cost: $12,000 + 15% OH&P ($1,800) = $13,800

The final cost of $13,800 includes a total of $2,447.50 in various overhead and profit add-ons, which represent 20 percent of the general contractors "cost." These multiple overheads and profits can be limited by inserting a clause in the contract agreement with the general contractor:

> *On Change Order work performed by the General Contractor's own forces, that Contractor shall receive an amount not to exceed 5 percent of direct costs for overhead and an amount not to exceed 10 percent for profit.*
>
> *Contractor shall receive a fee including overhead and profit to not exceed 5 percent for all subcontracted work for which the Subcontractor shall not apply a amount for overhead and profit to exceed 15 percent of that Subcontractor's direct costs.*
>
> *On Change Order work performed by one or more second- or third-tier subcontractors, the total allowable overhead and profit for each of these sub-subcontractors shall not exceed 15 percent, and the prime subcontractor shall apply an overhead and profit allowance not to exceed 5 percent.*
>
> *Notwithstanding the above, to the extent that Change Order in its aggregate results in a net increase in the Cost of the Work that exceeds 5 percent of the original GMP specified in (whatever section of the contract it is so stated), General Contractor shall be entitled to payment for Contractor's fee in connection with each subsequent Additive Change Order in the amounts specified above.*
>
> *In the case where a Change Order results in a decrease in costs, the same overhead and profit allowances specified above for additive Change Order will apply to those costs in the deduction.*

Another way to control the amount of overhead and profit that a general contractor applies to a change order is to specify a decreasing percentage as the value of the change-order work increases—in other words, an overhead and profit percentage on a sliding scale. As an example, a contract provision can be added as follows:

For the contractor for work performed by his own forces:

Up to an including $100,000	Allow 15%
$101,000 to and including $200,000	Allow 10%
$201,000 and over	Allow 5%

Of course, these dollar-value limits can be adjusted to present other reasonable amounts of value based on the size of the total construction contract. And this same sliding scale can be applied to subcontracted work; however, traditionally subcontractors add on 10 percent overhead and 10 percent profit or 15 percent overhead and profit and may well balk at a total of 5 percent, although a lower overhead and profit percentage than initially proposed by them might be arranged. And the general contractor's fee on all subcontracted work can be stated as a sliding scale as indicated previously.

INCLUDING A "DEFINITIONS" SECTION IN THE CONTRACT

Some attorneys will suggest including a list of definitions in the contract as a way to ensure that both contractor and owner have the same interpretation of a specific terminology. The contract seems straightforward, but some terms may need explaining, and a construction attorney usually has compiled such a list from previous contracts.

Consider the term *contractor*. Although we assume we know what this means, confusion may arise over whether it refers to a general contractor or a subcontractor. The term *contractor* in Specification Division 16—Electrical Work may refer to the electrical contractor, not the general contractor. When the contract states that the contractor is allowed a certain overhead and profit, for example, and the term *contractor* is vague and may apply to either a general contractor or a subcontractor, the need for a definitions article becomes more apparent.

What does *final completion* mean? We know that *substantial completion* means that the building is suitable for the purpose for which it has been designed, but maybe we need to define *final completion,* which can mean, in addition to substantial completion, that the work is 100 percent complete, all closeout procedures submitted and approved, all punch list items completed, all warranties and guarantees submitted and accepted, consent of surety (if job was bonded) provided, final lien release from the general contractor received, and the architect's certificate of final completion and acceptance executed. A comprehensive list of definitions is just another step to take to reduce any ambiguities that may occur between contractor and owner.

LIQUIDATED DAMAGES

When the completion of a project per contract time is essential, many contracts include a provision known as liquidated damages. Owners of commercial or retail buildings who have signed leases with tenants must have their buildings and tenant spaces completed per the contract schedule. College dormitories must be completed on time, or incoming students may have to be housed temporarily off-campus at considerable cost to the college or university.

A liquidated damages clause reinforces the need of adherence to a schedule, but the contract schedule will include any increases in the original schedule if and when change orders have been approved, increasing both contract sum and contract completion dates. A typical liquidated damages (LD) clause, inserted in the contract provision relating to "contract time," will read something like this:

> *If the Work is not substantially completed in accordance with the drawings and specifications, including any authorized changes, by the date specified above (the contract time is inserted here), or by such date to which the contract time may be extended, the contract sum stated in Article XX (whichever provisions contains the contract sum) shall be reduced by $_____ (the daily value of the*

LD) as liquidated damages for each day of delay until the date of substantial completion.

When the term *day* is used, the contract must be clear as to its meaning. Does it mean calendar day or workday? There is a difference.

The daily value of the LD is meant to represent the costs the owner estimates he will incur each day the building is unable to be occupied for the purpose for which it is intended. This clause is not meant to be a penalty clause, and many courts will rule that if deemed so, an incentive clause must also be added to the contract for early completion or else the penalty clause is void. Even though the contract states that the liquidated damages clause is not a penalty clause, certain criteria must be met for it to be enforceable:

- The amount established as the daily dollar amount of damages must be a reasonable assessment of the costs the owner will actual incur if delivery of the project is late.

- The breach of contract—the inability to finish on time—must have been difficult to establish at the time the contract was executed.

A contractor may still face damages for late delivery of a project even if a liquidated damages clause is not included in the contract; the contractor may be subject to "actual damages." When LDs are a part of the contract, the project owner calculates the reasonable amount of damages she will incur if the project is not delivered on time, which in some cases may be less than those actually incurred, since the former is an estimate and the latter an actual figure. Costs such as higher interest rates for construction loans, loss of revenue from leases, increased moving costs, and costs to remain in existing rented space will all be folded into the daily LD damages. But where "actual" costs are assessed to builder for late delivery, they represent an accounting of the actual costs, which may exceed the projected costs gathered prior to the start of construction.

When a contract includes an LD clause, a contractor will begin to document every potential delay that he or she may have otherwise overlooked if no LD clause had been included in the contract. If a query to the architect is not responded to in a day or two, the contractor will fire off a letter stating that the lack of a prompt response will cause a delay of a day or two or three. When a shop drawing submitted to the architect or engineer has not been processed in the time allotted in the contract, another letter will be sent off to the A/E, stating that the delay in returning the shop drawing will delay delivery of the material or equipment by a week or so. Any decisions required of the owner will be watched closely, and if the response is not received by the contractor in a reasonable period of time, it will be cause for another claim for delay.

The liquidated damages clause puts all parties to the process on notice that contract responsibilities will be carefully noted and documented, and thereby it places another burden on the owner to be responsive to obligations and also monitor the actions of the design team. Contracts with LDs usually end up with lots of paper flowing back and forth and a negotiated settlement.

ADDITIONAL CONTRACT PROVISIONS

In my experience working in the general contracting industry and later as a consultant to owners, some contract provisions may be beneficial to an owner and should be reviewed with the legal team. The following sections cover some of the more important ones.

Ensure that the contractor fully understands the nature of the work, particularly when the project involves some renovation or addition work. This clause will help:

> *As of the execution of this Agreement, Contractor has reviewed the plans, specifications, reports, and other information provided by the Owner, inspected the site, verified field conditions under which the work is to performed, familiarized himself with all known and observable conditions in which the Work is to be performed. Based on that review and the Contractor's experience and knowledge, and except as otherwise provided for in this Agreement, contractor acknowledges and agrees to the following conditions:*
>
> *1. The time provided in the Schedule for the performance of the Work is sufficient.*
>
> *2. That subject to subsequent changes in the work the lump sum (or GMP) is reasonable compensation for the work, including all reasonably foreseen and foreseeable risks, hazards, and difficulties in performing the work.*

One of a project manager's prime responsibilities is to review the shop drawings submitted by the subcontractors and vendors to ensure that they are complete and comply with the requirements of the plans and specifications. This careful review, however, is not always performed, and if it isn't, it places an added burden on the architect and engineer, who spend more time reviewing and rejecting these non-conforming submittals. And these added reviews can create delays in ordering materials and equipment.

Project managers are sometimes less than diligent in their review of these documents, and when obvious errors and discrepancies in the shop drawings are routinely overlooked or if the drawings don't even meet the plans or specifications, an owner may incur added costs for multiple reviews by the architect/engineers. The following provision will encourage the contractor to review these drawings carefully instead of merely passing them through from subcontractor and vendor to the A/E:

> *The Contractor shall submit complete and accurate submittal data in their first submission. If the submittal is returned requiring resubmittal, only one (1) additional submittal will be reviewed at the Owner's cost. Any additional submittals will be reviewed at a cost to the Contractor.*

From time to time, there will be unexpected delays in the construction process, and if these delays have not been created by the owner or their design consultants, the contractor will be held responsible for getting the project back on track. The following contract provisions may be helpful in advancing the owner's concern about keeping on schedule:

A. If any of the work is not on schedule, the Contractor shall immediately advise the Owner, in writing, of the proposed action to bring the Work back on schedule. In such event, the Owner will require the Contractor to work such additional hours, including any premium time such as extended work days, Saturdays, Sundays, and holidays, at no additional cost to the Owner, to ensure that the work is back on schedule.

B. If the contractor fails to take prompt and adequate remedial action to get the project back on schedule to the satisfaction of the Owner and the design consultants, the owner reserves the right to perform such work as it deems necessary to do so and will back-charge the cost thereof against payments to the Contractor.

(This may be a threat, but to actually carry it through may present a huge quagmire for the owner.)

Occasionally a contractor will be required to perform some time and material work (we examine this type of work in more detail in Chapter 8). Sometimes, the method of establishing costs and the documentation of "time" and "materials" can be murky and result in disagreements about what work was actually performed and how the costs were assembled. The following steps can be taken to clarify this process by adding suitable contract language:

All time and material work will be subject to compliance with the following provisions:

Any work performed on a time and material basis (T&M) shall be subject to audit by the Owner. Approval of all labor rates, overtime, and equipment costs charged by the contractor and the subcontractors shall be subject to audit by the Owner. Reimbursable labor costs shall include federal payroll taxes and other taxes and fringe benefits only to the extent that they have actually been paid by the Contractor or the applicable Subcontractor.

This is particularly important when it comes to fringe benefits, which may not apply in their entirety once certain worker salary levels have been reached and premium time work is authorized. For example, state and federal unemployment taxes (SUTA and FUTA) no longer apply once a base salary level of between $7,500 and $10,000 has been attained. Also, note that state unemployment tax limits vary from state to state. Contractors should be advised to alert their subcontractors to these limits and monitor their costs accordingly.

One general contractor issued a memo to the subcontractors as follows after being alerted to these costs by the owner's consultant:

Re: FUTA/SUTA taxes included in Hourly Rates

Please be advised that the consultant for (the Owner) has questioned the FUTA/SUTA hourly charges in the overtime billings submitted for October, November, and December.

The contention is that the limits for these taxes are reached within the first payroll quarter of the year for most full-time employees and by the last quarter of the year, every employees' limits should have been met.

The consultants are correct in asking for this reduction and they have cut our requisition accordingly. We will be reviewing each requisition submitted to us for overtime and premium time and will calculate credits by each sub-contractor for the FUTA/SUTA charges included in the rates.

In the coming months, the Owner will accept FUTA/SUTA charges as the new payroll year begins; however, we are very aware of the employee limits and the time frame in which those limits are reached. It is safe to anticipate that FUTA/SUTA rates will not be accepted after February, since these limits will be reached within an eight-week period.

This general contractor was a union contractor in a large metropolitan area, addressing subcontractors whose average wage rates were $35 to $50 per hour exclusive of fringe benefits, and that is why the eight-week period was considered the limit for deduction for these costs.

When collective bargaining agreements are in effect in the case of union workers, additional language can be added to the FUTA/SUTA limits:

Any benefit contributions included as fringe benefits in the hourly reimburs-able labor rates shall be limited to those required by the applicable collective bargain agreements. If requested by the Owner, the Contractor will provide copies of any collective bargaining agreement so requested to confirm the base wage rates and applicable benefit costs.

Equipment "costs" are not those arbitrarily assigned a value by the general contractor or subcontractor without regard for standard rental equipment costs. The general contractor or subcontractor equipment costs must be documented in detail to support a claim.

Another clause that is important, particularly when GMP contracts are in force, is one that segregates change-order costs from contract costs. To effect that condition, add the following:

Contractor shall maintain a separate account by job number or other suitable accounting practices, of all costs incurred under this T & M work and shall pre-pare report sheets in duplicate to identify the T&M work performed each day and shall furnish these reports to the Owner or a designated representative on the second day following the work. This daily report shall itemize all elements of cost such as names and classification of each worker, their corresponding hourly rate of pay, hours worked, and if equipment has been provided, the equipment type, size, and hours operated.

Material charges are to be documented with a copy of the vendor's invoice, and if not available at the time the daily ticket is presented, then the invoice is to be submitted within five (5) calendar days. If the vendor's invoice is not submitted within 60 days after delivery of said material or equipment, the Owner shall have the right to establish the cost according to the lowest wholesale price for the material or equipment based upon the quantities delivered to the construction site less any trade discounts.

All daily reports are to be signed by the general contractor's representative for work performed by that contractor's own forces and for subcontracted work; the signature of the subcontractor's supervisor as well as the general contractor's supervisor is required to validate and certify the items contained therein.

Quite often delays occur that are beyond the control of the contractor. Labor or transportation strikes and extended periods of inclement weather that vary significantly from normal weather patterns are two such occurrences. The contractor may request an extension of the project's completion date, and when the owner, A/E, and contractor agree on a reasonable extension of the schedule and it appears that no additional costs are involved, it is important to document that no additional costs will be presented to the owner. This can be covered by a no damages for delay clause in the contract similar to this one:

Any extension of time in which to complete the Work granted by the Owner for items beyond the Contractor's control shall be the sole remedy for any such delay, loss of productivity, hindrance in performance, impact damages or similar claims.

For weather delays, the following can be added:

The Contract time shall not be extended because of adverse weather conditions, nor shall the (lump sum or GMP) be increased due to adverse weather conditions, unless the Contractor establishes to the Owner's reasonable satisfaction that the experienced weather encountered was abnormal for the Contract time, exceeded those that have been encountered historically or may be reasonably expected to encounter at the project site and could not have been reasonably anticipated by the contractor.

Many contract provisions offer some control over change-order work, but nothing succeeds better than ensuring that an experienced, professional contractor and the A/E have established a fair and equitable relationship. But adding some contract language doesn't hurt. It sometimes is beneficial to sit in on the general contractor's change-order meeting with the subcontractor or vendor to learn more about the process and the negotiating skills of the general contractor. When the cost of the change-order work is in question, there are ways to proceed with the work while the actual value is resolved. In Chapter 8, we discuss the construction change directive (CCD), which simply states that if both parties can't agree on the cost of the changed work, a plan is put in place that is fair to all parties involved. A provision added to the construction contract can make this solution a little clearer and allow the work to proceed as cost evaluation continues:

> *If the Owner and Contractor are unable to agree on the amount of any cost or credit to the Owner resulting from a change in the work, the Contractor shall promptly proceed with and diligently prosecute such change in the Work and the cost or credit to the owner shall be determined on the basis of reasonable expenditure and savings.*

Another provision may avoid some change-order work, specifically those details that may have been missed by the architect or engineer but claimed as "extras" by the general contractor. If and when this situation occurs, the following simple statement in the contract may prove helpful in defusing claims for minor missing work items:

> *The Contractor has constructed several projects of this type and has knowledge of the construction and finished project.*

When change-order work involves estimates from subcontractors, participation in subcontractor negotiation meetings can be informative, and contract language can make provisions for such attendance:

> *The Owner at all times shall have the right to participate directly in the negotiations of change order proposals with subcontractors and material and equipment vendors.*

Although the definition of *substantial completion* is basically when the project has reached the stage of completion where it meets its intended use, some specific contract language may alert the contractor to additional obligations to fulfill. The definition of *final payment*, although spelled out in the contract, may also need some added clarification. Both of these terms come into play when items on the punch list remain incomplete. An unfinished punch list is a frustrating event for the owner wants to wrap up the contract and get on with business.

One way to ensure that punch list completion standards are made firm is to attach a provision such as this:

> *Unless otherwise agreed to in writing by the Owner, the project shall not be deemed substantially complete if the items on the Punch List would reasonably require more than two (2) calendar weeks to complete.*

When dealing with those nagging incomplete punch list items at the end of the project, there is another way to get the general contractor's attention, and that is to assign a value to each incomplete item and withhold double or triple that amount from the contractor's payment. For example, if some minor flooring repairs have not been made, obtain a firm, written quote from a reputable flooring contractor in the area and triple the amount. So if the repairs would normally cost $1,500, value them at $4,500, and advise the contractor that this amount will be withheld from the final payment until such time as the repairs are made. Set a deadline, and indicate that if the punch list work is not completed and approved by the architect by a given date, the other contractor will be notified to complete that flooring work and the cost will be deducted from the amount due the general contractor. This requirement can be converted to contract language:

> *When punch list items remain incomplete 14 days after the official final punch list review has been made by the Architect, all incomplete items shall be independently valued.*
>
> *That value will be tripled and withheld from the Contractor's Request for Payment.*
>
> *If, after 10 days' written notice to the Contractor, for each unfinished punch list item remaining incomplete or unacceptable, as deemed by the Architect, the Owner will engage another contractor to complete each item of said Work and deduct triple the actual cost of the work from any final payments due the Contractor.*

These suggestions are provided with the thought that an owner will discuss them with an attorney so proper language can be prepared for incorporation into the contract for construction.

Organizing for the construction process

7

Now that a construction contract has been fully executed and the start of the project is imminent, an owner must ramp up for the responsibilities he or she will face during construction. First and foremost, an owner's representative must be selected, one who will assume the role of the official owner contact for both contractor and architect. Although the architect's responsibilities may have been extended to include construction services, a fully engaged owner is necessary for a successful project. There are a number of responsibilities facing the owner's representative:

- Attendance at all project meetings

- The portal through which all contractor and architect/engineer correspondence will pass

- A reviewer of proposed change-order requests by the contractor, in consultation with the architect, and the originator of contract changes initiated by the owner

- Recipient of copies of requests for information from the contractor, for which a prompt response from the architect/engineer may need to be obtained

- Recipient of "information" copies of all correspondence among contractor, architect, and engineer

- Knowledge of the pace of construction and any problems that may arise with respect to contract compliance

- Approval of the architect's recommendation for progress payments to the contractor

ARCHITECTURAL FORMS

When an architect's role includes construction services, certain terms and forms are used as the job progresses, and an owner should be familiar with them. Here is a partial list of architect- and contractor-generated forms:
- Project meeting minutes
- Proposed change order (PCO)

© 2010 by Elsevier, Inc. All rights reserved.
Doi_No = 10.1016/B978-1-85617-548-7.00007-0

- Change-order form
- Change-order log
- Application and certificate for payment and continuation sheet
- Waiver of liens upon receipt of a progress payment
- Request for offsite storage of materials or equipment
- Affidavit that all taxes have been paid
- Contract CPM schedule and look-ahead schedules
- Transmittal forms
- Request for information (RFI)
- Response to a contractor's request for information
- Architect's supplemental instructions (ASI)
- Contractor's log of ASIs
- Architect's field report
- Contractor's field instruction log
- Shop drawing and sample log
- List of subcontractors
- Certificate of substantial completion
- Contractor's surety to final payment
- Waiver and release of lien upon receipt of final payment

THE PROJECT MEETING

Periodic project meetings on the construction site bring the architect/engineer, general contractor, subcontractors, and vendors together to discuss current and anticipated progress, manpower requirements, and any other project activities and concerns. Some architects prefer to conduct these meetings, while others may choose to have the general contractor take charge of the proceedings and prepare and distribute the meeting minutes. Either method is acceptable as long as the meeting is conducted in a professional manner and the meeting minutes accurately reflect the contents of that meeting.

These progress meetings provide a forum in which all parties associated with the project can raise questions, concerns, and interpretation of contract requirements and discuss all other matters of interest. These meetings are more than mere discussion sessions; policies and procedures are established and memorialized by minutes that become part of the project documentation. Meeting minutes can take on added importance when they are referenced as documentation during a dispute, claim, or litigation proceeding, so their accuracy is critical.

Figure 7-1 is a sample of two pages from the minutes of a project meeting. It exhibits a basic meeting format, which may vary slightly from contractor to contractor, but the basic elements are much the same:

1. A record of each attendee name, affiliation, phone number, and e-mail address, and list of those not in attendance who are to receive information copies

Meeting Minutes No. 14

Project:

Item	Description	Responsible	Open Date	Due Date	Closed	Closed Date
555-2	Upcoming Field Instructions		12/25/2007		☐	

1. Elevation 58 Skylight - Issued as ASK #043 - may change due to floor drain routing
2.   Exposed Ceiling - Release 1/7/08
3.   Exposed Brick - Release 1/7/08
4. Cooler Change - Aramark to make changes fit original layout
5. Bar/Concession #2 Bar dye and counter - Received 1/4/08
6. HC Seating Railing Design Changes - Received 1/4/08.
7.   Wall - Gate E change CMU wall to brick.

NOTE: IS PROCEEDING WITH ALL BFI'S UNLESS OTHERWISE DIRECTED.

Item	Description	Responsible	Open Date	Due Date	Closed	Closed Date
600-1	VACATION SCHEDULE		10/30/2007		☐	

1/29/08:
January 2008:
a.   1/25/08 – 2/2/08
February 2008:
b.   2/2/08 – 2/9/08

Item	Description	Responsible	Open Date	Due Date	Closed	Closed Date
650-1	ATTACHMENTS		10/30/2007		☐	

1/29/08: Sign In Sheet - Schedule updated 1/29/08
1. Open Issues List dated 1/28/08.
2. Open Submittal Log dated 1/29/08.
3. Open RFI Log dated 1/29/08.
4. Sketch Log dated 1/29/08
5. Drawing Log dated 1/29/08
6. BFI Log dated 1/29/08

Item	Description	Responsible	Open Date	Due Date	Closed	Closed Date
700-1	UPCOMING INTERIM PROJECT MEETINGS		10/30/2007		☐	

1. The following meetings are schedule for the week of 1/28/08:

- Monday (1/28)
   - Hot Water Heater & Boiler Exhaust Meeting
   - Subcontractor Meeting

- Tuesday (1/29)
   - MEP Coordination Sign-off Meeting
   - Folding Partition Meeting

- Wednesday (1/30)
   - Misc. Metals at
   - Stair LS-7 Meeting

- Thursday (1/31)
   - MEP Meeting
   -    Schedule Review Meeting
   - Hot Water Heater Follow-up Meeting

- Friday (2/1)
   -    Safety Meeting

FIGURE 7-1

A typical meeting minutes format.

```
┌─────────────────────────┐
│   Meeting Minutes No. 14 │
└─────────────────────────┘
```

Project:

Item	Description	Responsible	Open Date	Due Date	Closed	Closed Date
125-1	PERMITS		10/30/2007	02/05/2008	☐	

1. Fuel oil permit approval by BFD - Target 1/25/08.
2. · received information from that BFD still has not completed their review of documents.

Item	Description	Responsible	Open Date	Due Date	Closed	Closed Date
126-1	Testing & Inspections		11/06/2007	02/05/2008	☐	

1. Waterproofing Inspection advised that an initial meeting between and has already happened , to set up a meeting with to discuss expansion joint details by 2/1/08.

Item	Description	Responsible	Open Date	Due Date	Closed	Closed Date
130-1	SHIFT WORK		10/30/2007	02/05/2008	☐	

1/29/08:

1. advised the following subcontractors will be working extended days on the week of 1/28/08 as noted:
   a. laborers and carpenters (as needed)

Item	Description	Responsible	Open Date	Due Date	Closed	Closed Date
150-1	SITE SAFETY		10/30/2007	02/05/2008	☐	

1. No incidents reported the week of 1/28/08.

2. requested that additional storage containment cabinets be made available due to the increase of propane and gasoline containers.

3. stated that housekeeping is not acceptable.

Item	Description	Responsible	Open Date	Due Date	Closed	Closed Date
200-1	OPEN ISSUES LIST		10/30/2007	02/05/2008	☐	

1. reviewed the "Critical Open Items Tracker Summary" dated 1/28/08 (Copy Attached) regarding current outstanding issues for Roof Expansion (RF/LF). and ADA Seating. - ONGOING.

Item	Description	Responsible	Open Date	Due Date	Closed	Closed Date
250-1	SUBCONTRACTOR RELEASE LETTERS		10/30/2007	02/05/2008	☐	

1. to forward copy of Subcontractors to once executed by - ONGOING

2. to forward copies of all Rider B's and E's to as soon as possible. - ONGOING

Item	Description	Responsible	Open Date	Due Date	Closed	Closed Date
300-1	MEP COORDINATION	Ken Nobrega	10/30/2007	02/05/2008	☐	

FIGURE 7-1 (continued)

2. A review of previous "open" meeting minutes and actions required to resolve them

3. Discussion of current concerns, problems, questions, and responses; each item is numbered, described, actions by responsible person indicated, date when response is required, and date when the item was resolved

4. A review of outstanding requests for information, architect's supplemental instructions, and proposed change orders

5. A review of outstanding shop drawings are submitted to the architect for review and action.

6. The construction CPM schedule, highlighting any major deviations or problems; some contractors prepare a detailed schedule analysis that includes the percentage complete for each major construction component, the percentage of contract time elapsed, overall percentage of work completed, and activities planned for the next two weeks.

7. Site safety issues, if any, including accidents or safety violations that occurred during the period

8. Housekeeping issues relating to the site or the building

9. Contractor's report on the status of permitting; certain trades such as plumbing, HVAC, and electrical require the appropriate subcontractor to obtain permits before work can commence.

10. Review of all insurance certificates; each subcontractor has a specific insurance requirement, and a current certificate is required prior to working on the site.

11. Scheduling of the next progress payment to be submitted by the general contractor so the design team and the owner can be alerted to receive and review it

12. The date of the next meeting

13. Closing statement requiring all recipients of the meeting minutes to respond to any item that was not accurately stated in the minutes; absent any responses, the minutes will stand as an accurate assessment of what transpired.

Open items that remain "open" for an extended period of time may require an owner's intervention to resolve. When these types of issues remain unresolved, the owner, acting much like a mediator, must use his or her authority to settle them so they don't escalate into a delay claim or degenerate into other disputes or claims.

When questions or concerns are raised at these meetings, it is important for the responsible party to be clearly identified and the necessary response time to be noted. Some meeting minutes use the term BIC (ball in court) to identify the requisite respondent. Delays in responding to requests for information or clarification will be documented in the meeting minutes and can serve as confirmation if and when a delay-of-job claim is being considered by either party.

A statement made by an individual at the meeting may have been recorded incorrectly and should not go uncorrected. For example, the minutes may have included a statement that the architect was requested to respond within one week to the contractor's question regarding a structural steel shop drawing revision, when in fact the architect stated that she would review the contractor's shop drawing with the structural engineer in one week and would have a response in ten days. The timely review of structural steel shop drawings can result in delays in fabrication, so it is important that the architect correct that statement in the minutes.

Minutes are prepared after the meeting has concluded, and copies are sent out to not only all attendees but to other members of the construction team to keep everyone apprised of the items and events reviewed and discussed. An owner is well advised to read each meeting minute carefully and note any discrepancies. All too often we hear comments such as "Well, you never notified us of any changes, so we assumed that you accepted that statement." The last statement in most project meetings will include a clause similar to this:

> *These minutes are considered to be a fair and accurate account of this meeting. They will become record unless notification to the contrary is received within one week after the date of the meeting.*

Shop drawings

The term *shop drawing* can refer to a drawing, an equipment specification sheet, a diagram, or a material sample; these drawings are frequents topics of concern at project meetings. Although the contract plans and specifications are specific as to material and equipment requirements, these requirements need to be expanded and more detailed, and this is accomplished by the issuance of shop drawings from subcontractors and vendors. For example, the engineer's structural steel drawings and accompanying specifications don't reflect the exact length of each steel member, nor do they show a detailed diagram of the connection of one member to another. These details are developed by the steel subcontractor and submitted to the engineer for approval.

Although the specific manufacturer and model number of an air-conditioning unit is clearly spelled out in the specifications, the access panels for maintenance, the location and type of electrical connections, and the configuration of the equipment necessary to design its structural support will all be detailed on the shop drawing. The timely submission, review, and approval or rejection of shop drawings will take place via the project meetings, and the smooth flow of all required shop drawings is primary to an efficiently running project.

Material and equipment substitutions

As the shop drawing process begins, some contractors may offer substitutes for contract-specified materials or equipment. At times these substitutions may be submitted because of availability issues or because a contractor's prior experience with a product warrants consideration by the architect. Some architects are adamant that no substitutions will be permitted except in extenuating circumstances, while other architects are more open to product substitution as long as it meets certain criteria.

At the first project meeting, a statement regarding substitutions may be helpful in guiding the contractor as he or she prepares material or equipment shop drawings. The following guidelines can be established:

- For products where more than one manufacturer has been specified, no substitutions will be allowed.

- Where specifications require a matching of materials via a sample, the architect will be the final judge of acceptability.

- When a substitute product is submitted by the contractor, he or she must certify that it meets or exceeds the quality level of the originally specified product.

- The same or a better warranty will apply to the substitute product.

- The contractor will coordinate the installation and make any other changes required to accommodate the product at no additional cost to the owner.

- The contractor waives any claim for additional costs or time extensions that may subsequently become apparent.

- The contractor will reimburse the owner for any additional time for the architect/engineer to review and approve the product, including any redesign services.

To accomplish these requirements, the contractor is to submit (1) three copies of the request for substitution and (2) three copies of the shop drawing, product data, certified test results if applicable, and other documentation attesting to the proposed product's equivalence.

Value engineering

The U.S. General Services Administration (GSA) defines *value engineering* as "an organized effort at analyzing designed building features, systems, equipment, and material selections for the purpose of achieving essential functions at the lowest life-cycle cost consistent with required performance, quality, reliability, and safety." A contractor may offer value engineering suggestions based on past performance with a product he or she deems equal or superior to a product specified in the contract documents. And quite often an architect, being familiar with the product being offered, will accept this substitution.

A contractor who is submitting value engineering proposals must thoroughly review the proposed change to ensure that quality, performance, and costs associated with that suggested change are equal to the specified product. When a contractor doesn't closely examine all the consequences of the value engineering proposal, he is not serving the client well. Consider the following example of a VE proposal that was actually submitted but, fortunately, rejected.

A contractor had proposed substituting two roof-mounted air-handling units for one large unit and offered the owner a credit to do so, claiming that the purchase and installation of the two smaller units cost somewhat less than the one large unit and that the owner could apply the savings toward some hardware upgrades, thereby avoiding a change order. Although the contractor was correct in the assumption based on the cost to purchase and install the two smaller units, he failed to add the extra cost for the roofer to create an additional reinforced roof opening and to flash in the opening, as well as the added costs for the electrical contractor to install an additional conduit run, connection, and circuitry. Together with the additional energy costs for these smaller units as opposed to the more efficient large one, the owner would have not fared well by accepting this VE proposal.

When a VE proposal is submitted, the architect and engineer should review that proposal thoroughly with the contractor, and if accepted, accept it only on the condition that any added costs incurred by the owner will be borne by the contractor.

Project schedule

Another important item for discussion at the project meeting is the construction schedule. Is the project on track for the completion date required by the contract? Are there any problems that may delay completion? Are materials, equipment, and manpower sufficient to maintain an orderly flow of work? Most construction schedules today are CPM (critical path method) schedules.

Bar chart schedules are frequently used for short-term, small projects where there are relatively few activities and where each activity is expressed as a horizontal bar with a start date and finish date.

A CPM schedule is more complex, encompassing many activities, often in the hundreds, that will take place during the entire construction project and the order in which each will start and finish, commencing with site clearing and site work and ending with the punch list and the certificate of occupancy.

The schedule is a dynamic document that will change as some construction activities occur sooner than anticipated and others occur later than scheduled. When this orderly flow of activities and work tasks is disrupted, the CPM schedule will be revised to record the changes that may or may not impact the overall project completion date.

Minor changes to the CPM may also occur as additional tasks are added, as when a change order increases the project's scope or when deleted items reduce the scope of work. Some changes in one trade's activity can cascade onto others, creating a series of delays that may not be so minor. For example, a significant delay in starting interior metal wall framing because the contractor did not have sufficient materials or labor will push ahead the start and completion of drywall, painting, and even installation of other finish operations.

An important ingredient in a CPM schedule is "float"—the extra days that are added to the schedule to be used for unanticipated minor delays caused by inclement weather, slight delays in material or equipment deliveries, or temporary manpower shortages. All of these minor delays, if and when they occur, will decrease the number of float days available to the contractor. A CPM schedule contains the following basic elements:

- Activity—specified by an ID number and a brief description

- Activity duration—the number of days required to complete the activity

- Early start—the anticipated date when this activity will take place, expressed as an actual date, such as July 15, 2009

- Early finish—anticipated completion of this planned activity, also expressed as an actual date

- Graphic display of the start and completion of each activity, expressed in a solid horizontal bar

The contractor should distribute a current, updated CPM schedule at each project meeting and review the most salient points, including any changes from the last schedule, whether the project is still on track for contract completion, or any concerns about potential delays.

Some contractors will provide even more information regarding the schedule, as reflected in Figure 7-2, which includes percentage of contract time elapsed and percentage of work complete and provides the actual percentage complete of some important components. This presentation also includes what is referred to as the two-week look-ahead, an announcement of the work scheduled to start or complete within the following two weeks.

When it appears that the project may be falling behind in the scheduled completion of certain events, an architect or owner may request the contractor to prepare a summary of manpower requirements for the coming week to project the number of workers in each trade that will be required to either maintain the schedule or accelerate the work. This summary can then be used to check actual versus projected work in the week that follows to verify that the contractor performs as planned. The summary of workforce (Figure 7-3) was necessary to ensure that sufficient manpower was being employed to meet the project's completion date because of a series of delays experienced by the contractor.

Importance of the owner's role

The owner will be deeply involved in all of the correspondence passing back and forth among contractor, architect, and engineer and should be copied on all such matters relating to contract cost and contract time. The degree of involvement for

Project No. 24205.00

SCHEDULE:

Notice to Proceed	6/10/2008
Start Date:	6/16/2008
Contract Finish Date:	6/01/2009
Anticipated Finish Date	
Days Elapsed:	149
% of Time Elapsed:	42%
% of Work Completed:	25%
Weather Days:	17

WORK IN PROGRESS/SCHEDULE:

Parking Lot Fill	100%
Underslab Plumbing	98%
On Site Water	100%
Slab prep	90%
Site lighting underground	60%
Duct bank	100%
Elevator shaft	100%
Curb and gutter prep	60%
Offsite curb and gutter prep	15%
Slab pour	60%
Framing	5%
Paving	60%
Curb and gutter	60%
Rain leaders	60%
Offsite Sediment control	50%

NEXT TWO WEEKS:

Complete site utilities (rain leaders)
Complete underslab plumbing
Continue utilities
Complete slab prep
Complete curb, gutter and asphalt base at housing
Install stone parking base
Install site light bases
Continue framing.
Phase II sediment controls
Phase II rough grade

FIELD REPORT:

Weather: overcast 40's

OLD BUSINESS

FIGURE 7-2

A narrative schedule presentation, including a two-week look-ahead attached to the meeting minutes.

Project: _____ Report Number: _____141_____

Location: _____ Date: _____

Project No.: ____11937____ Prepared By: _____

SUMMARY OF WORKFORCE FOR WEEK   7/15/2006   THRU   7/21/2006

SUMMARY OF WEEK'S	SAT	SUN	MON	TUE	WED	THU	FRI
WEATHER CONDITION							
COMPANY/TRADE	SAT	SUN	MON	TUES	WED	THURS	FRI
Cente	36		56	65	67	67	60
Heary	4	6	9	13	12	8	9
Fidel	42	23	76	81	81	77	80
MB	33	56	46	52	54	51	49
Scrib			2	2		2	4
Simplex			7	6	7	6	14
Elevator			3	2	3	3	3
CH			5	5			
TS							
ISE	6		6	9	10	9	8
Fews	13	7	6	9		10	9
Business	9						
Manol	11		13	13		7	9
Norm							
ED							
Heid							
Champion							
Benfield							
Davenport							
Ryan Restor			4	4	4	5	4
Cherry			7	7	7	7	8
Mach	6	6	5	4	6	6	8
TOTAL	160	98	245	272	251	258	265

FIGURE 7-3

A summary of past weekly workforce activities.

each owner will vary, but an owner who is apprised of the current status, progress of the project, and any looming problems will be more apt to make the right decision if and when the time comes.

Many of the shop drawings, samples, specifications, and architectural and engineering sketches that pass from subcontractor to general contractor, architect, and engineer do so via a form called a letter of transmittal (Figure 7-4). This form not only documents what is being transmitted but establishes the date of the transmission, to whom it was sent, a description of each document in the package, the person sending the transmittal, what action is required, and a remarks section.

Nowadays most A/E sketches (referred to as SKs) are transmitted electronically, and only bulk shop drawings, product samples, and multiple drawing reproductions use this transmittal form. When an owner is copied on any of these transmittals, she should take at least a cursory glance at the contents being transmitted, since the contractor or the design consultants are in effect telling the owner, "This is something you need to know."

The owner is the final approver of change orders. The process usually begins when the contractor submits a proposed change order (PCO) to the architect. The PCO can be generated by a request from the architect, as shown in Figure 7-5, where the contractor was requested to submit a price to add painting and wallpaper: "Add wall coverings per HRC Architect RFQ#54, dated January 5, 2009."

After the proposed change order is reviewed by the owner and the architect and approved, a change order will be prepared that, in this case, will increase the contract sum by $8,330. The formal change order will not only note the increase in the contract sum but the increase in contract time if applicable. Adding such items as painting or wall coverings rarely requires an increase in contract time.

Most contractors prepare a change-order log, which is more accurately called a proposed-change-order log, to track the review process. The log will include all PCOs issued to date by PCO number, date of issuance, status, and, if approved, the change order in which they were included. When a PCO has been withdrawn or not accepted, it will be so noted. This PCO log is another document that will be reviewed at each project meeting so resolution of all open items can be obtained.

Contractors become rightfully concerned when PCOs remain "open" for an extended period of time, since their costs are based on performing the work within a reasonable time after issuance and not having to do the work out of sequence. An owner's review and prompt resolution of all "open" PCOs will be much appreciated by the contractor.

Requests for information

Requests for information (RFI) are questions that are raised by either the general contractor, one of the subcontractors, or a vendor and are passed on to the architect's team for a response. Each RFI is numbered sequentially so it can easily be tracked and monitored.

LETTER OF TRANSMITTAL

	DATE		JOB NO.

ATTENTION

RE:

TO _____

_____

_____

WE ARE SENDING YOU   ☐ Attached   ☐ Under separate cover via _____ the following items:

☐ Shop drawings   ☐ Prints   ☐ Plans   ☐ Samples   ☐ Specifications

☐ Copy of letter   ☐ Change order   ☐ _____

COPIES	DATE	NO.	DESCRIPTION

THESE ARE TRANSMITTED as checked below:

☐ For approval   ☐ Approved as submitted   ☐ Resubmit _____ copies for approval

☐ For your use   ☐ Approved as noted   ☐ Submit _____ copies for distribution

☐ As requested   ☐ Returned for corrections ☐ Return _____ corrected prints

☐ For review and comment   ☐ _____

☐ FOR BIDS DUE _____ 19 _____ ☐ PRINTS RETURNED AFTER LOAN TO US

REMARKS _____

_____

_____

_____

_____

_____

_____

_____

COPY TO _____

SIGNED: _____

If enclosures are not as noted, kindly notify us at once.

FIGURE 7-4

A typical letter of transmittal format.

Enfield Construction, Inc.
Contractors and Construction Managers
Since 1954

582 North Calvert Street
Baltimore, Maryland

| Proposed Change Order |

To: Mr. James Smith
Statewide Real Estate Holdings
1287 South Rampart Street
Alexandria Virginia 22024

Re: Add wall coverings per HRC Architect RFQ #54, dated January 5, 2009

Attachments:
1. SOEP Painting PCQ No. 1231–813 in the amount of $7,900.00 dated 1/24/08.

Cost Code	Description		Amount
09901.0010	SOEP - Painting & Wallcovering		7,900.00
		Subtotal:	7,900.00
17400.0010	Permits (1.1%)		88.00
18000.0010	Subguard (1.25%)		99.00
19000.0010	Fee (3.0%)		243.00
		Subtotal:	430.00
The total amount to provide this work is			8,330.00

We will submit for your signature three (3) AIA G701 Forms upon receipt of a signed copy of this letter, signature should be placed where indicated below.

_____   _____
Owner   General Contractor

_____   _____
Date   Date

FIGURE 7-5

A proposed change order (PCO), also referred to as a change order request (COR). This one is in response to an architect's Request for Quotation (RFQ).

While some RFIs may not have a critical time frame for resolution, delays in processing and responding to them are often cause for disagreements and disputes during the project. Most contractors will prepare a log of requests for information to track the passage from originator to the architect and its return back to the originator. This RFI log is usually reviewed at every project meeting, and response dates are discussed. These requests for information represent another document that should be responded to promptly by the architect or engineer.

Owners should also take an active role in monitoring RFI responses; if the architect has taken an unreasonably long time in responding to the contractor's query, this concern should be voiced (quietly) with the architect after the meeting is over.

Architect's supplemental instructions and field instructions

An architect may issue an architect's supplemental instruction (ASI) to clarify some part of a contract plan or specification that does not result in a change in contract sum or contract time. An ASI document will usually contain the following caveat:

The work or instructions included in this document are issued in accordance with the contract documents and represent no change in either Contract Sum or Contract Time. By proceeding with this work or instructions, the contractor acknowledges that there will be no change in the contract sum or the contract time.

After the contractor has responded to the content in the ASI, it will be closed and require no further action except if the contractor takes issue with the architect, claiming that the ASI did indeed impact the project's cost or schedule. The architect and contractor need to quickly resolve these matters, and this is also a time when an owner could intercede to assist in the resolution.

An architect's field instruction (FI) is very similar to an ASI and can be generated when the architect makes a tour of the project and responds to a question from the field or makes a note of some item or issue that should be clarified or corrected. The architect will generally follow up on this field observation or directive with a written response or a clarification sketch via a field instruction. A contractor may have also pointed out the need for a minor improvement or a suggestion that the architect will take under advisement, after which an FI will be issued.

To monitor what may be a long list of either ASIs or FIs, the contractor will prepare a log similar to the one shown in Figure 7-6. Figure 7-7 is another method by which a contractor can track ASIs. And, once again, an owner can intercede if the ASI or FI process begins to break down so the information continues to flow.

The submittal log

The prompt processing of shop drawings is critical to the smooth flow of construction. Shop drawings and submittals refer to a multitude of items, from full-size drawings prepared by the fire protection subcontractor to detailed manufacturer's specification sheets on plumbing fixtures to a sample brick or paint color charts. Each one of these items may require a shop drawing to be submitted to the architect

Contractors and Construction Managers Since 1954

582 North Calvert Street
Baltimore, Maryland

Field Instruction pro	Title/Description	Date issued	Date issued Sp	Dwg's issued	SK's issued	Space issued	Proceed/ Net Proceed
A	Eloctial Value Engineering	11/29/07	11/30/07	Sex Summary Steel dated 10/24/07	N/A	N/A	Proceeding with work
1	Crutiose Revisous	11/12/07	11/12/07	C3 AI 1. CI.A2 1, CI.A3,I CI. A41	CLSK M1. CISK. M2. CLSK M3 CLSK P1. CLSK P2. CLSK E12	N/A	Proceeding with work
2	Eloctial Value Engineering	11/8/07	13/9/07	RESI 01B. SI.1B. SI. 2B. SI 3B. SI 4B KE A3 4	SSK-004. SSK-005. SSK-006. SSK-007 SSK-008. SSK-009	N/A	Proceeding with work
3	Eloctial Value Engineering	11/12/07	11/12/07	A6.16	ASK-0024. ESK-16 AND 17. PSK-3	N/A	Proceeding with work
4	Eloctial Value Engineering	11/13/07	11/13/07	NA	ASK-022 AND 023	N/A	Proceeding with work
5	Eloctial Value Engineering	11/13/07	11/13/07	NA	ASK-025	N/A	Proceeding with work
6	Eloctial Value Engineering	11/14/07	11/14/07	NA	ASK-026	N/A	Proceeding with work
7	Eloctial Value Engineering	11/19/07	11/19/07	NA	ASK-028, SSK-010	N/A	Proceeding with work
8	Eloctial Value Engineering	11/29/07	11/29/07	NA	ASK-030	N/A	Proceeding with work
9	Eloctial Value Engineering	11/29/07	11/29/07	NA	ASK-029A AND 0318	N/A	Proceeding with work
10	Eloctial Value Engineering	12/4/07	12/4/07	NA	SKE-010 Thin SKE-028	N/A	Proceeding with work
11	Eloctial Value Engineering	12/5/07	12/5/07	REAXI3B, KEAVI3. RE.AV62	N/A	N/A	Proceeding with work
12	Eloctial Value Engineering	12/6/07	12/6/07	NA	SKE-028, 029, SKE-030	N/A	Proceeding with work
13	Eloctial Value Engineering	12/6/07	12/6/07	NA	ASK-031A AND 031B	N/A	Proceeding with work
14	Eloctial Value Engineering	12/7/07	12/7/07	NA	SSK-010, 011, 012	N/A	Proceeding with work
15	Eloctial Value Engineering	12/17/07	12/18/07	NA	ASK-011, 035, 036, 033, SKP7 AND 8. SKM4 Thin 15 SKE 31 Thry 40	N/A	Proceeding with work
16	Eloctial Value Engineering	12/18/07	12/18/07	NA	ASK-033	N/A	Proceeding with work
17	Eloctial Value Engineering	N/A	N/A	NA	N/A	N/A	Void
18	Eloctial Value Engineering	N/A	N/A	NA	N/A	N/A	Void
19	Eloctial Value Engineering	12/20/07	12/21/07	REFPI.1A Rev 1	N/A	N/A	Proceeding with work
20	Eloctial Value Engineering	1/2/08	1/2/08	NA	ASK 050	N/A	Void
21	Eloctial Value Engineering	1/6/08	1/4/08	HC.A2	N/A	N/A	Pending
22	Eloctial Value Engineering	1/7/08	1/7/08	NA	ASK-048, 049, 094 SKM22, 23, 24, EPSK-01	N/A	Pending
23	Eloctial Value Engineering	1/4/08	1/4/08	NA	ASK-051	N/A	Pending
24	Eloctial Value Engineering	1/14/08	1/15/08	AVS-BLS1, AV4 2 BSL1	N/A	N/A	Pending
25	Eloctial Value Engineering	1/21/08	1/21/08	NA	ASK-051	N/A	Pending
26	Eloctial Value Engineering	1/21/08	1/21/08	NA	ASK-055	N/A	Pending

FIGURE 7-6

A field instruction (FI) log that allows the contractor to monitor the flow of field instructions and the status of each one: authorization to proceed, pending, or void.

Exhibit C – Schedule of ASI's Within Contract
9/22/05

ASI	Description	Issue Date
001	Foundation Design Clarification	10/20/2004
002	Building Grid Locations	11/5/2004
003	Misc Revisions BP 2 & 3	11/15/2004
004	Misc. Revisions BP 2 & 3	12/6/2004
005	Foundation Redesign & Low Roof	12/28/2004
006	Column Misc. Structural Revisions	1/21/2005
007	Geometric Plan Revisions	2/11/2005
008	Civil Changes	2/4/2005
009	Structural & Arch Revisions	2/18/2005
010	Finish Sill Elevations	3/3/2005
011	Loading Dock Columns	3/15/2005
012	Enlarged Plans and Details	3/17/2005
013	St Paul Bld Slab Edge	3/22/2005
014	Slab Edge - Plans and Details	3/22/2005
015	Slab Edge - Plans and Details	3/23/2005
016	St Paul Bld Slab Edge	3/25/2005
017	Charles Bldg Slab Edge	3/28/2005
018	Misc. Site / Utility Revisions	4/7/2005
019	St Paul Bldg - Partial Foundation	4/6/2005
020	Misc.	4/8/2005
021	Misc	4/22/2005
021R	Stud Gauge Revisions	9/12/2005
022	Plan - Second Floor Mechanical	4/22/2005
023	VE Priority 1	4/26/2005
024	Misc.	4/22/2005
025	Misc.	5/4/2005
026	Electrical Clarifications	5/10/2005
027	Misc.	5/19/2005
028	RFI #299, 273, & 260, Charles Building Framing Modifications	6/24/2005
029	Community Kitchens(MEP) and other Clarifications	5/31/2005
030	VE Lighting Modifications, Window Washing Equipment , Structural Changes & Value Engineering	6/10/2005
031	2nd Floor Charles Building - ADA	6/20/2005
032	Elevator, Walls & Electrical Modifications	7/1/2005
033	VE Lighting, Revised Electrical Panels, & Dimming	7/25/2005
034	Plumbing and Electrical Revisions Bookstore Café	7/27/2005
035	Structural Revisions - Charles	7/27/2005
036	ADA Modifications -Charles Building & Remedial Column Reinforcement	8/12/2005
037	Not issued	N/A
038	Stair Rail VE	8/12/2005

FIGURE 7-7

Tracking architect supplementary instructions (ASI) that have no contract sum or time impact.

for review, comment, and approval. The architect's review will be in the form of a stamp with one of these comments:

- Approved—no exceptions taken

- Approved as noted—the architect will have made minor changes to the information furnished by the contractor and will expect the contractor to include those changes in the product or material to be supplied

- Revise and resubmit—the architect takes exception to one or more portions of the shop drawings, and the contractor must revise those sections and resubmit the entire set

- Rejected

- No review or not reviewed—a comment when the shop-drawing submission is so deficient that the architect will perform only a cursory review and go no further; this is basically a kind of contractor rebuke.

Neither the general contractor nor the subcontractors can order any materials or equipment without receiving the architect's approval via a stamp on the shop drawing.

As one can imagine, the number of shop drawings can be staggering, numbering in the hundreds, and passing through many hands: the supplier, the general contractor, a subcontractor, the architect and/or engineer, and back down the path through which they came.

A delay in preparing a shop drawing by the general contractor or the subcontractor for review by the architect should be closely monitored by the architect. These delays can signal a problem much worse than the resultant delay in delivery of that product. A general contractor may have delayed submission of a specific shop drawing because he or she may not have awarded a subcontract agreement or a purchase order to the company furnishing that shop drawing. Perhaps the general contractor is having trouble making this award because all the quotes are considerably over budget and he is searching for a company that can meet the budget. The initial estimate may have been too low, and in an effort to "make budget," he ends up with a company that could perform poorly.

When a subcontractor fails to submit key shop drawings in a timely fashion, that should also be looked at askance. The subcontractor may have a cash-flow or credit problem, and the supplier of that product may be hesitant to accept the order.

For all of these reasons, the review of shop drawing submissions should have an important place at the progress meeting. A submittal log (Figure 7-8) is the key to tracking shop drawings. This log should contain at least the following:

- The specification manual's applicable section as a reference point

- The revision number (If the first submission by the general contractor is not approved and has been returned to be modified, the subsequent submissions are numbered: 1, 2, 3, and so on. Several revisions may indicate that the general contractor is not reviewing submissions carefully before sending them on to the architect/engineer and must focus on complying with the requisite specification requirements.)

Submittal Log by Days Held

Date: 2/4/2008

Spec Sect/ Submittal	Rev	Description	Received Date	Sent Date	Days Held
05120-017-0 5	0	- LS-7: Stair Framing Sections (Dwg. E2002)	1/28/2008	1/28/2008	7
05120-017-0 6	0	Stair - LS 7: Stair Support Frames (Dwg 2000)	1/28/2008	1/28/2008	7
05120-017-0 7	0	Stair - LS-7: Stair Support Frames (Dwg 2001)	1/28/2008	1/28/2008	7
05120-017-0 8	0	Stair - LS-7: Stair Support Frames (Dwg 2002)	1/28/2008	1/28/2008	7
05120-017-0 9	0	Stair - LS-7: Stari Framing (Dwg 2003)	1/28/2008	1/28/2008	7
05120-017-1 0	0	Stair - LS-7: Stair Framing (Dwg 2004)	1/28/2008	1/28/2008	7
05120-017-1 1	0	Stair - LS-7: Stair Framing (Dwg 2005)	1/28/2008	1/28/2008	7
05120-017-1 2	0	Stair - LS-7: Stair Framing (Dwg 2006)	1/28/2008	1/28/2008	7
05120-017-1 3	0	Stair - LS-7: Stair Framing (Dwg 2007)	1/28/2008	1/28/2008	7
05120-017-1 4	0	Stair - LS-7: Stair Framing (Dwg 2008)	1/28/2008	1/28/2008	7
05120-017-1 5	0	Stair - LS-7: Stair Framing (Dwg 2009)	1/28/2008	1/28/2008	7
05120-017-1 6	0	Stair - LS-7: Stair Framing (Dwg 2010)	1/28/2008	1/28/2008	7
05120-017-1 7	0	Stair - LS-7: Stair Framing (Dwg 2011)	1/28/2008	1/28/2008	7
05120-017-1 8	0	Stair - LS-7: Stair Framing (Dwg 2012)	1/28/2008	1/28/2008	7
05120-018-0 1	0	Framing at Stair LS-1 Left Field (Dwg. E25)	1/22/2008	1/25/2008	10
05120-018-0 2	0	Tube Beams at LS-1 Stair (Dwg. 1035)	1/22/2008	1/25/2008	10
05120-018-0 3	0	Tube Beams at Stair LS-1 (Dwg. 1036)	1/22/2008	1/25/2008	10
05500-021	1	Stair LS-1 (J) Drawings E6501 and E6502	2/1/2008	2/1/2008	3
07542-007	1	Samaclad	1/8/2008	1/11/2008	24
08216-001	0	Gate E Cargo Doors - Shop Drawing & Frame/HDWE Schedule	1/22/2008	1/22/2008	13
08360-003	1	Zap Controls - Sectional Door Drive and Control	2/1/2008	2/1/2008	3
08411-08	0	Modern - A410 Wall Panels	1/18/2008	2/1/2008	3
09510-004	0	ACT-1 Ceiling Panel and Grid product data and samples	1/28/2008	1/28/2008	7

FIGURE 7-8

A submitted log maintained by the contractor to monitor the architect's review of shop drawings and the number of days the architect spent reviewing each drawing.

- A brief description of the contents of the submission

- The date when the contractor received the submission from the vendor or subcontractor, to be compared to the date that this submission was passed on to the architect/engineer (It is the responsibility of the general contractor to review the shop drawing for compliance with the contract requirements before sending it on to the architect.)

- Shop drawing logs, to include a column indicating when the drawing was returned and its status: approved, approved as noted, rejected

The log in Figure 7-8 indicates the number of days the shop drawing remained under review before being returned by the architect. In this log 13- and 24-day lag times should be highlighted with an explanation.

This shop-drawing log will be reviewed at each project meeting; if the owner observes that the architect is not responding promptly, after the meeting he should confer privately to determine the reason for any delays.

Pace of construction

The pace of construction is of interest to the owner for any number of reasons. First and foremost is the assurance that the project completion date will be met and all other contract requirements appear to be under control. When a contractor falls behind on the schedule, it must be noted quickly and not allowed to continue. There may be a valid reason for a delay, and the contractor may have already planned to accelerate the project to compensate for that delay. But some contractors may not want to acknowledge a delay, thinking that somehow they will make up the time without making it obvious, and quite often these contractors don't alert the architect and owner until the delay has serious consequences.

Participation in the weekly or biweekly construction meetings provides an owner with scheduling information, a key element in the project's path to completion. An owner or architect should not hesitate to ask questions if there is a concern that progress is not what it should be. Are there money problems? Are there personnel problems? Is there a problem with a particular vendor or subcontractor?

An owner should approach an unwarranted delay by the contractor as the serious matter it is and put the contractor on notice of the consequences of the delay. If the owner's attorney must become involved, a letter or two may be all that's needed to jog the contractor to pick up the pace. An owner who takes an active role in the project can determine when to act and what force is required early on to correct a design consultant or contractor delay.

Contractor's application for payment

At the first project meeting, a protocol for the submission of the contractor's requisition should be established so the contractor has clear direction regarding the preparation of the initial application for payment. Procedures for the submission and processing of payment requests vary somewhat from job to job, so it is best to review the current owner's requirements.

Prior to that first project meeting, both owner and architect should meet and agree on these submission procedures. It may be helpful to prepare a formal memo to be distributed to the general contractor explaining these procedures to avoid any misunderstandings. A sample memo can include these guidelines:

- Application form—AIA G702/G703 or similar (Figures 7-9a and b) to be used for all payment requests

- The schedule of values presented in the initial application for payment, not to be changed unless authorized by the owner, in writing

- Line items in the schedule of values, not to be invoiced for more than 100 percent; this would increase the contract sum, which can only be accomplished by issuance of a change order by the architect.

- Materials or equipment stored offsite at the date of the payment request, not to be requisitioned without written permission from the owner beforehand (Specific documents for offsite storage will be required if that request is approved, including a bill of sale, certificate of insurance, and a bailment agreement, copies of which can be given to the general contractor.)

- Subcontractor list are submitted at the first project meeting and updated as additional subcontractors are hired by the general contractor (This list is important for several reasons: An owner should be aware of all companies working on the site and to whom payments are to be made and from whom lien waivers are required.)

- Lien waivers, to be submitted by the general contractor and each subcontractor requesting payment during the application period (The general contractor will also be required to submit a release of lien with every progress payment, shown in Figure 7-10, and at the completion of the project; Figure 7-11 shows a final payment lien release form.)

- Change orders (Although the protocol for the preparation and submission of change orders may already have been established, the way in which they are to be listed in the application for payment requires clarification. The owner may elect to pay for that portion of the change-order work that has been completed if the work is considerable in nature and is still in progress or wait until the change order work has been 100 percent completed.)

- No change orders that have not been formally accepted by a fully executed change order submitted to the general contractor with owner and architect signatures affixed

- All owner change-order work to be clearly separated by a line item for each change order, not lumped into the schedule of values; the additions to or deductions from the initial contract amount are thus clearly delineated.

APPLICATION FOR PAYMENT

HAND CARRY or OVERNIGHT DELIVERY TO:

FROM:

Contractor:

Remittance Address:

Phone No.

Fax No.

PAGE 1 OF 3 PAGES

PROJECT:

CONTRACT NO:

APPLICATION NO:

PERIOD TO:

INVOICE DATE:

APPLICATION:

Application is made for Payment, as shown below, in connection with the Contract.

Schedule of Values,

1 ORIGINAL CONTRACT SUM $
2 Net change by Change Orders $
3 CONTRACT SUM TO DATE (Line 1 +/- 2) $
4 TOTAL COMPLETED & STORED TO DATE $
(Column G)

5 RETENTION

a ___ 10 % of Completed Work $ 0
(Column D + E)

b ___ 0 % of Stored Material $ 0
(Column F)

c ___ 0 % of Contract Sum $ 0

Total Retention (Lines 5a + 5b, or Line 5c)

6 TOTAL EARNED LESS RETENTION $ 0
(Line 4 less Line 5 Total)

7 LESS PREVIOUS CERTIFICATES FOR PAYMENT $ 0

8 CURRENT PAYMENT DUE $ 0

CHANGE ORDER SUMMARY

	Additions	Deductions	Net Change
Previously Approved	$ -	$ -	$ -
Approved this Month	$ -	$ -	$ -
Totals	$ -	$ -	$ -

CERTIFICATION:

The undersigned Contractor certifies that to the best of the Contractor's knowledge, information and belief the Work covered by this Application for Payment has been completed in accordance with the Contract Document, that all amounts have been paid by the Contractor for Work for which previous Certificates for Payment were issued and payments received from ___ and that the current payment shown herein is now due.

CONTRACTOR:

By: ___ Date: ___

___ Printed Name and Title

State of: ___ County of: ___

Subscribed and sworn to before me this ___ day of ___

Notary Public: ___

My Commission expires: ___

EARLY PAYMENT DISCOUNT OPTION:

Payment terms shall be in accordance with the above referenced Contract No. In the event that elects to provide for shorter payment terms, we agree to do so for the following discount structure:

___ % Discount, net 10 days, if paid by ___ (Date)
___ % Discount, net 15 days, if paid by ___ (Date)
___ % Discount, net ___ days, if paid by ___ (Date)

JOB NO.		SUB JOB NO.		CMS CONT. NO.		
		ITEM NO.	ITEM NO.	ITEM NO.	ITEM NO.	
GROSS TO DATE						
RETENTION TO DATE						
TOTAL DUE TO DATE						
LESS PREV. APPLS						
CURRENT PAY						
VENDOR	APPV	DATE	APPV	DATE	APPV	VP

FIGURE 7-9

An application for payment. (a) Summary sheet. (b) Detail sheet.

SCHEDULE OF VALUES

FROM:

Contractor: 0
Remittance: 0
Address: 0

PROJECT:
CONTRACT NO:
APPLICATION NO:
PERIOD TO:
INVOICE DATE:

PAGE 2 OF 3 PAGES

0
0
0

Application for Payment, containing Contractor's signed Certification is attached.
In tabulations below, amounts are stated to the nearest dollar.

A	B	C	D		E	F	G		H	I
			Work Completed			Materials				Retainage
Item #	Description of Item	Scheduled Value	From Previous Application	This Period		Presently Stored (Not in D or E)	Total Completed And Stored to Date (D+E+F)	% (G/C)	Balance to Finish (C-G)	Held 10%
1		$ —	$ —	$ —		$ —	$ —	0%	$ —	$ —
2		$ —	$ —	$ —		$ —	$ —	0%	$ —	$ —
3		$ —	$ —	$ —		$ —	$ —	0%	$ —	$ —
4		$ —	$ —	$ —		$ —	$ —	0%	$ —	$ —
5		$ —	$ —	$ —		$ —	$ —	0%	$ —	$ —
6		$ —	$ —	$ —		$ —	$ —	0%	$ —	$ —
7		$ —	$ —	$ —		$ —	$ —	0%	$ —	$ —
8		$ —	$ —	$ —		$ —	$ —	0%	$ —	$ —
9		$ —	$ —	$ —		$ —	$ —	0%	$ —	$ —
10		$ —	$ —	$ —		$ —	$ —	0%	$ —	$ —
11		$ —	$ —	$ —		$ —	$ —	0%	$ —	$ —
12		$ —	$ —	$ —		$ —	$ —	0%	$ —	$ —
13		$ —	$ —	$ —		$ —	$ —	0%	$ —	$ —
14		$ —	$ —	$ —		$ —	$ —	0%	$ —	$ —
15		$ —	$ —	$ —		$ —	$ —	0%	$ —	$ —
16		$ —	$ —	$ —		$ —	$ —	0%	$ —	$ —
17		$ —	$ —	$ —		$ —	$ —	0%	$ —	$ —
18		$ —	$ —	$ —		$ —	$ —	0%	$ —	$ —
19		$ —	$ —	$ —		$ —	$ —	0%	$ —	$ —
20		$ —	$ —	$ —		$ —	$ —	0%	$ —	$ —
21		$ —	$ —	$ —		$ —	$ —	0%	$ —	$ —
22		$ —	$ —	$ —		$ —	$ —	0%	$ —	$ —
23		$ —	$ —	$ —		$ —	$ —	0%	$ —	$ —
24		$ —	$ —	$ —		$ —	$ —	0%	$ —	$ —
25		$ —	$ —	$ —		$ —	$ —	0%	$ —	$ —
Summary of Changes		$ —	$ —	$ —		$ —	$ —	0%	$ —	$ —
Totals:		$ —	$ —	$ —		$ —	$ —	0%	$ —	$ —

FIGURE 7-9 (continued)

Exhibit J

Standard

Waiver and Release of Lien Upon Progress Payment

OWNER:

GENERAL CONTRACTOR: _____

PROJECT NAME: _____

COMMONWEALTH OF MASSACHUSETTS
COUNTY OF SUFFOLK

     The undersigned hereby acknowledges receipt of payment in full, and hereby waives and releases any claim, lien or right to claim a lien, for labor, services, or materials furnished through _____ (*date of last previous application for payment*) under contract with _____ on the job of _____

(*Owner*) to the following Property: (*Name and Address of Project*). This waiver and release does not cover any retention or labor, services or materials furnished after the date specified.

     Any and all contractors, subcontractors, laborers, suppliers and materialmen that have provided labor, material or services to the undersigned for use or incorporation into the construction of the improvements to the Property have been paid and satisfied in full, and there are no outstanding claims of any character arising out of, or related to, the undersigned's activities on, or improvements to, the Property.

     This Waiver constitutes a representation by the undersigned signatory, for and on behalf of the firm or company listed below, that the payments received by said firm constitute full and complete payment for all work performed, and all costs or expenses incurred (including, but not limited to, costs for supervision, field office overhead, home office overhead, interest on capital, profit, and general conditions costs) relative to the work or improvements at the Property through the date specified above, except for the payment of retainage. The undersigned hereby specifically waives, quitclaims and releases any claim for damages due to delay, hindrance, interference, acceleration, inefficiencies or extra work, or any other claim of any kind it may have against Owner, any Owner Affiliate (as defined in the Contract for the Project), the General Contractor (if this Waiver is signed by a subcontractor or supplier), or any other person or entity with a legal or equitable interest in the Property, as of the date of this Waiver, except as follows:

_____

_____

_____

Exhibit J – Page 1

FIGURE 7-10

Waiver and Release of Lien form to be submitted with the general contractor's and subcontractor's progress payments.

This Waiver is specifically made for the benefit of Owner and Owner's lender, any tenant and lender of any tenant, and any other person or entity with a legal or equitable interest in the Property.

In Witness Whereof, the undersigned signatory, acting for and on behalf of the firm or

company listed below and all of its laborers, subcontractors, and suppliers, has placed his hand

and seal this ____ day of _____, 20__.

FIRM OR COMPANY:

_____

By:_____

Sworn to and subscribed   Print Name:_____
before me this _____ day of
_____, 20___.   Its:_____

_____
Notary Public

(NOTARY SEAL)

My Commission Expires:

_____

Exhibit J – Page 2

FIGURE 7-10 (continued)

Standard

Waiver and Release of Lien Upon Final Payment

OWNER:

GENERAL CONTRACTOR: _____

PROJECT NAME: _____

The undersigned, in consideration of the final payment in the amount of $_____, hereby waives and releases its lien and right to claim a lien for labor, services, or materials furnished to _____ on the job of Limited Partnership (*Owner*) to the following described property:
(*Name and Address of Project*).

Any and all contractors, subcontractors, laborers, suppliers and materialmen that have provided labor, material or services to the undersigned for use or incorporation into the construction of the improvements to the Property have been paid and satisfied in full, and there are no outstanding claims of any character arising out of, or related to, undersigned's activities on, or improvements to, the Property. This Waiver is specifically made for the benefit of Owner and Owner's lender, any tenant and lender of any tenant, and any other person or entity with a legal or equitable interest in the Property.

This Waiver constitutes a representation by the undersigned signatory, for and on behalf of the undersigned, that the payment referenced above, once received, constitutes full and complete payment for all work performed, and all costs or expenses incurred (including, but not limited to, costs for supervision, field office overhead, home office overhead, interest on capital, profit, and general conditions costs) relative to the work or improvements at the Property. The undersigned hereby specifically waives, quitclaims and releases any claim for damages due to delay, hindrance, interference, acceleration, inefficiencies or extra work, or any other claim of any kind it may have against Owner, any Owner Affiliate (as defined in the Contract for the Project), the General Contractor (if this Waiver is signed by a subcontractor or supplier), or any other person or entity with a legal or equitable interest in the Property, relative to the work or improvements at the Property.

Exhibit K – Page 1

FIGURE 7-11

Waiver and Release of Lien form to be submitted by the general contractor with the final application for payment.

(a)   waive any and all liens and right of line on such real property for labor or materials, or equipment, appliances or tools, performed or furnished through the following date: _____(payment period, except for retainage, unpaid agreed or pending change orders, and disputed claims as stated above; and

(b)   subordinate any and all liens and right of lien to secure payment for such unpaid, agreed or pending change orders and disputed calms, and such further labor or materials, or both labor and materials, or rental equipment, appliances or tools, except for retainage, performed or furnished at any time through the twenty-fifth day after the end of the above payment period, to the extent of the amount actually advanced by the above lender/mortgagee through such twenty-fifth day.

Signed under the penalties of perjury this _____ day of _____, 200___.

Contractor:

By:_____

its:_____

COMMONWEALTH OF MASSACHUSETTS

_____, SS.

_____
(DATE)

Then personally appeared the above named _____ and acknowledged the foregoing instrument to be his/her free act and deed and the free act and deed of [INSERT] before me,

_____

Notary Public
My Commission Expires:_____

FIGURE 7-11 (continued)

- Equipment summary (If the project is a cost-plus or cost-plus with a guaranteed maximum price, any equipment purchased during construction may have residual value at the close of the project. The type of equipment on this list can be established at the first project meeting by either a minimum dollar value or other mutually acceptable criterion. The equipment list can be included in the current request for payment, along with the initial cost and brief description. Additions to the list can be made with each subsequent application for payment.)

- Unless otherwise stated, general conditions, which are time-related, to be billed in equal proportions over the length of the project; no deviation is permitted unless approved in writing by the owner.

- Weather-related general conditions costs included in the contract as an "allowance" to be included in the requisition with sufficient documentation to support the charges; a cost-to-date document would also be helpful.

- Subcontractor requisitions and vendor invoices (Subcontractors should use the same requisition form as the general contractor. The amount requisitioned is to be based on percentage complete, as for the general contractor.)

- Vendor copies of receiving tickets signed by either the owner's representative or the contractor's designated authority

- Rental-equipment reimbursements, including a signed ticket verifying the date and time that the equipment was picked up at the site

- Lien releases/lien waivers (Lien waivers must be attached to all requests for payments from the general contractor and the subcontractors. This requirement will occur only on the second and subsequent applications for payment. With the first payment, the subcontractor will make his or her own first payment for all labor, materials, and equipment and can only certify on the second request that the first payment was made appropriately.)

Offsite material and equipment storage

A contractor may purchase equipment and materials and store them offsite for a number of reasons. A long lead time for delivery of vital materials and equipment requires the contractor to place orders for those items far in advance so they are available when needed at the job site. With just-in-time delivery not guaranteed, the contractor may elect to receive these materials and equipment and store them in an offsite facility so they are protected against damage or potential theft. The contractor has to pay for the materials when received and is justified in requesting reimbursement by the owner. But, while justified, payment for these offsite materials and equipment presents a few concerns for the architect and owner:

1. Do the materials or the equipment meet the contract specifications?

2. Are the materials and/or equipment actually stored in the quantity and the space offsite as described by the contractor?

3. Are the materials and/or equipment stored in an environment that will maintain their quality and integrity?

4. Is the storage area secure and protected against fire and theft?

5. Will the contractor provide insurance for the materials or equipment while in transit from storage to the construction site?

6. Once payment is made, will the contractor provide a paid bill of sale, transferring ownership from the contractor to the owner?

When materials are stored offsite, where quality, content, and quantity are questionable, the contractor must provide the purchase order and a receiving ticket or invoice from the supplier verifying that the materials or equipment have in fact been delivered to the warehouse so stated.

If there is any concern about the offsite storage facility or the quality of the product stored there, the owner may wish to have the contractor accompany the architect and inspect the product prior to approving payment. The owner, in the written approval for offsite storage, should also include this inspection provision, adding that any costs associated with the inspection may be required to be paid by the contractor.

Although the stored product was ordered by the contractor, an owner should also consider the fact that the contractor did so with the owner's interests in mind to maintain job progress, so the owner may relent on reimbursement for the entire travel expense.

Three documents should be furnished by the contractor prior to approving payment for offsite items: a bailment agreement, a bill of sale, and an insurance certificate. The bailment agreement (Figure 7-12) is a document that includes the description of the items stored offsite, the exact location where they are stored, and the "bailee's" (vendor or subcontractor who furnished the item) agreement to be responsible for the safe storage of these items.

The bill of sale (Figure 7-13) transfers title to the owner when payment is made to that vendor or subcontractor by the general contractor. This is important because in the event that the vendor's or subcontractor's business fails and the company declares bankruptcy, without evidence of ownership of the offsite stored material, the bankruptcy judge may impound that material or equipment during the entire bankruptcy proceedings.

Finally, the vendor or subcontractor must provide a certificate of insurance for coverage not only during storage but also as the product is transported from the storage facility to the job site.

PROJECT CLOSEOUT

An otherwise successful project can end not so successfully as the contractor's project manager and project superintendent turn their sights on that new job to which they have recently been assigned by their boss. As the completion of the project begins to unfold, the manifold requirements of project closeout come into play, which, if completed properly, will provide a successful handoff from the contractor to the owner.

In preparation for project closeout, the owner, with assistance from the architect, can become more familiar with the process by carefully reviewing the contract specifications and listing all required tests, inspections, and warranties. Each division in the specification manual, where applicable, will contain a section on tests, warranties, extra materials, and special tools to be provided by the contractor. Most of these requirements will be found in the electrical, mechanical, fire-protection, and plumbing sections of the specifications, but they are also scattered throughout

Agreement No. _____

Bailment Agreement

This Bailment entered into as of the _____ day of _____, 20____, between
_____, herein called "Bailor" and
_____, LLC, herein called "Bailee."

Bailor hereby loans to Bailee the property described below, herein called "bailed property."

1. The bailed property consists of:

2. Bailee may use the bailed property for the following purposes:

3. The bailed property will be shipped f.o.b. _____ plant to
_____
_____, Attention: _____,
and returned to Bailor f.o.b. _____ plant on or before _____.

4. Bailee shall have no liability for loss of or damage to the property except to the extent of insurance coverage provided for hereinafter. Bailee will carry fire and extended coverage perils insurance on the property in the amount of not less than $_____ which Bailor has advised is adequate to cover the property.

5. Unless otherwise provided herein Bailee may modify the property provided that it is substantially restored to its original condition, ordinary wear and tear excepted, prior to its return to Bailor. Bailee shall, to the extent reasonably required, be responsible for normal maintenance of the bailed property, exclusive of replacement parts.

6. Bailor shall be liable for all ad valorem taxes assessed against the bailed property while in the possession of Bailee.

FIGURE 7-12

A Bailment Agreement that is required when a contractor requests payment for materials or equipment that is stored offsite.

the project manual. With the assistance of the architect and engineer the following closeout activities can be monitored:

- The punch list—the contractor must be directed to diligently pursue completion of the punch list. As we discussed previously, items that remain unresolved after a reasonable period of time can be completed by the owner by

BILL OF SALE OF PERSONAL PROPERTY

KNOW ALL MEN BY THESE PRESENTS THAT._____

(Vendor/Subcontractor) for, and in consideration of the sum of $_____
and other goods and valuable consideration, upon receipt of payment of which the Undersigned ("Seller") does by these present GRANT, BARGAIN AND SELL into (Contractor) ("Purchaser"), the goods and chattels located at (place where stored-street, city, state address) as described in Schedule "A" attached hereto and by this reference, made a part of hereof the Property.

IN CONSIDERATION OF THE FOREGOING AND THE COVENANTS HEREIN CONTAINED, SELLER AGREES AS FOLLOWS:

   1. Seller does hereby covenant and warrant to the Purchaser that Seller is the lawful owner of the Property; that the Property is free from all liens and claims whatsoever; that Seller has good right to sell the same; that Seller will warrant and defend same against the claims and demands of all persons.

   2. Seller will provide safe and proper storage for the Property and will cause to be placed conspicuously and securely on the Property a sign or signs which will show that the Property is the property of the Purchaser.

   3. The Property shall be held at Seller's risk and shall be kept insured against fire, theft and all other hazards by Seller at Seller's expense while in its custody or control in an amount equal; to the replacement cost thereof, with loss payable to the Purchaser. Copies of insurance certificates evidencing such insurance are to be furnished to the Purchaser.

   4. The Purchaser shall have the right to inspect the Property at any time during normal business working hours at the storage facility of the Seller. The failure to inspect shall not be deemed a waiver of any of the rights of the Purchaser, and if the Property is found to be defective, in materials or workmanship, stolen or lost, in whole or in part, the Seller shall replace the same at its own cost.

   5. The Property shall be subject to removal by Purchaser at any time upon Purchaser's instructions.

   6. Seller does hereby warrant to Purchaser that value of the property described herein is $_____

FURTHER- IN WITNESS WHEREOF, The undersigned has set his hand on

This_____ (day) of _____(Month)

SELLER:_____

   Title_____

WITNESS: _____

State of:_____

County of:_____

(To be notarized by a Notary Public)

FIGURE 7-13

A bill of sale that is required when a contractor requests payment for materials or equipment that is stored offsite and is submitted when payment has been received.

withholding triple the amount of the cost to correct from any funds due the contractor. The price for the work should be based on a firm price from a reputable specialty contractor. A general contractor may also argue that a punch list item is actually a warranty item, and therefore he or she has an extended period of time in which to correct. As an example, a light fixture may be blinking on and off because of a bad ballast. If such an occasion arises, have the architect render an opinion whether it is punch list or warranty, and resolve the matter.

- Submission of the record drawings, otherwise known as the as-built drawings. These drawings should have been reviewed by the architect and engineer for accuracy from time to time as they were prepared by the subcontractor and verified by the general contractor. For example, before some underground utility lines were backfilled, the architect or engineer could verify the "as-built" condition by a field visit as the contractor-inserted line and grade on the as-builts. Variances in work in place as opposed to design line and grade of these underground utilities should be incorporated into these drawings. If these as-built conditions are not accurate, if at some future date they must be uncovered for repair, relocation, damage, or otherwise, extra costs in locating the underground utilities will undoubtedly occur. Any changes in the structure of the building and its internal systems are also important when future additions or modifications to that structure are required. The new contractor, relying on the accuracy of those "as-built" drawings, will have based the estimate of the work on them only to find errors in dimensions, elevations, and location. Location of all concealed mechanical and plumbing valves, dampers, piping, and ductwork are important not only when future modifications to those components are required but also for routine maintenance.

■ Reports of all test reports and other inspections by local officials or as required by the contract. These may include the following:
   — Earth-compaction tests performed during the site-work stage of the project
   — Infiltration or exfiltration tests for underground storm or sanitary lines to ensure the integrity of the pipe joints against leakage
   — Concrete compression tests for cast-in-place concrete and/or pre-stressed or posttensioned concrete component tests
   — Mill reports if the building's structure is steel to confirm the quality levels of steel produced by the manufacturer and ensure conformance with the contract specifications
   — If the building has structural steel framework, weld, bolt-up, shear-stud test reports
   — For all masonry work, mortar cube compression tests and flashing inspection reports if required
   — Reports on HVAC test and balance procedures, known as TAB; start-up and commissioning checklists (Figure 7-14) if used; video of boiler fire-up if made
   — Operating and maintenance manuals (O&M) for all equipment, as specified in the contract documents, reviewed and approved by the engineer
   — Fire protection inspections witnessed by the local authorities verifying the required flow volume and pressure

■ All required extra materials, referred to as "attic stock," to be turned over to the owner. This can include extra boxes of ceramic tile, additional boxes of acoustical ceiling pads, extra hardware parts, extra cans of paint for each color, or carpet and other flooring materials. Another section in the specifications near the attic-stock requirement pertains to special tools, which are also turned over to the owner. These can include the special wrenches needed to replace sprinkler heads or testing devices for various HVAC controls.

Start-Up checklist For Rooftop Equipment

Equipment Identification:
Type; _____
Manufacturer:_____
Model No.: _____
Serial No; _____
Location: _____

Electrical Component: Y   N
Electrical connections complete with disconnect ___   ___
Wire sizing/starter per specifications   ___   ___
Terminations and panel circuit identified/labeled ___   ___

Equipment Installation
Vibration isolation- if required   ___   ___
Piping complete   ___   ___
Seismic restraints, if required, in place   ___   ___
Manufacturer's O&M manuals complete   ___   ___
Manufacturer's start-up sheets attached   ___   ___
Warranty certificate(s) received   ___   ___

Distribution
Distribution piping complete and tested   ___   ___
Piping test report attached   ___   ___
Gas piping complete and tested   ___   ___
Gas piping test report attached   ___   ___
Insulation complete   ___   ___
Confirm installation per plans and specs   ___   ___
Test, adjust balance liquid flow   ___   ___

Automatic Controls
Temperature control wiring complete   ___   ___
Control point connections verified   ___   ___
Control set point verified   ___   ___
Electrical system interlocks completed   ___   ___
BAC system interfaced with Owner's computer   ___   ___

_____   _____
Manufacturer Rep/Subcontractor   Date

_____   _____
Engineer   Date

_____   _____
Owner's Representative   Date

FIGURE 7-14

A typical checklist for HVAC equipment start-up when this equipment is being commissioned in the presence of the manufacturer's representative and the engineer.

AFFIDAVIT THAT ALL TAXES HAVE BEEN PAID

RE:

Date: _____

TO:

The undersigned certifies that all federal, state and local taxes (including sales, consumer, use and excise taxes) applicable to the work and services performed and materials and equipment incorporated into the work, in each case pursuant to the contract referred to above, have been paid in full.

SUPPPLIER/CONTRACTOR: _____

By: _____

Date: _____   Title: _____

STATE OF   )

COUNTY OF   )

On this _____ day of _____, ____, before me personally came _____

_____, to me known who, being by me duly sworn, did

depose and say that he resides at _____ that

he is _____ of _____

The corporation that executed the foregoing instrument and that he signed his name

thereto by order of the Board of Directors of said corporation.

_____
Notary Public

My Commission Expires: _____

FIGURE 7-15

An affidavit that all taxes have been paid that the general contractor must submit when the work is completed.

SAMPLE CONSENT OF SURETY

State Construction Engineer
North Carolina Department of Transportation
1543 Mail Service Center
Raleigh NC 27699-1543

Dear Sir:

The Contractor, _____, for North
Carolina Contract Number: _____, in _____ County,
whose performance we have guaranteed by our Bond Number _____, has requested
that we give our consent to the payment, at your option, of all monies due on his final estimate
according to the provisions of his contract.

We hereby give our consent to the payment of the final estimate and agree that such action on
your part will not operate to qualify or invalidate the Bond.

Sincerely,

_____

By:_____

Seal of Surety

The Consent of Surety should be prepared on the surety's official stationery and it must be signed
by a general officer of the corporation or by an attorney-in-fact. If signed by an attorney-in-fact,
a power of attorney must be attached giving the attorney-in-fact specific authority to write
Consent authorizing the release of monies and it must also bear the corporate seal.

FIGURE 7-16

A sample Consent of Surety form.

- Warranties and guarantees as required by the contract documents. Although
  the contractor, as a standard provision, includes a 1-year warranty, many individual product warranties offer extended service. Roofing warranties vary
  from 10 to 30 years. Insulated glass usually carries with it a 10-year guarantee
  against defects. Compressors used in air-conditioning equipment have a 5-year

or greater warranty period. Some mechanical and electrical warranties include a no-cost 24-hour service contract during the builder's 1-year guarantee period.

■ Material safety data sheets (MSDS). These individual sheets, which contain product safety warnings—whether they contain hazardous materials—were sent to the general contractor prior to the product arriving on site. The GC would have collected those MSDS sheets in a binder and should turn them over to the owner at the closeout.

When allowances and a contingency were included in the construction contract, both accounts should be reconciled. If the allowance(s) had not been reconciled and closed out by change order, now is the time to do so. If the contingency was to be shared by both contractor and owner, that, too, can be reconciled and closed out via a change order.

If the contract was in the form of cost-plus not to exceed a guaranteed maximum price, an accounting will be required to arrive at the agreed-upon final contract sum. Architects and owners frequently request that the general contractor prepare interim "costs to date" reports during the project; if this has been the case, the final accounting will be easier to reconcile.

If there was an equipment list that included contractor purchases of items having a residual value at the end of the project, the contract should either advise the contractor that the owner will take possession of the item or items or settle on a value, allowing the contractor to retain possession and issue an appropriate credit.

Along with the general contractor's final release of lien, some architects require an affidavit that all federal, state, and local taxes have been paid (Figure 7-15).

When payment and performance bonds have been furnished by the general contractor, that contractor will provide the owner with a consent of surety form (Figure 7-16) in which the bonding company certifies that all bond provisions and requirements have been met.

Change orders

8

The project that has no change orders is a rarity, and the change-order process is often a contentious affair. Change orders can occur for the following reasons:

- Changes in the scope of work requested by the owner
- Changes in the work due to conditions unknown at the time of contract signing, which neither the design consultants nor the contractor could have reasonably anticipated
- Claims by the general contractor or subcontractors for errors, omissions, or inconsistencies in the contract documents that can be corrected only by extra work that is beyond the scope included in or intended by the contract documents

Each of these occurrences can create an adversarial relationship between owner and contractor, especially when the change-order process addresses a shortfall in the plans and specifications. The key to a successful change-order system depends on an understanding of why the change-order work is required and a clear, concise explanation of the cost and the impact, if any, that these proposed changes will have on the contractor's workforce demands and the project schedule.

The owner, architect, and contractor can take steps to diminish the conflicts the change-order process causes and inject a sense of reasonableness into that process. One such approach is for each team member—owner, architect, and contractor—to try to view the change order from the perspective of the other team members.

When change-order proposals are initiated by the contractor and they claim that the added costs are the result of missing details in the project's plans or specifications, or of a design error that requires correction at additional cost, an owner's response might well be "This is not my problem. I bargained for a complete project, and it looks like this is something that should have been included in my contract with either the architect or contractor or both." Most contractors willingly correct errors

© 2010 by Elsevier, Inc. All rights reserved.
Doi_No = 10.1016/B978-1-85617-548-7.00008-2

or design issues that are minor and require little or no added expense. They have a professional responsibility to correct them as part of their contract obligation. But larger sums raise other concerns.

A few less-than-scrupulous contractors have used the change-order process to their own advantage and exploited any attempt at reasonableness to gain a few quick bucks; they don't care about upsetting an owner they will most likely never see again. Fortunately, the overwhelming majority of general contractors do not share this view.

Most contractors would rather not issue a single change order because they are very aware of how disruptive it can be. When hard, competitive bidding is the process used in contractor selection, the contractor is obliged to estimate only that portion of work specifically shown on the drawings—no more, no less. Although it may have been incumbent upon the bidders to point out any plan or specification shortcomings, they may, rightfully so, be hesitant to yield any competitive edge to their competition. Although some incomplete or missing items can be absorbed by the contractor at no added cost to the owner, larger ones cannot be treated so casually. The prebid conference discussed in Chapter 2 is the forum to filter out many of these sorts of problems, assign a value to them if necessary, adjust the bidder's price prior to executing a construction contract if an adjustment is warranted, and get on with the job.

Owners would be wise to direct their design consultants to conduct these kinds of prebid meetings as part of the change-order-avoidance process. The preparation of a set of plans and specifications for today's complicated projects is a mighty task. The details required can be staggering, and it is nearly impossible for that perfect, error-free set of drawings to emerge from the architect's and engineer's desks. Two concepts frequently govern the accepted quality standards for drawings: standard of care and best practices; both cut the architect and engineer a little slack that they rightfully deserve.

The standard of care is a critical element in an architect's professional portfolio; it states in essence that the architect is required to perform a reasonable and prudent job in the preparation of the contract documents commensurate with what he or she would perform in the geographic community in the same time frame and under the same circumstances when provided with the same facts.

The best-practices standard, applied to the design consultant, can equally be applied to a contractor's performance. Best practices, as defined by the American Institute of Architects, are those that encompass the collective wisdom of AIA members and related professionals. An architect's performance can be measured by the creative design and the high quality of the design documents, and architects are very much aware of the problems that arise if those design documents contain numerous errors and omissions.

The contractor views a non-owner-generated change order as relief from the costs associated with events that are not his or her responsibility. As stated previously, most contractors would rather not have any change-order work because they are acutely aware of an owner's view that the change order may not be fully warranted and that the contractor should have anticipated some of this extra work as they estimated the project price. The contractor is also aware that the owner is assuming that the cost of the proposed change will be higher than it should be.

When change orders originate from subcontractors and the proposed changes are created by deficiencies in the plans or specifications, both general contractor and subcontractor are merely trying to recover costs that could not have reasonably been anticipated. General contractors often rely on their subcontractors to do the really in-depth review of the plans and specifications because these subcontractors have much more in-depth knowledge of their trades. The terms *general contractor, subcontractor,* and *specialty contractor* convey the image of progressive specialization, and that is usually the case with identifying problems in the plans and specifications.

During the bidding process, general contractors will be soliciting bids from several subcontractors as they are preparing their estimates. The really in-depth review of each vendor and subcontracted component begins after a contract with the owner has been executed and the negotiation process with the vendors and subcontractors commences. This period of time, during detailed scrutiny of the plans and specifications, is when many errors, inconsistencies, or omissions will be discovered, and that's when any problems with the plans and specs that arise must be dealt with. Does the general contractor or the subcontractor absorb some of the costs or request a change order to repair any of these problems? Major plan/specification errors, omissions, or inconsistencies that have a significant cost implication will most likely be presented to the architect at that time.

When these problems arise, it is the responsibility of the general contractor to provide sufficient details in the proposed change-order request to allow the architect and owner to make a rational decision. The change-order process works best when this information is furnished to the owner promptly and the contractor receives a quick response. If the architect or engineer requires more information, it will be furnished, and the architect or engineer should make those requests promptly. When contractor-generated proposals are viewed from the perspective of each party, they are more likely to be resolved quickly and amicably.

CHANGE-ORDER PROVISIONS IN THE AIA GENERAL CONDITIONS DOCUMENT

Several articles in the AIA General Conditions A201 document deal with the submission and processing of change orders:

- Article 3.2.1 requires the contractor to study and compare various drawings for the purpose of "facilitating construction by the contractor and are not for the purpose of discovering errors, omissions, or inconsistencies." If any of these deficiencies are found, they are to be reported to the architect for clarification.

- Article 3.2.2 states, again, that any errors noted by the contractor should be promptly reported to the architect and that this review is made in the contractor's capacity as a contractor and not as a licensed professional.

- Article 3.2.3 states that if the contractor believes that any clarifications received from the architect during this discovery period involve extra cost or time, the contractor will make a "claim,"—that is, submit a proposed change order. But 3.2.3 also states that if the contractor fails to perform the obligations of subparagraphs 3.3.1 and 3.2.2, the contractor "shall pay such costs and damages to the owner as would have been avoided if the contractor had performed such obligations." This subparagraph also stipulates that the contractor will not be liable to the owner or architect unless the contractor recognized such error, inconsistency, omission, or difference and knowingly failed to report it to the architect. It appears that the contractor's approach to a disclaimer may be that they "knowingly" failed to report it.

- Article 4.3 defines a claim as a "demand or assertion by one of the parties seeking adjustment or interpretation of contract terms, payment of money, extension of time."

- Article 4.3.4 deals with claims for concealed or unknown conditions—a difficult issue for all parties involved.

- Article 4.3.7 relates to claims for additional contract time. Subparagraph 4.3.7.2 describes how claims of adverse weather conditions affecting project completion are to be prepared.

- Article 4.4 describes the procedures to resolve claims and disputes, and Articles 4.5 and 4.6 describe the process for resolution, first mediation, and, if that fails, binding arbitration.

Article 7 of the AIA General Conditions A201 document details change-order procedures; one section discusses the construction change directive, referred to as a CCD. The CCD process is a method by which disagreements over the cost of a change order can potentially be resolved by employing those conditions up front.

The CCD permits an architect or owner to direct a change in the work in cases where a contractor doesn't have enough time to prepare a cost proposal or when extra work must be performed but the cost has not yet been determined. A CCD breaks the logjam that can occur in these types of situations because it allows the work to continue with a set of guidelines to be met as costs are accumulated and presented. The general contractor is provided direction to commence the work and provide the owner and architect with all documented costs, including the following:

- The cost of labor, including all fringe benefits

- The cost of materials and equipment, including delivery charges to the job site

- Rental cost of machinery and equipment, whether rented from the contractor or other sources; hand tools, hammers, and screwdrivers are excluded as reimbursable costs, since the worker, not the employer, in most cases has an obligation to provide them.
- Cost of any bonds, insurance, permits, and fees
- Costs of supervision and field office personnel directly attributable to the work involved in the change
- The contractor's overhead and profit

This procedure varies from the normal contractor's practice of preparing a lump-sum price for the work. Given authorization, the contractor proceeds to perform the work accordingly. If the work actually costs less than the estimate, the contractor will reflect an increase in the stated profit, but if the work costs more, the contractor will absorb any losses.

With the CCD, the cost of the work is based on actual costs, not an estimate of the work, and this cost-plus approach appears to be fair to all parties. It is still incumbent upon either the architect or the owner's representative to ensure that all costs assembled are valid and justified. Copies of invoices for materials, rental equipment, and signed tickets documenting all labor costs are equally important.

The CCD approach should be readily accepted by all parties because it allows the work to proceed and provides an acceptable method of establishing costs. The provision that allows for supervision and field office personnel, while usually rejected by an owner when a regular cost proposal is submitted, is included in the CCD process. Other lump-sum types of change orders usually do not include field office personnel and supervision costs, since these costs are assumed to be part of the contractor's overhead and profit add-on, so the contractor is perfectly willing to proceed with a constructive change directive.

REVIEW PROCEDURES FOR CHANGE ORDERS

The general contractor has the responsibility to review all change orders received from vendors or subcontractors before passing them on to the architect. The owner may well ask the contractor, "Are these changes being submitted actually changes or merely the subcontractor's or vendor's interpretation of a change in scope? Are the costs associated with this change presented in sufficient detail for us, as owner and architect, to analyze? Are all of the costs justified? Is a credit due for deleted work?" These questions should have been answered by the project manager *before* the cost proposal was submitted to the architect and owner, but many project managers are remiss in doing these kinds of reviews.

Time and time again we find the general contractor's project manager merely passing through change orders without reviewing them for accurate content, applicability, or even correct numbers. Lack of proper screening places an undue burden on the architect, engineer, and owner. If an architect or owner receives several proposed change orders that are insufficient for proper review, they should be returned to the project manager posthaste with a strong note that the PCO(s) is not acceptable in the form in which it was submitted and that no claim for delay or extra cost will be tolerated due to the delay in receiving proper, detailed information.

Early in the administration of the project—perhaps at the first or second project meeting—the contractor should be advised that all proposed change orders, if and when they occur, are to be thoroughly reviewed by the project manager prior to submission to the design consultants for review. It is a good idea to have the contractor prepare a stamp that reads, "This proposed change order no. ___ has been reviewed by (name of the project manager) for applicability, content and pricing." When someone will be held accountable in such a manner, the likelihood of rejection will generally prompt them to do a proper review.

Quite often a proposed change order not only adds costs to the project but the proposed changes may also have generated a credit. For example, if vinyl wall covering is to be added to the conference room walls, a second coat of paint is not necessary, so the additional cost to install the wall covering should include a credit for the finish coat or coats of paint in terms of both labor and materials. This seems to be an obvious credit, but it does not always appear on the "cost" sheet. It is the responsibility of the project manager to consider a prior condition or the contract requirements before assembling the costs for the changed work.

Although generic contract language relating to change-order submission, review, and approval may be used, it is a good idea to establish some basic ground rules with the contractor in the opening stages of the contract administration. These ground rules can relate to some of the more basic situations where change-order work may occur:

- Documentation to substantiate proposed change-order work
- Time and material work authorized by the owner
- Premium time work authorized by the owner
- Winter or hot weather conditions that may require additional work

DOCUMENTATION OF CHANGE-ORDER WORK

Your architect can prepare and present a protocol for the submission of change-order work so all parties are aware of the format in which these requests are to be submitted, the procedures to be used for pricing, and the documentation required for review and comment. A simple plan can be presented at a project meeting and incorporated into the project meeting minutes, or a protocol can be transmitted to the general contractor in a separate letter from the architect.

Change-order protocols might include the following:

1. Each proposed change order is to contain a brief explanation of the nature of the change and whether it is initiated by the contractor or the owner. All supporting documentation is to be attached, such as a letter from the owner/ architect, general contractor, subcontractors, or vendors requesting the change and architect/engineer-furnished sketches or drawing revisions.

2. If the scope of the work is being increased or decreased, the prior condition is also to be listed by either the architect or engineer. This should not deter the contractor from analyzing the effect of the change but may make it somewhat easier to highlight those changes. In the absence of an architect's statement to that effect, the contractor is obliged to perform that analysis.

3. All costs submitted by the general contractor for work performed with his or her own forces, or work to be performed by subcontractors should be broken down into labor (number of hours and hourly rate), materials (quantity or square or cubic feet if applicable), and equipment with accompanying proposals from the vendor. All costs are to be net of overhead and profit. Subcontractor's costs are to be net and include as a separate line item the actual addition of OH&P. The same will apply to the general contractor costs if applicable.

4. Rental equipment costs are to be designated as general-contractor-owned or from a rental agency not affiliated with the general contractor. A request for a rate differential for idle equipment and operated equipment should be included. Although the actual hourly equipment rates may be the same, idle equipment costs for contractor-owned equipment should not reflect the operator's hourly rate, fuel, and routine maintenance costs. Idle rental equipment will have no operator or fuel costs, which are normally added by the contractor renting the equipment. Equipment, either rented or contractor-owned, will have different rates for daily, weekly, and monthly periods, with weekly and monthly rates lower than hourly and daily rates.

5. If the change-order work is being authorized as time and material instead of as a lump-sum cost, the T&M protocol indicated following should be used.

6. If requested by the owner or the architect/engineer, the owner's representative can attend the meeting when the general contractor negotiates change-order work with the subcontractors or vendors.

Time and material (T&M) work protocols might include the following:

1. The contractor's supervisor is to provide daily tickets for all T&M work that is self-performed by the contractor's workers or by any of the subcontractors. The ticket should include the worker's name or trade designation (apprentice, journeyman, foreman), hours worked, and task performed. This daily ticket should be signed by the contractor's supervisor, attesting to the accuracy of the information. A copy will be sent to the owner's representative each day unless the owner requests otherwise. Receiving tickets for all materials and equipment are also to be attached to the daily ticket. These tickets should be signed by the general contractor's superintendant or foreman, substantiating receipt. (The project manager will review and either accept these costs or request a modification from the subcontractor before sending this on to the architect for review and comment or approval.)

2. For all subcontracted work, follow the preceding procedures by submitting the subcontractor's daily tickets to the general contractor's supervisor for review and acceptance. Receiving tickets for all materials and equipment are to be attached to the daily ticket. When second-tier subcontractors are employed, the prime subcontractor is to receive the daily tickets signed not only by the second-tier subcontractor but by the prime subcontractor, certifying to its accuracy.

3. In a case where the T&M work is being performed for an extended period of time, the owner may require the general contractor to prepare weekly tallies of "costs to date" and "projection of costs to complete."

Premium-time costs can be documented as follows:

1. For the general contractor's self-performed work, follow the preceding procedures for T&M work, but include the reason for the premium time work—for example, weather delays, request by owner, delays in receiving response to queries to the owner/architect. Daily tickets are to be prepared as indicated previously and signed for verification by the contractor's supervisor.

2. For the subcontractor, follow the procedures for T&M work as previously indicated. Include the reason for the premium time work on the ticket. Labor hours are to be verified by the subcontractor's supervisor.

3. An account of hours for extended work is to be presented to the owner on a weekly basis, or more often when requested, and costs tabulated if so requested by the owner.

Winter conditions or weather delay claims can be documented as follows:

- On all daily tickets, provide all information listed following; all tickets are to be certified by the general contractor's onsite supervisor.

- Indicate the operation taking place and why the weather-protection work was required.

- When winter conditions are involved, provide temperature readings, one at 7:00 A.M., one at noon, and a third at 2:00 P.M.

- When winter conditions are being documented, provide receiving tickets for all types of fuel consumed.

- Attach tickets for any other materials or equipment employed and why these materials and/or equipment were required.

- If extra work related to unusual weather conditions is being undertaken, the general contractor is to provide proof of the nature of those unusual conditions—for example, 100-year storm, excessive rainfall, hurricane-force winds, and so on.

COST ISSUES

We've discussed including in the contract for construction a list of costs to be reimbursed and costs not to be reimbursed, and change-order work usually involves two basic types of costs: direct costs, those costs required to complete the work, and indirect costs, costs associated with the contractor's field and home-office expense. Another type of cost, referred to as impact cost, will generally apply when a delay or series of delays is experienced by the contractor and a delay claim is being considered. Any of these costs could be incorporated into a proposed change order, but that doesn't necessarily mean that they are justified; some may need to be challenged by the design consultants and the owner.

Direct costs include the following:

- Contractor-owned equipment operation and fuel, whether active or idle, at rates no higher than published rates; frequently these costs are limited to 75 percent of market rate rental rates.

- Rental equipment, including replaceable accessories such as jackhammer points and diamond cutting blades.

- Insurance premiums, which are based on the cost of work; as costs increase, so do insurance premiums.

- Bond premiums; as the cost of work increases, so does the cost of the bond.

- Building permit costs in those jurisdictions where building officials track costs added to the initial permit cost via change orders. (If a change order increases the original cost, the general contractor may be required to add a proportionate amount to the building permit.)

- Labor, fringe benefits, and payroll taxes

- Material costs, including costs to deliver or pick up and to distribute the material to the area within the building where it will be used

- Fasteners; stainless-steel expansion bolts, for example, are expensive and will be included in the cost.

- Photographs, express deliveries, reproduction costs (plans and/or architect sketches)

- Safety equipment and personal protective equipment (hard hats, goggles, gloves, foot protection, and respirators)

- Subcontractor and vendor costs

- Additional temporary heat or cooling in excess of the amount required by the base contract work

- Travel expenses of the project manager and project superintendent, including parking, tolls, and so forth

- Added utility costs for the building under construction if weekend work or after-hours work is required

Indirect costs include the following:

- Project management
- Project engineer
- Project superintendent
- Field office operating expenses
- Estimating costs to prepare the proposed change order

Impact costs include the following:

- Lost productivity due to trade stacking and other inefficiencies

- Idle equipment and its maintenance

- Lack of availability of skilled tradespeople due to increased demand to complete the work and remain on schedule

- Cost of disruption to the orderly flow of work

- Cost to work out of sequence

- Cost of extended warranties on equipment installed during the project that may expire before the standard one-year guarantee period

CHANGES IN SCOPE OF WORK

One of the major complaints voiced by owners is that the contractor fails to provide sufficient detail to allow for a comprehensive review of their change-order work. This back-and-forth is a waste of time for all concerned. Even if there is an agreement that extra costs are involved, without a clear statement by the contractor of each of the components in the proposed change-order request and its respective costs, a prompt review and response from the owner cannot be expected. As we discussed previously, specific requirements for change-order submissions can be included in the contract for construction, and such requirements are not unreasonable:

1. Each proposed change order should include a brief narrative of the work and the party initiating the request. For example, if the architect has issued a letter or request for pricing and has attached a sketch or description of the extra work, a statement referring to this letter or sketch should be included. If the owner has issued a verbal request directly to the contractor (not a good policy; the request should be in writing and via the architect), the date, place, and exact request must be included in the contractor's response.

2. If the scope of the work has increased or decreased, the contractor should reference the prior condition either with a descriptive narrative or a portion of the plan or specification that reflects the original condition. For example, if the species for wood doors changed from red oak to mahogany, the contractor should

so note and attach the specification section or door schedule that includes the oak doors. This difference is often referred to by the contractor as the "delta."

3. When the scope of work has either decreased or increased, the value of the prior condition must be included so the difference between the original costs and the new costs can be discerned. The contractor should make sure that the original costs are what they should be and not less.

4. All labor costs submitted for either the general contractor's forces or the subcontractor's are to include the number of hours for each type of trades-person involved and whether they are journeymen or foremen. The rate for each trade times the number of hours is the proper way to list each labor component.

5. All material costs proposed by the general contractor can be documented with quotes from the related suppliers or, in the case of a subcontractor, the supplier's quote for major components such as air-conditioning units, electrical fixtures, doors, and frames.

6. Equipment costs for use of a general contractor's equipment should be documented with comparable rates from a rental agency.

Figure 8-1 lists the equipment's T&M rate, operator per-hour costs, fuel, and OH&P, along with a local equipment rental company rate.

ABC Excavation Co.	T&M Rate	Operator	Fuel	OH&P	Bare Cost	DEF Rental-Hourly Daily	Weekly
345 Exc	$ 175.00	$ 38.00	$ 13.66	$ 22.25	$ 101.09	$ 191.25	$ 114.75
330 Exc	$ 154.00	$ 38.00	$ 10.86	$ 18.48	$ 86.66	$ 151.25	$ 90.25
963 Loader	$ 126.00	$ 38.00	$ 7.05	$ 18.90	$ 62.06	$ 130.00	$ 78.00
D-4 Dozer	$ 110.00	$ 38.00	$ 4.66	$ 16.50	$ 50.84	$ 60.00	$ 35.75
Roller	$ 80.00	$ 38.00	$ 2.50	$ 12.00	$ 27.50	$ 52.50	$ 31.25
735 Site Dump	$ 140.00	$ 38.00	$ 13.45	$ 19.00	$ 69.55	$ 178.75	$ 106.75
IT28 Loader	$ 120.00	$ 38.00	$ 4.49	$ 18.00	$ 59.51	$ 75.00	$ 44.75
Mini Exc.	$ 96.00	$ 38.00	$ 2.81	$ 14.40	$ 40.79	$ 31.25	$ 21.25

FIGURE 8-1

A list of equipment with bare rental rates, an add for operator, fuel, and contractor's overhead and profit. The last two columns on the right offer a comparison between rental rates of contractor owned equipment and those of a construction equipment rental company.

2008 Alban Rents Book Rental Rates

Category/Class			2008	Daily Rate	Weekly Rate	4 Week Rate	Daily Excess Rate	Weekly Excess Rate	4 Week Excess Rate
3010160SST	Caterpillar	416D 2W/ST	78 hp Backhoe Loader w/cab	250	750	2,250	15.63	8.52	6.39
				250	750	2,250	15.63	8.52	6.39
3010160EST	Caterpillar	416D 2W/EX	78 hp Backhoe Loader w/cab	290	870	2,600	18.13	9.89	7.39
				290	870	2,600	18.13	9.89	7.39
3010162SST	Caterpillar	416D 4W/ST	78 hp Backhoe Loader w/cab	300	890	2,650	18.75	10.11	7.53
				300	890	2,650	18.75	10.11	7.53
3010162EST	Caterpillar	416D 4W/EX	78 hp Backhoe Loader w/cab	320	960	2,870	20.00	10.91	8.15
				320	960	2,870	20.00	10.91	8.15
3010202	Caterpillar	420D	88 hp Backhoe Loader w/cab	330	990	2,970	20.63	11.25	8.44
				330	990	2,970	20.63	11.26	8.44
3010264	Caterpillar	426C	88 hp Backhoe Loader w/cab	320	960	2,870	20.00	10.91	8.15
				320	960	2,870	20.00	10.91	8.15
3010302	Caterpillar	430D/436C	98 hp Backhoe Loader w/cab	400	1,200	3,590	25.00	13.64	10.20
				400	1,200	3,590	25.00	13.64	10.20
3010460	Caterpillar	446C	110 hp Backhoe Loader w/cab	640	1,910	5,730	40.00	21.70	16.28
				640	1,910	5,730	40.00	21.70	16.28
3018105	CTI	CL 4000	Backhoe Fork Attachment	40	120	340	2.50	1.36	0.97
				40	120	340	2.50	1.36	0.97
4370403	Fermac	640B	90 hp Tractor 3 point hitch	260	780	2,330	16.25	8.86	6.62
				260	780	2,330	16.25	8.86	6.62
3035405	Caterpillar	904B	0.8 yd3 Mini Wheel Loader	290	850	2,550	18.13	9.66	7.24
				290	850	2,550	18.13	9.66	7.24
3030060	Caterpillar	906	1.04 yd3 Mini Wheel Loader	300	890	2,650	18.75	10.11	7.53
				300	890	2,650	18.75	10.11	7.53
3030080	Caterpillar	908	1.3 yd3 Mini Wheel Loader	360	1,060	3,180	22.50	12.05	9.03
				360	1,060	3,180	22.50	12.05	9.03
3040140	Caterpillar	914G	1.8 yd3 Rubber Tire Loader	430	1,280	3,840	26.88	14.55	10.91
				430	1,280	3,840	26.88	14.55	10.91
3040240	Caterpillar	924 GZ	2 yd3 - 2.7 yd3 Rubber Tire Loader	480	1,440	4,310	30.00	16.36	12.24
				480	1,440	4,310	30.00	16.36	12.24
3040280	Caterpillar	928G	2.8 yd3 - 6.5 yd3 Rubber Tire Loader	500	1,790	5,370	37.50	20.34	15.26
				600	1,790	5,370	37.50	20.34	15.26
3050380	Caterpillar	938G	2.75-2.65 yd3 Rubber Tire Loader	710	2,120	6,360	44.38	24.09	18.07
				710	2,120	6,360	44.38	24.09	18.07
3050500	Caterpillar	950G	3.25-4.5 yd3 Rubber Tire Loader	920	2,750	8,250	57.50	31.25	23.44
				920	2,750	8,250	57.50	31.25	23.44
3050620	Caterpillar	962G	3.5-5.0 yd3 Rubber Tire Loader	990	2,970	8,910	61.88	33.75	25.31
				990	2,970	8,910	61.88	33.75	25.31
3050660	Caterpillar	966G	4.25-5.25 yd3 Rubber Tire Loader	1,320	3,960	11,880	82.50	45.00	33.75
				1,320	3,960	11,880	82.50	45.00	33.75
3070140	Caterpillar	IT14G	Integrated Tool Carrier w/Forks	490	1,470	4,410	30.63	16.70	12.53
				490	1,470	4,410	30.63	16.70	12.53
3070240	Caterpillar	924G	Integrated Tool Carrier w/Forks	560	1,660	4,960	35.00	18.86	14.09
				560	1,660	4,960	35.00	18.86	14.09
3040300	Caterpillar	930G	Integrated Tool Carrier w/Forks	580	1,740	5,210	36.25	19.77	14.80
				580	1,740	5,210	36.25	19.77	14.80
3080380	Caterpillar	IT38G	Integrated Tool Carrier w/Forks	640	1,910	5,730	40.00	21.70	16.28
				640	1,910	5,730	40.00	21.70	16.28
3086625	Caterpillar	IT28/IT38	IT Fork Attachment - Loose	50	150	430	3.13	1.70	1.22
				50	150	430	3.13	1.70	1.22
4985356	Misc.	IT28/IT38	IT Fork Extentions	40	100	290	2.50	1.14	0.82
				40	100	290	2.50	1.14	0.82
3180120	Caterpillar	12H	12' Motor Grader	1,000	2,990	8,950	62.50	33.98	25.43
				1,000	2,990	8,950	62.50	33.98	25.43
3180142	Caterpillar	143H	12' Motor Grader/All Wheel Drive	1,110	3,320	9,960	69.38	37.73	28.30
				830	2,480	7,440	51.88	28.18	21.14
3140301	Caterpillar	D3G	70 hp Track Dozer	380	1,140	3,420	23.75	12.95	9.72
				380	1,140	3,420	23.75	12.95	9.72
3140311	Caterpillar	D3G LGP	70hp Track Dozer Low Gnd Press	400	1,190	3,570	25.00	13.52	10.14
				400	1,190	3,570	25.00	13.52	10.14
3140452	Caterpillar	D4G XL	80 hp Track Dozer	420	1,260	3,780	26.25	14.32	10.74
				420	1,260	3,780	26.25	14.32	10.74
3140451	Caterpillar	D4G LGP	80hp Track Dozer Low Gnd Press	480	1,430	4,280	30.00	16.25	12.16
				480	1,430	4,280	30.00	16.25	12.16

FIGURE 8-1 (continued)

Some general contractors own several basic pieces of excavating equipment, such as bulldozers, excavators, and compaction equipment, while others subcontract all or a major portion of this type of work to excavating contractors. The excavating subcontractors generally own basic equipment and rent additional specialized equipment to augment their inventory. Equipment costs will appear in many types of change orders ranging from site work or hoisting to supplying heaters or fans for cold- or hot-weather protection, and all need some scrutiny when included in change orders to ensure that the rates are fair and equitable. Costs are significant for some large excavators: up to $250 per hour or more, including operator and fuel costs.

Equipment, whether contractor-owned or rented from an equipment rental company, is charged either at an hourly, daily, weekly, or monthly rate, depending on the requirements of the operation. And, of course, the hourly rates decrease as the length of time employed increases. Figure 8-1 reflects the difference in rental rates. When equipment costs are included in a change order, the general contractor should determine if the correct rates are being charged. In the chapter on contracts, we discussed inserting language into the contract with the general contractor that limits the hourly and weekly rates to be applied, depending on the number of hours or days the equipment was used. If the equipment was used more than three hours, the daily rate in lieu of the hourly rate should be charged, and if the equipment was operating for three days or more, the weekly rate should apply.

If the owner-contractor contract did not include such restrictive provisions, certain guidelines must still be followed when equipment rates are included in a proposed change order. The first matter to consider is how long the equipment was working and whether it was active all of that time.

Equipment rates vary quite a bit. An hourly or daily rate of, say, $85 per hour will be reduced to about $52 per hour, or between 30 and 40 percent, when a weekly rate is applied, and that initial rate will be reduced further to $39 per hour (a 45 percent reduction) when rented on a monthly basis. This doesn't include the operator's cost or fuel costs, which have to be added along with the appropriate contractor overhead and profit.

A call or two to a local construction equipment rental company will yield information on comparable rental rates in the area. The architect or engineer can make the calls. Rental companies usually follow these guidelines:

- The rental rate is based on using the equipment for one 8-hour shift, 40 hours per week, or 176 hours per month.

- Overtime rates would be about 12 percent more than the daily rate, 2.5 percent over the weekly rate for hours in excess of 40 hours, and just over 1 percent more for the hourly rate for hours exceeding the monthly maximum operating time.

- Delivery charges are usually added to the rental rate to transport the equipment to the site and pick it up.

- The lessee (operator) is responsible for the cost of fuel and for any routine maintenance costs, such as greasing, oil changes, and operator costs.

The *AED Green Book* is a nationally recognized manual of equipment rental rates. Updated yearly, it includes the average rental rates for a wide range of construction equipment, provided as monthly, weekly, and daily rates. Geographic cost adjustments to account for regional differences are provided (called modifiers) that allow a user to adjust a rate for a specific area of the country. There are ten such regions, ranging from New England and mid-Atlantic areas to the South, Southwest, and Northwest, as well as Alaska and Hawaii. As an example, a national rental rate, adjusted for New England and the mid-Atlantic areas and designated as Region 1, would appear as follows:

National average rental rate for loader	$690 per week
Multiplier for Region 1	× 0.98
Adjusted weekly rental rate	$676 per week

There is a distinct difference between an active rate for equipment that is working and requires an operator, fuel, and some routine maintenance and equipment that sits idle for extended periods of time—say, a day or two. During this idle or down time, operator, fuel, and maintenance costs should accrue. Although idle rented equipment will still incur the same rental costs because it was rented on a daily, weekly, or monthly basis, the hourly costs will be lower because nobody is using it and it does not require fuel or routine maintenance costs. The same condition is not true for contractor-owned equipment. Contractors are not bound by a rental agreement, and their downtime or idle costs are substantially less.

There are no hard and fast rules for establishing idle or downtime rates, but a figure of 50 percent of the active rate should be the starting point. If the contractor anticipates that the equipment will be idle for an extended period of time, he or she may decide to move it to another project where it can continue to generate income. In that case no downtime charges are warranted.

Some court decisions support idle time and downtime costs. In the court decision for the *Appeal of Dillon Construction, Inc.*—ENGBCA No. PCC—101, November 21, 1995, the court ruled that if contractor-owned equipment was idle, the contractor could claim 50 percent of the equipment costs as listed in an equipment manual. The court may have been referring to the *AED Green Book*, the nationally recognized listing of rental equipment costs.

In *C.L. Fairly Construction Co.*, ASBCA No. 32,581 90-2, BAC (CCH) par. 22,665 (1990), the court ruled that the contractor must establish that the equipment in question could have otherwise been productively employed before he or she can claim any downtime costs. In other words, can the contractor show evidence that this equipment could have been shipped to other jobs and generate revenue? Most likely, an invoice from a rental company substantiating the fact that the contractor had to rent equipment rather than employ his own, idled equipment would be sufficient proof for the claim.

Thus, when the cost of major pieces of equipment is included in a change order, you should scrutinize as follows:

1. What rates are applied, daily, weekly, or monthly, and are these time frames appropriate?

2. Are the rates appropriate when the cost of the operator and a reasonable cost for fuel and maintenance are added or deducted in the case of idle equipment?

3. Should the active rate be applied over the entire period, or should some time be charged at downtime or idle rates?

CHANGE ORDERS AND SCHEDULING

With the signing of a construction contract, the builder will generally submit a schedule prepared by the critical path method (CPM). This schedule reflects the timeline for starting and completing the various work tasks and segments of the project. A critical path is established for the most important milestone events; as work progresses, the critical path changes. The first critical event may be the completion of the building's foundations, after which the critical path may shift to the erection of steel if that is the building's structural system. When delays interfere with the critical path, the contractor may incur a delay. Quite often the delay of one work task can impact other subsequent work tasks and cascade, resulting in a delay of more consequence than the initial one and thereby extending the completion date of the entire project.

Some changes in the work may impact the construction schedule, and when an extension of completion time is anticipated, the contractor will advise the architect and owner accordingly. This notification in itself can become an issue: Is there truly a delay, or is the contractor "crying wolf"? Will that changed work really impact the schedule? The owner may ask, "How can the contractor with any certainty at this early stage in the project determine that additional time will be required to incorporate the work in question?" No doubt some changes could be cause for an extension of time, but others will not. The time to settle this issue is when the claim has been presented, not at a later date.

Some contractors may state that any increase in time will be determined at some future date—say, if they miss the contract completion date by "X" days. At that point, they could present a claim that the delay was due to the changes that took place previously and that were documented properly. In one respect, this may be a fair assessment of establishing the delay, but if actions by the contractor unrelated to the change extend the project completion date, how can those initial delays be separated from the contractor-generated delays?

When a contractor requests additional time to complete the project, he or she may be considering a request for extra costs if in fact they do finish late. By extending the completion date, he will have extended their general conditions costs, since he will have to pay the superintendent to remain on the jobsite, and the various rental costs will continue for the office trailer, phone service, and toilets; the contractor would expect to be reimbursed for those costs. These general conditions costs are not inconsequential and can be upward of $20,000 to $40,000 per month, Recapturing all or a portion of these costs is important to the general contractor when justifiable delays are encountered.

One way for an owner to deal with contractor-proposed extended-schedule issues is to advise the contractor that the owner recognizes that the changes may increase the completion time by "X" number of days; however, the contractor will proceed with the work, and the owner reserves the right to challenge any time extensions. If both parties agree, then that statement should be added to the formal change order. When the formal change order is issued in this case, some forms list in small print at the bottom three options to choose from, one of which is to be checked off: the change order will increase, decrease, or have no effect on the contract end date.

When signing a change order, owners must be aware of this provision if it is included on the change-order form. Another approach to potential delay claims is to provisionally sign a cost proposal, acknowledging the work but reserving the right to challenge not only the cost of the work but the time required to perform the work, which could subsequently impact the contract completion date. Figure 8-2 is part of a contractor's cost proposal for an increase in which they claim not only an additional cost of $3,816 but, by the nature of this work, a possible extension of the contract completion time. As both parties, owner and contractor, sign this agreement, there is room to negotiate a settlement later on with respect to additional time issues.

Float and the schedule

In all contractor- or architect-prepared schedules, extra days are included on the assumption that not all activities will go as planned, and added time—a cushion—is added for minor unanticipated disruptions to the smooth flow of work. There will always be days when it rains, when the equipment does not show up on time, or when a particular subcontractor did not bring sufficient crews to the site. These extra days included in the schedule are known as float and are used by the contractor to make adjustments to the schedule if and when those unanticipated delays occur.

Unless specifically stated in the contract for construction, this float is for the sole use of the general contractor. If the owner wishes to add float days for unanticipated events, he or she must so state when the contract for construction is being prepared. The owner may be planning to purchase some special materials, custom carpeting, wall coverings, or custom-made electrical fixtures, all of which are to be installed by the general contractor at a specific time frame in the schedule. If no owner float days are included and that carpet, wall covering, or electrical fixture arrives on the site later than the installation schedule indicates, the contractor may request an "extra" to go back and install the item. If the owner had included his or her own float days to cover such contingencies, these delays will not impact the contractor's schedule.

Other cost issues must also be explored. Payroll costs should not be overlooked when reviewing costs. Not only may the appropriate wage rate need to be reviewed, but the other "add-ons" for fringe benefits and applicable overhead and

SHORT DESCRIPTION:

REASON: Design Change

SCOPE OF WORK:

CONTRACTORS AFFECTED:

REFERENCED DOCUMENTS:

WT REQUESTS AUTHORIZATION TO PROCEED WITH THE CHANGES DESCRIBED ABOVE.
COST / SCHEDULE IMPACT WILL BE AS FOLLOWS:

COST: $3,816.00   For GMP   Funding: Other: _____
                  Contracts:

SCHEDULE: ☐ NO IMPACT
              ☐ POSSIBLE IMPACT
              ☐ IMPACT TO SCHEDULE AS FOLLOWS   _____

OWNER DIRECTION:

☐ PRICE ONLY - DO NOT PROCEED UNTIL PROPOSAL IS APPROVED
☐ PROCEED AND CONFIRM COST   ☐ CANCEL
☐ PROCEED FOR NTE COST INDICATED ABOVE
☒ PROCEED T&M
☐ PROCEED FOR LUMP SUM COST ABOVE
Other: _____

DISTRIBUTION:
☐ ORIGINAL FILE
☐ OWNER REP.
☒ ARCHITECT

Other: _____

_____ _____
OWNER'S REPRESENTATIVE   DATE

Owner acknowledges receipt of this Change Notification 22, and hereby authorizes Contractor to proceed with the work on a T &M basis. However, Owner reserves the right to challenge the cost of the work and extent of additional time requested for reasonableness.

FIGURE 8-2

Owner statement on contractor's proposed change order, authorizing work but not agreeing to a cost or time extension.

profit percentages are to be considered. Labor represents a significant portion of each construction component, and it plays a major part in any change-order work.

Hourly rates for union workers include many employee benefits that increase those rates. Merit-shop and nonunion workers also include some fringe benefits, but generally not to the same degree and amount as for workers with collective-bargaining agreements. Figure 8-3 shows a page from a collective-bargaining agreement that lists all of the union benefits to which the contractor or subcontractor will add his or her own: FICA, unemployment insurance, travel costs, and so forth.

When premium-time rates are included in the cost of the work—for work that exceeds the standard 40-hour week or holiday and Sunday work—sometimes the general contractor is not as diligent as he should be as he reviews the labor rates

ARTICLE III

Wages

Section 1.

The hourly rate of wages and fringe benefit contributions to be paid by each Employer for all Employees covered by this Agreement who work within the territorial jurisdiction of each Chapter as stated in Article II, shall be as follows:

A. CHAPTERS:

Effective Date	Total Package	Wages	H/W	P	IPF	AF	ATF	IMI/AT	IMI	MCAM
8/1/07	$63.91	$41.79	$7.38	$7.32	$1.50	$4.20	$.27	$.17	$1.18	$.10
2/1/08	$64.65	$42.48	$7.38	$7.32	$1.50	$4.20	$.27	$.17	$1.23	$.10
8/1/08	$66.40	$44.23	$7.38	$7.32	$1.50	$4.20	$.27	$.17	$1.23	$.10
2/1/09	$67.14	$44.92	$7.38	$7.32	$1.50	$4.20	$.27	$.17	$1.28	$.10
8/1/09	$68.94	$46.72	$7.38	$7.32	$1.50	$4.20	$.27	$.17	$1.28	$.10
2/1/10	$69.88	$47.61	$7.38	$7.32	$1.50	$4.20	$.27	$.17	$1.33	$.10
8/1/10	$71.78	$49.51	$7.38	$7.32	$1.50	$4.20	$.27	$.17	$1.33	$.10
2/1/11	$72.82	$50.50	$7.38	$7.32	$1.50	$4.20	$.27	$.17	$1.38	$.10
8/1/11	$74.92	$52.60	$7.38	$7.32	$1.50	$4.20	$.27	$.17	$1.38	$.10
2/1/12	$75.96	$53.59	$7.38	$7.32	$1.50	$4.20	$.27	$.17	$1.43	$.10

B.   Deducted from net wages after taxes in all Chapters -- D, BACPAC, IUD

   D -   Local Union Dues Deduction
   BAC/PAC - $.01 per hour -- BAC Political Action Committee
   IUD -   International Union Dues Deduction

C.   Definitions of abbreviations used:

   H/W -   Health and Welfare Fund
   P -   Local Pension Fund
   IPF -   International Pension Fund
   AF -   Annuity Fund
   ATF -   BAC Local 3 Apprenticeship and Training Fund
   IMI/AT -   International Masonry Institute Apprentice Training Fund
   IMI -   International Masonry Institute Industry Marketing & Promotion Fund
   MCAM -   Mason Contractors Construction Advancement Program

FIGURE 8-3

Typical union collective bargaining agreement statement of wage rates and union benefits.

before passing them on to an owner. The first step in a comprehensive review of labor costs is for the owner or architect to request a complete breakdown of the chargeable hourly rates for regular time, time and one-half, and double time at the beginning of the project as the contractor negotiates each subcontractor agreement. When received, these costs should be reviewed by the general contractor and, after any adjustments have been made, passed on to the owner for review, comment, and/or approval. These rates would then apply throughout the entire job, unless new collective-bargaining or other such labor agreements have been instituted or updated in the interim period. By agreeing to labor rates early on, another potential problem will have been anticipated, and any changes to the work will be a little easier to resolve.

A matrix similar to Figure 8-4 can be distributed by the general contractor to all subcontractors to obtain current wage rates that will form the basis for future extra work. Certain elements of the basic wage structure will not apply to premium-time work, whether time and one-half or double time, and this form will be used to sort out those elements. The N/A (not applicable) in each such category in Figure 8-4 addresses those categories in case the subcontractor needs to be reminded.

This particular matrix, as completed, represents a union worker's hourly rate, including regular-time hourly rates, time and one-half, and double time for a foreman and a journeyman in this trade. Each of the fringe benefits is listed for straight time, and corresponding increases when premium time rates are applied. Note that insurance rates such as state health and welfare and some forms of insurance remain unaffected by premium-time work. Table 8-1 shows another example of an actual subcontractor labor rate breakdown that the owner questioned.

The increase applied to premium-time costs relating to small tools, travel, and parking in this wage breakdown was questioned by the owner and architect. The small tools category is meant to reimburse the worker for hand tools that wear out and for which they, not the company, are responsible. On the basis of the value for that item in the labor breakdown, $1.25 per hour would amount to $50 per week for regular-time work; if that carpenter worked eight hours on Saturday, an additional $14.96 was requested. One owner, when presented with those rates, indignantly replied, "That buys an awful lot of screwdrivers," and, of course, it was deleted. That same owner's representative questioned how small tools and travel compensation can increase after an 8-hour day: Isn't it still the same per hour? And what about parking: Did it increase 50 percent after 8 hours? All of these items were subsequently adjusted, but not before the general contractor was advised that he appeared to be performing less than a due-diligence review of change-order costs.

All labor cost breakdowns include employer costs for contributions to federal and state unemployment taxes. In total, these two taxes can add anywhere from $2.50 to $4.50 per hour to the cost of labor and, to a point, are certainly justified. However, should these costs increase by 50 percent for time and one-half work and by 100 percent for double-time work? Unless the subcontractor can substantiate those increases for premium time, they should not be approved.

Furthermore, each of these tax contributions is limited to a specific level of wages earned. In Massachusetts, for example, state unemployment (SUTA) taxes are paid on the first $10,800 of wages; the federal tax (FUTA) is paid only on the first $7,000 in wages. After these limits have been reached, no further contributions to either fund are required. In New Hampshire, SUTA is paid on the first $8,000 of wages.

Subcontractor: _____
Trade: Installation
Effective Date: 3/1/2007
Expiration Date: 8/31/2007

Description	Foreman			Journeyman		
	Straight Time	Half Time	Double Time	Straight Time	Half Time	Double Time
Base Hourly Rate	$ 36.88	$ 18.44	$ 36.88	$ 33.88	$ 16.94	$ 33.88
Health and Welfare	$ 7.78	N/A	N/A	$ 7.78	N/A	N/A
Pension Fund	$ 4.65	N/A	N/A	$ 4.65	N/A	N/A
Annuity Fund	$ 7.26	N/A	N/A	$ 7.26	N/A	N/A
Education Fund	$ 1.18	N/A	N/A	$ 1.18	N/A	N/A
Local Supplemental Pension*		N/A	N/A		N/A	N/A
Industry Promo Fund*		N/A	N/A		N/A	N/A
Holidays*	$ 1.37	N/A	N/A	$ 1.37	N/A	N/A
LMCT Fund*		N/A	N/A		N/A	N/A
NFT/NEMI Fund*		N/A	N/A		N/A	N/A
SASMI Fund*		N/A	N/A		N/A	N/A
Equality Fund*		N/A	N/A		N/A	N/A
		N/A	N/A		N/A	N/A
Sub Total of Union Rates	$ 59.12	$ 18.44	$ 36.88	$ 56.12	$ 16.94	$ 33.88
FICA	$ 2.82	$ 1.41	$ 2.82	$ 2.59	$ 1.30	$ 2.59
Federal Unemployment Ins.	$ 0.13	$ 0.07	$ 0.13	$ 0.12	$ 0.06	$ 0.12
State Unemployment Ins.	$ 3.13	$ 1.57	$ 3.13	$ 2.90	$ 1.45	$ 2.90
Workman's Compensation	$ 3.55	N/A	N/A	$ 3.34	N/A	N/A
Public Liability (Bodily Injury)	$ 1.38	N/A	N/A	$ 1.35	N/A	N/A
Property Damage	$ 1.21	N/A	N/A	$ 1.11	N/A	N/A
Umbrella/Auto	$ 0.24	N/A	N/A	$ 0.22	N/A	N/A
Sub Total of Ins. and Taxes	$ 12.46	$ 3.04	$ 6.08	$ 11.63	$ 2.81	$ 5.61
Small Tools		N/A	N/A		N/A	N/A
Travel	??			??		
Sub Total of Other	$ -	$ -	$ -	$ -	$ -	$ -
Total Union/Ins/Other	$ 71.58	$ 21.48	$ 42.96	$ 67.75	$ 19.75	$ 39.49
Overhead and Profit at 15%	$ 10.74	$ 3.22	$ 6.44	$ 10.16	$ 2.96	$ 5.92
Total Labor Rate	$ 82.32	$ 24.70	$ 49.40	$ 77.91	$ 22.71	$ 45.41

N/A	Not applicable
*	If Applicable

FIGURE 8-4

Matrix to be used for reporting of subcontractor wage rates and associated hourly costs.

Table 8-1 Labor Rate Breakdown

Description	Straight Time	Time & One-Half	Double Time
Base hourly rate	$36.88	$55.32	$73.76
Health & Welfare	$15.67	$23.50	$31.34
FICA	$2.82	$4.23	$5.64
Fed Unemployment	$0.30	$0.45	$0.60
Travel	$1.10	$1.52	$2.20.
Small Tools	$1.25	$1.87	$2.50
Parking	$1.00	$1.50	$2.00

Let's take as an example a carpenter who earns $25 per hour. After she has worked seven 40-hour weeks and has earned $7,000, her FUTA contribution has been fully paid; after eleven 40-hour weeks, she earned $11,000 and her SUTA has been paid. So it would appear that if this carpenter had been fully employed since the first of the year, certainly all unemployment tax contributions have been fulfilled by mid-March, and any time and material work performed after that period should not include these contributions. The general contractor should overview these contributions, but the owner and architect must be prepared to step if he or she doesn't.

QUANTUM MERUIT AND UNJUST ENRICHMENT

As we traverse from the practical to the legal, an owner should become familiar with two certain terms if he or she intends to dispute certain types of change orders. Just as there are some contractors who are looking to take advantage of some owners, so are there some owners who will deny a contractor a reasonable payment for the additional work they have requested. I have experienced situations in which an owner has requested the contractor to proceed with extra work either on the basis of time and material or a lump sum, and upon completion of that work, the owner has either refused to pay the initially agreed-upon price or attempted to reduce an otherwise reasonable cost. Legal action to recover what may not be a substantial sum is an option, and the contractor, if litigation is pursued, will have several court decisions in his favor. One is *quantum meruit,* and the other is *unjust enrichment.*

Quantum meruit (pronounced "mare-o-it") is referred to as a "quasicontract" method of recovery of costs, associated in this case with construction change orders. When confronted with a need to justify the claim for extra work, the contractor needs to prove that the owner has benefited by the work that he or she directed and accepted. This is the thrust of quantum meruit.

If an owner had directed the contractor to change ten wood doors from oak veneer to cherry veneer, this change would most likely result in an increased cost to the owner; if the agreed-upon price was not honored by the owner after the upgraded doors were installed, the contractor could clearly show that the owner certainly benefited by this change.

Unjust enrichment is similar to quantum meruit but is based on the concept that an owner can't get something for nothing. In the case of the cherry doors, the owner has been enriched by the higher value of the cherry veneer. A contractor who threatens to institute a claim in this situation will be on very firm ground, and an owner should think twice before denying such a claim.

Another term that is similar to unjust enrichment is *betterments and enrichments;* it should be filed in the owner's memory bank in case such a demand is made by the contractor. It refers to the betterments and enrichments an owner received due to the added value of the contractor's change-order work.

The change-order process can proceed without too many problems if the following steps are taken:

- Establish as many costs as possible at the beginning of the project.

- Provide the general contractor with clear guidelines for submitting change orders, the detail required, and the documentation necessary for a reasonable review.

- Impress upon the general contractor the responsibility to review and critique vendor and subcontractor proposals for change-order work before passing them on to the architect/engineer and owner.

- Fair dealings from the owner's team should be rewarded with fair dealings from the contractor's team.

Green and sustainable buildings

Green building construction is based on designs that are more environmentally friendly, and it employs materials and equipment that result in energy efficiency and the reduction of waste. The U.S. Environmental Protection Agency (EPA) simply defines *sustainability* as "meeting the needs of the present without compromising the ability of future generations to meet their needs."

THE IMPACT OF CONSTRUCTION ON THE ENVIRONMENT

The U.S. Department of Energy reports that there are more than 4.5 million commercial buildings in the United States, occupying as much as 67 billion square feet of space. These buildings make enormous demands on the supply of materials used for construction and the environment in which they exist:

- Commercial and institutional buildings consume one-sixth of the world's freshwater supply.
- They consume one-half of all virgin wood harvested.
- These buildings require 36 percent of our country's total energy use and 65 percent of all electrical consumption.
- Buildings produce 30 percent of all U.S. waste output—approximately 136 million tons annually.
- These buildings are responsible for 30 percent of all greenhouse gas emissions.

These statistics drive the green building and sustainable movements in this country, and both concepts have moved from the "tree hugger" image to mainstream design and construction. More and more federal and state agencies are mandating

© 2009 by Elsevier, Inc. All rights reserved.
Doi_No = 10.1016/B978-1-85617-548-7.00009-4

new buildings to incorporate energy-efficient heating and cooling systems; reduction in all forms of waste, both solid and liquid; and environmentally friendly work spaces.

WHOLE-BUILDING DESIGN

The focus of green and sustainable building design is directed to achieve the following goals:

- Reduce capital and building maintenance costs

- Reduce energy costs

- Reduce the environmental impact of not only the structure but the site on which it will be placed

- Increase the comfort, health, and safety of building occupants

- Provide an environment conducive to increasing productivity

Whole-building design encompasses the building's envelope (exterior walls) and roof, its interior components, the mechanical and electrical equipment, the orientation of the building on its site, the materials of construction, energy consumption, indoor air quality, acoustics, and preservation of the site's natural state—all from a holistic approach as opposed to being considered as separate components.

Another aspect of whole-building design is one that can be practiced by both green building and nongreen building owners: waste reduction. We are known as the "disposable generation," and so many things we buy are packaged, sometimes in packages 500 percent larger than the item inside, adding to the disposal problem. Reducing waste is more than just cutting back on what we discard; it also involves purchasing materials that are more durable and easier to repair and maintain and using reclaimed materials such as recycled concrete or bricks reduced to aggregate size for placement under new asphalt or concrete paving. We also can consider using materials that have dual functions, such as exposed HVAC ductwork or stained concrete floors to approximate the look of tile. And both new and existing building owners can initiate a program to recycle day-to-day materials.

Reducing waste has several benefits:

- Reduction in transportation costs from the waste generator to the waste-disposal area

- Reduction in energy (fuel) costs with a concurrent reduction in air pollution

- Freeing up land previously set aside as landfill

A research study prepared by George Goldman and Aya Ogishi at the University of California at Berkeley in 2001 dramatically displayed the impact of waste disposal and waste reduction in California. In 1999, California estimated that its annual waste amounted to 33 million tons, 21 million of which were generated by nonresidential

Table 9-1 Average Statewide Impact for Waste Disposal vs. Diversion

	Disposed	Diverted	Additional Gain from Diversion (Difference)
Total Sales ($/ton)	$119	$254	$135
Output Impact ($/ton)	$289	$564	$275
Total Income Impact ($/ton)	$108	$209	$101
Value-added Impact ($/ton)	$144	$290	$146
Jobs Impact (Jobs/1,000 tons)	2.46	4.73	2.27

Source: State of California Sustainable Building Task Task Force Study (October 2003)

buildings. Green building attempts to reduce waste and refocus on recycling and reuse had significant economic impact. Not only were the costs to dispose of waste materials reduced, but the recycling process actually created more jobs. Table 9-1 reflects the impact of diverting rather than disposing of waste materials.

In an April 2007 press release, North Carolina reported that the amount of waste generated within the state during fiscal year 2005–2006 amounted to 1.36 tons per person per year—a 5 percent increase over the previous fiscal period and a 27 percent increase from the baseline figure set in FY 1991–1992. Industrial and commercial construction and demolition accounted for more than 70 percent of the state's waste stream. Stepped-up state purchases of recycled products and recovery of plastic, glass, and metal containers by recycling have been positive steps in reducing waste generation.

The U.S. Green Building Council

The U.S. Green Building Council (USGBC) is a nonprofit organization committed to expanding sustainable and green building practices. With 15,000 organizations through the country, USGBC penetrates all geographic regions, working with architects, engineers, contractors, building systems and materials manufacturers, and government agencies to achieve a mission: to "transform the way buildings and communities are designed, built, and operated, enabling an environmentally and socially responsible, healthy, and prosperous environment that improves the quality of life."

USGBC developed a trademark-protected rating system, which includes a program of standards and certification for accreditation known as LEED—Leadership in Energy and Environmental Design. This LEED program encompasses all types of construction projects:

- LEED-NC—new construction
- LEED-EB—existing buildings

- LEED-CI—commercial interiors
- LEED-C&S—core and shell
- LEED-H—home and residential
- LEED-ND—neighborhood development

LEED certification programs are also available for schools, retail, and health care facilities.

Each certification level has six credit areas, with points awarded for the degree of compliance in such areas as sustainable sites, energy and atmosphere, water efficiency, indoor environmental quality, materials and resources, and innovation in design. These are the four progressive levels of certification:

- Certified (the lowest level): 26–32 points
- Silver: 33–38 points
- Gold: 39–51 points
- Platinum: 52 points or more

The LEED approach focuses a great deal of attention of the building's site and looks at 14 areas of concern:

1. Site selection
2. Erosion and sedimentation control during construction
3. Urban development
4. Brownfield redevelopment
5. Alternative transportation: public transportation access
6. Alternative transportation: bicycle-friendly
7. Alternative transportation: alternative fuel-refueling stations
8. Alternative transportation: parking reduction
9. Reducing site disturbance: protecting and restoring open spaces
10. Reducing site disturbance: maximizing open space
11. Storm water management: flow reduction
12. Storm water management: flow treatment
13. Landscape and exterior design to reduce heat islands: nonroof areas
14. Landscape and exterior design to reduce heat islands: roof surfaces

Figure 9-1 shows a LEED checklist for new construction projects that includes sustainable site points, water efficiency, energy and atmosphere, materials and resources, indoor environmental quality, and innovation and design-process standards.

LEED 2009—or LEED 3.0, as it may be called—is restructuring and revising its point system based on science-based allocations. This update will include LEED Online, an Adobe Life Cycle–based technology that will allow project teams to submit 100 percent of their documentation for certification online.

This updated version uses a 100-point rating system that weights credits. No longer will one credit necessarily equal one point. The weighting systems take into account the environmental and human benefits of each LEED credit. LEED 2009 will also address the shortcomings present in their early life cycle assessment (LCA) by

LEED for New Construction v 2.2
Registered Project Checklist

Yes	?	No			
			Energy & Atmosphere		**17 Points**
Yes			Prereq 1	**Fundamental Commissioning of the Building Energy Systems**	Required
Yes			Prereq 1	**Minimum Energy Performance**	Required
Yes			Prereq 1	**Fundamental Refrigerant Management**	Required

*Note for EAc1: All LEED for New Construction projects registered after June 26, 2007 are required to achieve at least two (2) points.

			Credit 1	**Optimize Energy Performance**	1 to 10
			Credit 1.1	10.5% New Buildings / 3.5% Existing Building Renovations	1
			Credit 1.2	14% New Buildings / 7% Existing Building Renovations	2
			Credit 1.3	17.5% New Buildings / 10.5% Existing Building Renovations	3
			Credit 1.4	21% New Buildings / 14% Existing Building Renovations	4
			Credit 1.5	24.5% New Buildings / 17.5% Existing Building Renovations	5
			Credit 1.6	28% New Buildings / 21% Existing Building Renovations	6
			Credit 1.7	31.5% New Buildings / 24.5% Existing Building Renovations	7
			Credit 1.8	35% New Buildings / 28% Existing Building Renovations	8
			Credit 1.9	38.5% New Buildings / 31.5% Existing Building Renovations	9
			Credit 1.10	42% New Buildings / 35% Existing Building Renovations	10
			Credit 2	**On-Site Renewable Energy**	1 to 3
			Credit 2.1	2.5% Renewable Energy	1
			Credit 2.2	7.5% Renewable Energy	2
			Credit 2.3	12.5% Renewable Energy	3
			Credit 3	**Enhanced Commissioning**	1
			Credit 4	**Enhanced Refrigerant Management**	1
			Credit 5	**Measurement & Verification**	1
			Credit 6	**Green Power**	1

FIGURE 9-1

LEED certification for New Construction Checklist.
Source: U.S. Green Building Council, Dupont Circle, Washington, D.C.

incorporating a multiattribute evaluation within the rating systems that will replace credits in the materials and resources section.

Another key change is the recognition of regional differences—for example, water conservation in Las Vegas, Nevada, is somewhat different from water conservation issues in upstate New York. A regionally based weighting system will assign bonus points when environmental issues specific to a region are addressed. USGBC has also tapped into the U.S. Environmental Protection Agency's TRACI (Tool for the Reduction and Assessment of Chemical and Other Environmental Impacts) program and has weighted the new rating systems according to this standard.

LEED for New Construction v 2.2
Registered Project Checklist

Yes	?	No			
			Materials & Resources		**13 Points**
Yes			Prereq 1	**Storage & Collection of Recyclables**	Required
			Credit 1.1	**Building Reuse,** Maintain 75% of Existing Walls, Floors & Roof	1
			Credit 1.2	**Building Reuse,** Maintain 95% of Existing Walls, Floors & Roof	1
			Credit 1.3	**Building Reuse,** Maintain 50% of Interior Non-Structural Elements	1
			Credit 2.1	**Construction Waste Management,** Divert 50% from Disposal	1
			Credit 2.2	**Construction Waste Management,** Divert 75% from Disposal	1
			Credit 3.1	**Materials Reuse,** 5%	1
			Credit 3.2	**Materials Reuse,** 10%	1
			Credit 4.1	**Recycled Content,** 10% (post-consumer + 1/2 pre-consumer)	1
			Credit 4.2	**Recycled Content,** 20% (post-consumer + 1/2 pre-consumer)	1
			Credit 5.1	**Regional Materials,** 10% Extracted, Processed & Manufactured	1
			Credit 5.2	**Regional Materials,** 20% Extracted, Processed & Manufactured	1
			Credit 6	**Rapidly Renewable Materials**	1
			Credit 7	**Certified Wood**	1

Yes	?	No			
			Indoor Environmental Quality		**15 Points**
Yes			Prereq 1	**Minimum IAQ Performance**	Required
Yes			Prereq 2	**Environmental Tobacco Smoke (ETS) Control**	Required
			Credit 1	**Outdoor Air Delivery Monitoring**	1
			Credit 2	**Increased Ventilation**	1
			Credit 3.1	**Construction IAQ Management Plan,** During Construction	1
			Credit 3.2	**Construction IAQ Management Plan,** Before Occupancy	1
			Credit 4.1	**Low-Emitting Materials,** Adhesives & Sealants	1
			Credit 4.2	**Low-Emitting Materials,** Paints & Coatings	1
			Credit 4.3	**Low-Emitting Materials,** Carpet Systems	1
			Credit 4.4	**Low-Emitting Materials,** Composite Wood & Agrifiber Products	1
			Credit 5	**Indoor Chemical & Pollutant Source Control**	1
			Credit 6.1	**Controllability of Systems,** Lighting	1
			Credit 6.2	**Controllability of Systems,** Thermal Comfort	1
			Credit 7.1	**Thermal Comfort,** Design	1
			Credit 7.2	**Thermal Comfort,** Verification	1
			Credit 8.1	**Daylight & Views,** Daylight 75% of Spaces	1
			Credit 8.2	**Daylight & Views,** Views for 90% of Spaces	1

FIGURE 9-1 (continued)

LEED for New Construction v 2.2
Registered Project Checklist

Yes	?	No			
			Innovation & Design Process		5 Points
			Credit 1.1	**Innovation in Design:** Provide Specific Title	1
			Credit 1.2	**Innovation in Design:** Provide Specific Title	1
			Credit 1.3	**Innovation in Design:** Provide Specific Title	1
			Credit 1.4	**Innovation in Design:** Provide Specific Title	1
			Credit 2	**LEED® Accredited Professional**	1

FIGURE 9-1 (continued)

New also are credits for reducing energy consumption that are linked to benefits in other categories such as greenhouse gas emissions and fossil fuel use. This weighting of impact categories is not only science based but also includes value judgments. Although this new point system uses some science-based allocations, USGBC recognizes that this hybrid approach cannot factor in all of the health and environmental benefits that still need to be considered.

Indoor water consumption was given more importance in the update by changing the point system to require a 30 percent reduction instead of last year's 20 percent savings and providing a reward of 10 points instead of 5. Credits for access to transportation systems also increased from 1 point to 6, and credits for energy efficiency and renewable energy increased from 13 points to 26 points.

PROMOTING GREEN BUILDINGS

Other states are promoting green buildings differently, such as providing tax credit incentives to owners. New York led the way with the country's first green building tax credit. Baltimore County, adjacent to Baltimore City, Maryland, in 2006 provided a full property tax abatement for 10 years for buildings achieving at least a LEED Silver certification. Chatham County in Georgia offers developers a property tax abatement for Gold certificate green buildings. In Nevada, no sales tax is levied on building materials used in projects with a LEED certification of Silver or higher, while other cities—Boston and Washington, D.C.—use a different approach, requiring owners of new buildings to adopt LEED standards for all buildings over a certain size.

Retrofitting existing buildings to green standards has been picking up speed, particularly as reports come in citing the cost savings that accrue to a building owner. Figure 9-2 shows an existing building operations and maintenance checklist reflecting the key items to be addressed, such as site, water efficiency, energy and atmosphere, and materials and resources. Upgrading existing mechanical systems

Project Name: _____

Project Address: _____

Yes	?	No		
			Project Totals (Pre-Certification Estimates)	**92 Points**
			Certified: 34–42 points Silver: 43–50 points Gold: 51–67 points Platinum: 68–92 points	

Yes	?	No		
			Sustainable Sites	**12 Points**
			Credit 1 — **LEED Certified Design and Construction**	1
			Credit 2 — **Building Exterior and Hardscape Management Plan**	1
			Credit 3 — **Integrated Pest Mgmt, Erosion Control, and Landscape Mgmt Plan**	1
			Credit 4 — **Alternative Commuting Transportation**	1 to 4
			Credit 4.1 10% Reduction	1
			Credit 4.2 25% Reduction	2
			Credit 4.3 50% Reduction	3
			Credit 4.4 75% Reduction or greater	4
			Credit 5 — **Reduced Site Disturbance**, Protect or Restore Open Space	1
			Credit 6 — **Stormwater Management**	1
			Credit 7.1 — **Heat Island Reduction**, Non-Roof	1
			Credit 7.2 — **Heat Island Reduction**, Roof	1
			Credit 8 — **Light Pollution Reduction**	1

LEED for Existing Buildings: Operations & Maintenance
Registered Project Checklist

Yes	?	No		
			Water Efficiency	**10 Points**
Yes			Prereq 1 — **Minimum Indoor Plumbing Fixture & Fitting Efficiency**	Required
			Credit 1.1 — **Water Performance Measurement**, Whole Building Metering	1
			Credit 1.2 — **Water Performance Measurement**, Submetering	1
			Credit 2 — **Additional Indoor Plumbing Fixture and Fitting Efficiency**	1 to 3
			Credit 2.1 10% Reduction	1
			Credit 2.2 20% Reduction	2
			Credit 2.3 30% Reduction	3
			Credit 3 — **Water Efficient Landscaping**	1 to 3
			Credit 3.1 50% Reduction	1
			Credit 3.2 75% Reduction	2
			Credit 3.3 100% Reduction	3
			Credit 4.1 — **Cooling Tower Water Mgmt**, Chemical Management	1
			Credit 4.2 — **Cooling Tower Water Mgmt**, Non-Potable Water Source Use	1

FIGURE 9-2

LEED certification for Existing Buildings Checklist.

Source: U.S. Green Building Council, Dupont Circle, Washington, D.C.

LEED for Existing Buildings: Operations & Maintenance
Registered Project Checklist

Yes	?	No			
			Energy & Atmosphere		**30 Points**
Yes			Prereq 1	**Energy Efficiency Best Management Practices**	Required
Yes			Prereq 1	**Minimum Energy Efficiency Performance**	Required
Yes			Prereq 1	**Refrigerant Management,** Ozone Protection	Required

*NOTE for EAc1: All LEED for Existing Building projects registered after June 26th, 2007 are required to achieve at least two (2) points under EAc1.

			Credit 1	**Optimize Energy Efficiency Performance**	1 to 15
				ENERGY STAR Rating: 65 / Alternative Score: 15% Above Nat'l Average	Required
			Credit 1.1	ENERGY STAR 67 / Alternative Score: 17% Above Average	1
			Credit 1.2	ENERGY STAR 69 / Alternative Score: 19% Above Average	2
			Credit 1.3	ENERGY STAR 71 / Alternative Score: 21% Above Average	3
			Credit 1.4	ENERGY STAR 73 / Alternative Score: 23% Above Average	4
			Credit 1.5	ENERGY STAR 75 / Alternative Score: 25% Above Average	5
			Credit 1.6	ENERGY STAR 77 / Alternative Score: 27% Above Average	6
			Credit 1.7	ENERGY STAR 79 / Alternative Score: 29% Above Average	7
			Credit 1.8	ENERGY STAR 81 / Alternative Score: 31% Above Average	8
			Credit 1.9	ENERGY STAR 83 / Alternative Score: 33% Above Average	9
			Credit 1.10	ENERGY STAR 85 / Alternative Score: 35% Above Average	10
			Credit 1.11	ENERGY STAR 87 / Alternative Score: 37% Above Average	11
			Credit 1.12	ENERGY STAR 89 / Alternative Score: 39% Above Average	12
			Credit 1.13	ENERGY STAR 91 / Alternative Score: 41% Above Average	13
			Credit 1.14	ENERGY STAR 93 / Alternative Score: 43% Above Average	14
			Credit 1.15	ENERGY STAR 95+ / Alternative Score: 45%+ Above Average	15

LEED for Existing Buildings: Operations & Maintenance
Registered Project Checklist

			Energy & Atmosphere, continued		
			Existing Building Commissioning		
			Credit 2.1	**Investigation and Analysis**	2
			Credit 2.2	**Implementation**	2
			Credit 2.3	**Ongoing Commissioning**	2
			Performance Measurement		
			Credit 3.1	**Building Automation System**	1
			Credit 3.2-3.3	**System Level Metering**	1 to 2
			Credit 3.2	40% Metered	1
			Credit 3.3	80% Metered	2
			Other		
			Credit 4	**Renewable Energy**	1 to 4
			Credit 4.1	On-site 3% / Off-site 25%	1
			Credit 4.2	On-site 6% / Off-site 50%	2
			Credit 4.3	On-site 9% / Off-site 75%	3
			Credit 4.4	On-site 12% / Off-site 100%	4
			Credit 5	**Refrigerant Management**	1
			Credit 6	**Emissions Reduction Reporting**	1

FIGURE 9-2   (continued)

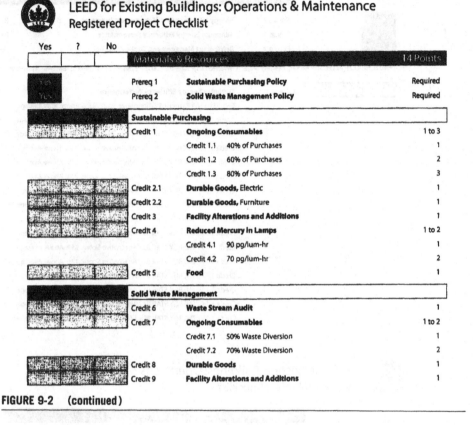

FIGURE 9-2 (continued)

to energy-efficient equipment can result in long-term paybacks, as witnessed by the annual heating, cooling, and air-conditioning costs for various types of buildings compiled by APS, a division of Pinnacle West Capital Corporation (Figure 9-3).

Waste recycling, green cleaning, and environmentally sensitive purchasing are less complex tasks for a company to assume; water and energy improvements are key areas in which to concentrate but are more sophisticated in their approach. Building owners can start with some basic questions: Do we need that light? If not, take it down. If we do need it, what wattage do we need, and what type of bulb—incandescent or fluorescent?

One research organization surveyed 14 LEED-EB building owners and found that an investment of $2.43 per square foot for some green modifications would pay for itself in a couple of years. Designers have found that simple changes such as the subtle use of an overhang and sunscreens can be adapted to increase natural lighting conditions and reduce energy burdens in a typical office building by as much as 30 percent.

LEED for Existing Buildings: Operations & Maintenance
Registered Project Checklist

Yes	?	No				
			Indoor Environmental Quality			**19 Points**
Yes			Prereq 1	**Outdoor Air Introduction and Exhaust Systems**		Required
Yes			Prereq 2	**Environmental Tobacco Smoke (ETS) Control**		Required
Yes			Prereq 3	**Green Cleaning Policy**		Required
			IAQ Best Management Practices			
			Credit 1.1	**IAQ Management Program**		1
			Credit 1.2	**Outdoor Air Delivery Monitoring**		1
			Credit 1.3	**Increased Ventilation**		1
			Credit 1.4	**Reduce Particulates in Air Distribution**		1
			Credit 1.5	**Facility Alterations and Additions**		1
			Occupant Comfort			
			Credit 2.1	**Occupant Survey**		1
			Credit 2.2	**Occupant Controlled Lighting**		1
			Credit 2.3	**Thermal Comfort Monitoring**		1
			Credit 2.4–2.5	**Daylight and Views**		1 to 2
			Credit 2.4	50% Daylight / 45% Views		1
			Credit 2.5	75% Daylight / 90% Views		2
			Green Cleaning			
			Credit 3.1	**High Performance Cleaning Program**		1
			Credit 3.2–3.3	**Custodial Effectiveness Assessment**		1 to 2
			Credit 3.2	Score of ≤ 3		1
			Credit 3.3	Score of ≤ 2		2
			Credit 3.4–3.6	**Sustainable Cleaning Products and Materials**		1 to 3
			Credit 3.4	30% of Purchases		1
			Credit 3.5	60% of Purchases		2
			Credit 3.6	90% of Purchases		3
			Credit 3.7	**Sustainable Cleaning Equipment**		1
			Credit 3.8	**Entryway Systems**		1
			Credit 3.9	**Indoor Integrated Pest Management**		1

LEED for Existing Buildings: Operations & Maintenance
Registered Project Checklist

Yes	?	No				
			Innovation in Operations			**7 Points**
			Credit 1.1	**Innovation in Operations:**	Provide Specific Title	1
			Credit 1.2	**Innovation in Operations:**	Provide Specific Title	1
			Credit 1.3	**Innovation in Operations:**	Provide Specific Title	1
			Credit 1.4	**Innovation in Operations:**	Provide Specific Title	1
			Credit 2	**LEED® Accredited Professional**		1
			Credit 3	**Documenting Sustainable Building Cost Impacts**		2

FIGURE 9-2 (continued)

Annual Heating, Cooling and Ventilating Electricity Costs	
Building Type	Annual Cost Per Square Foot
Large office	$0.70
Small office	$0.50
Large retail	$0.50
Small retail	$0.25
Sit-down restaurant	$1.10
Quick service restaurant	$2.45
Large grocery	$0.65
Small grocery	$1.70
In-patient healthcare	$0.95
Out-patient healthcare	$0.70
Primary school	$0.35
Secondary school	$0.45
College/University	$0.60
Hotel/Resort	$0.80

FIGURE 9-3

Heating, cooling, and ventilating costs per building type on an annual basis.

Source: APS division of Pinnacle West Capital Corporation, Phoenix, Arizona.

GREEN BUILDING COMPONENTS

One of the prime goals of green design and construction is to produce a net-zero energy building, one that produces as much energy onsite as it consumes. While this is the ultimate goal and is rarely achieved, it sets the basis for reducing energy consumption by investigating the ways heating and cooling are affected by the building's envelope: its exterior walls and roof.

The surface of a building's roof is a major absorber of heat in the summer. A number of new roofing surfaces that lower interior temperatures during hot weather have been developed. These products range from highly reflective roofing membranes to a green roof, literally and figuratively, that creates a cooler surface. A green roof is one where a vegetation layer is grown on the surface; on hot summer days, this can significantly reduce the temperature on that roof surface. The estimated cost of a green roof ranges from $10 to $25 per square foot, with annual maintenance costs of $0.75 to $1.50 per square foot. Initial costs of a green roof are higher than conventional membrane-type roof coverings and, due to the increased weight, may require reinforcing an existing roof's structural support system.

Other methods that can be used for a green roof involve selecting roofing membranes that are highly reflective. The measure of solar reflectance, expressed as a percentage, is *albedo*; the higher the percentage, the higher the albedo. Fresh snow has an albedo rating of 90 percent, and black surfaces are in the less than 10 percent range. Roofing materials with the EPA's ENERGY STAR® label have higher solar reflectance, and qualified roof products with the ENERGY STAR label can help reduce the peak demand for air conditioning in buildings by 10 to 15 percent. About 25 percent of manufacturers of commercial roofing products sell these materials. The ENERGY STAR website provides a roofing comparison calculator to help building owners estimate how much energy and money can be saved by installing an ENERGY STAR–labeled roof product.

Energy-efficient HVAC equipment is also at the top of the list of green equipment. Annual heating, cooling, and ventilating costs represent a major portion of a building's operation expenses. Federal law mandated a minimum efficiency rating of 13 SEER (seasonal energy efficient ratio) for unitary equipment with less than 65,000 Btu capacity. Upgrading from a 13 SEER to a 14 SEER may reduce cooling costs by 7 percent, and updating from a 13 to a 15 SEER may reduce cooling costs by 13 percent. Designers should consider a base SEER equipment rating and several options for upgrading to a more efficient unit.

EPA'S ENERGY PERFORMANCE RATINGS

The Environmental Protection Agency has developed a suggested list of guidelines for design consultants and contractors to further green energy conservation. For owners considering incorporating some of the energy-efficiency and energy-conservation measures in their new buildings, the EPA suggests pursuing the following strategies:

- Establish an energy performance target with your design consultants. The EPA's Building Design Guidance Checklist (Figure 9-4) can be helpful in that respect.

- During predesign the owner and the design consultants should investigate energy-related design concepts that consider the geographic environment and local climate, building orientation, and other features that will impact performance in the future, such as the effect of building overhangs and landscaped areas.

- Develop a scope of work, project budget, and schedule that include energy-efficient strategies and performance goals.

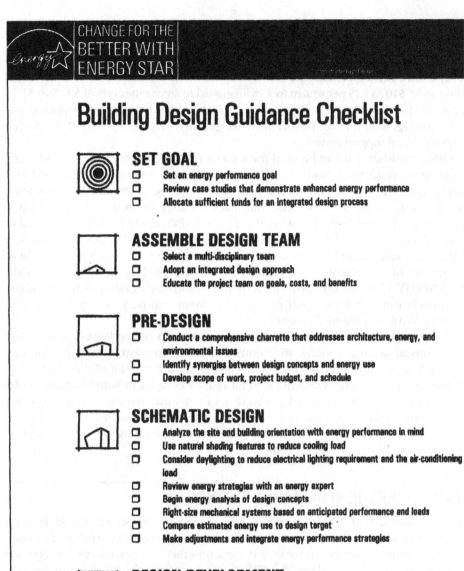

FIGURE 9-4

Energy Star® Building Design Guidance Checklist.

Source: U.S. Environmental Protection Agency.

CONSTRUCTION & BID DOCUMENTS

- ☐ Include Statement of Energy Design Intent (SEDI)
- ☐ Specify design team participation during construction
- ☐ Include approval process for change orders to methods and materials
- ☐ Encourage building owner to hold all parties accountable for achieving the energy performance goal
- ☐ Seek incentives for meeting the energy performance goal
- ☐ Include design team's summaries of energy-efficient features and anticipated functions
- ☐ Select qualified manufacturers
- ☐ Place the "Designed To Earn The ENERGY STAR" graphic on final drawings that achieve 75 or better in Target Finder

COMMISSIONING THE BUILDING

- ☐ Specify detailed commissioning activities in project contracts
- ☐ Seek separate funding and hire specialists
- ☐ Include the commissioning firm as part of the design team early in the project
- ☐ Communicate your energy performance goal
- ☐ Encourage energy-use tracking over time

TRACKING, MEASUREMENT & VERIFICATION

- ☐ Communicate the energy performance target to the M&V team
- ☐ Document how energy performance compares to the design intent
- ☐ Use EPA's Portfolio Manager to track and rate annual energy performance

THE ENERGY STAR FOR COMMERCIAL BUILDINGS

- ☐ Compare the building's actual energy consumption to the industry benchmark using Portfolio Manager
- ☐ Apply for the ENERGY STAR if the building achieves a 75 or higher
  - ☐ Complete the application letter and the Statement of Energy Performance (SEP)
  - ☐ Have a licensed Professional Engineer certify indoor environmental criteria
- ☐ Communicate the success of the building design team/firm and their energy design strategies

FIGURE 9-4 (continued)

- At the schematic design stage, perform various simulations for various energy options and technologies, and compare them to the initial performance goal.

- During the design development stage, prepare energy performance specifications for materials and equipment. Gather manufacturers' technical information, which can be included in the bid documents.

- A building automation system included in the design criteria will allow owners to track actual energy consumption over time and provide the necessary automated controls to tune equipment performance.

- When the contractor-selection process begins and contractor qualification forms are issued, include a section for the contractor's previous experience with green buildings and energy-conservation activities.

- The commissioning process at the completion of construction and prior to building turnover is critical to ensure that the design energy performance has been attained.

- EPA suggests that once an owner's building has been operating for 12 months, log on to their Web-based portfolio manager to track and rate annual energy performance.

The EPA national energy performance rating helps energy managers assess how efficiently their building uses energy compared to a nationwide rating. The system's scale ranges from 1 to 100, with high efficiencies having a higher number; a rating of 50 percent would be considered average. Factors used in determining efficiency include the size of the building, its location, and the number of occupants. The rating system compares actual energy data entered by the building owner to an estimate, thereby reflecting the owner's building rank relative to his or her peers'. This EPA performance estimate is gleaned from a statistical analysis of data collected by the Department of Energy's Energy Information Administration during a quadrennial commercial building energy consumption survey that tests this model with real buildings.

THE U.S. DEPARTMENT OF ENERGY BUILDING TECHNOLOGIES PROGRAM

The Department of Energy is a major source for information offered to design consultants and owners to improve building performance. The Department's Building Technologies Program contains a database of buildings from not only the United States but from around the world, providing information about green building performance and other notable environmental features. The database can be tapped from within the EPA website by entering a specific project name, building type and size, location, or owner. A sample of one of the companies included in the database is the Herman Miller Company, a manufacturer of high-quality office furniture and equipment (Figure 9-5). Its new building, located in Zeeland, Michigan, and

U.S. Department of Energy
Energy Efficiency and Renewable Energy

EERE Home

Building Technologies Program

Search Help ‖ More Search Options ‖
EERE Information Center

‹ Buildings Database Home
About the Database
Using the Database
Search by Project Name
Search by Owner
Search by Location
Search by Building Type & Size
List All Projects
Detailed Search
Submit a Project

Herman Miller MarketPlace
› Overview
› Process
› Finance
› Land Use
› Site & Water
› Energy
› Materials
› Indoor Environment
› Images
› Ratings & Awards
› Lessons
› Learn More

Herman Miller MarketPlace

Overview
- Location: Zeeland, MI
- Building type(s): Commercial office
- New construction
- 95,000 sq. feet (8,830 sq. meters)
- Project scope: 2-story building
- Suburban setting
- Completed January 2002
- Rating: U.S. Green Building Council LEED-NC, v.2/v.2.1--Level: Gold (39 points)

The goal of the Herman Miller MarketPlace was to create a prototype office environment that supports progressive business-place thinking within a sustainable framework. The expectation of this 95,000-square-foot speculative office was nothing less than that it be a great place to work.

Environmental Aspects
Located in a commercial development next to a McDonald's restaurant, MarketPlace demonstrates that green buildings can be woven successfully into the commercial landscape.

Photo credit: Kevin Beswick: People, Places and Things Photographic

The entire design team worked to create an effective building without losing track of the initial cost. The target was an HVAC system that requires 40% lower energy cost than what is budgeted in the baseline model ASHRAE 90.1-1999. Current operating records show this goal is being met.

Owner & Occupancy
- Owned by The Granger Group, Corporation, for-profit
- Occupants: Corporation, for-profit
- Typically occupied by 430 people, 40 hours per person per week; and 20 visitors per week, 40 hours per visitor per week

Building Programs
Indoor Spaces: Office (93%), Public assembly (17%), Cafeteria (3%), Other (2%), Lobby/reception (2%), Circulation (1%), Restrooms

Outdoor Spaces: Restored landscape (50%), Drives/roadway (25%), Parking (22%), Patio/hardscape (2%), Garden—decorative (1%)

Keywords
Integrated team, Design charrette, Green framework, Simulation, Green specifications, Contracting, Commissioning, Performance measurement and verification, Transportation benefits, Indigenous vegetation, Stormwater management, Efficient irrigation, Insulation levels, Glazing, HVAC, Efficient lighting, Durability, Recycled materials, Local materials, Certified wood, C&D waste management, Occupant recycling, Connection to outdoors, Daylighting, Natural ventilation, Moisture control, Thermal comfort, Low-emitting materials, Indoor air quality monitoring

FIGURE 9-5

The Herman Miller LEED building project.

Source: U.S. Department of Energy.

completed in 2002, as listed in the program, provides a thumbnail sketch of the building program. The keyword section of each of these case studies presents an opportunity to click on various aspects of the building's design and construction to learn more about the measures each of these building owners took to achieve his or her individual goals. Topics such as performance measurement, insulation levels, and efficient irrigation can be explored in more depth.

Toyota Motor Sales South Campus Office in Torrance, California, constructed a 624,000-square-foot building that was awarded the USGBC Gold certification in 2003. The building was constructed of a tilt-up concrete exterior wall system with thermally insulated double-pane windows and a highly reflective "cool roof." By recycling water for cooling, landscape irrigation, and fecal-free bathroom flushing, they saved millions of gallons of potable water each year. The HVAC system is 58.6 percent more efficient than the state's Title 24 Energy Code. Diverting construction waste along with concrete panel waste and temporary onsite concrete casting slabs, the company was able to divert 95 percent of its waste from being disposed of in local landfills. With a goal of a 10 percent return on investment, Toyota is realizing a savings of $400,000 annually on these systems.

ENERGY MODELING

The U.S. Department of Energy (DOE) initially funded development of an energy simulation software program known as DOE-2, which analyzed the building's envelope—its roof, walls, and windows—in response to internal loads and the external climate for every hour in a year. These studies included gains from solar heat, interior equipment, lights, and the building's occupants. This software program also included mechanical equipment performance and local utility company electrical costs.

This software program has been updated as of 2008 and has incorporated more "user-friendly" interfaces for architects and engineers; DOE calls it EnergyPlus. The agency initiated a program known as eQUEST (doe2.com/equest) that allows an architect or engineer to tap into DOE-2's analytical power. The private sector has also joined the crowd in developing energy-efficiency software. Mechanical equipment manufacturer Trane developed Trane Trace 700, a software model that compares the energy and economic impact of building alternatives such as architectural features and HVAC. Over 30 air-side systems, advanced chiller plant configurations, and water source and ground source heat pumps are all reviewed on Trane Trace 700. Building Information Modeling (BIM) and estimating and design software producer Revit®, a division of AutoDesk, are developing an integration interface with some of these green analytical programs. Bentley®, another architecture, estimating, and project management software provider, has created multidisciplinary applications that tie EnergyPlus, Trane Trace 700, and other such software together.

Building information modeling is the process of creating a three-dimensional design platform that allows the architect and the engineering consultants to participate in the design process on a real-time, multidiscipline basis. Energy modeling differs from this architectural design process by focusing on building performance rather than design, so the architect is provided a choice of materials and equipment and can tweak the design to obtain more energy efficiency.

Energy modeling allows the design team to optimize building design and prioritize investment in those areas that will return the best value from the building's energy source. Modeling of this type can commence during the building's conceptual design, during which a skilled modeler can assemble a simple model of the proposed building to test the effects of site location, building orientation, wall section design, and heating and cooling plants on capital and operating costs. These various options can be reviewed and weighed as design proceeds to schematic design before finally being firmed up during design development.

GREEN INFRASTRUCTURE

An equally important part of green design and construction is the building's site infrastructure. One of the first steps to consider when creating an energy-efficient building is to look for ways to work with the existing site to take advantage of possible savings in heating, cooling, and lighting costs for proper building orientation. Site design is also important.

Although it may be slightly more expensive for a site excavating contractor to work around existing trees on a site, the advantages of keeping some of the more mature trees can provide payback by offering cooling shade and reducing some of the impact of the site's heat islands.

Low-impact development

One major problem today is water quality, which began deteriorating years ago when factories dumped toxic wastes into our rivers. In 1972, the National Pollutant Discharge Elimination System was created under the Clean Water Act to regulate the discharge of pollutants into the nation's waters. The Clean Water Act also regulates storm water management; construction projects with more than a one-acre site fall within the purview of this Act.

As the green movement emerged in the 1990s, attention was also focused on the site and discharge of surface water. The result was a strategy known as low-impact development. The passive approach to storm water management is to allow water to meander as it travels across the unpaved portions of the site, where it can filter down without picking up the gasoline, oil, and other toxic particles that collect on a hard-paved area and subsequently discharge into a storm water system. Natural drainage basins, existing vegetation, and preserved open space can all act as water collection and filtration areas and recharge underground aquifers. These other design considerations can be used to develop micro-watersheds on the building site:

- French drains, perforated pipe installed in a bed of small stone, which permit collected water to dissipate back into the soil

- Swales and shallow ditches, which can direct water to open space filtration areas

- Sidewalks pitched to open spaces or vegetation areas

- Pervious paving areas, such as paving blocks through which grass can grow, porous-concrete paved areas, or pavers set in a gravel base

- Development of a rain garden where hardy vegetation can be planted to absorb or distribute downspout-collected rain water

Contrary to the passive water management system just described, an active system would include pumps and storage tanks to collect and store storm water for reuse. By collecting site runoff before it picks up any surface pollutants and storing it, the relatively clean rainwater can be reused for lawn and planting irrigation, fountains, ornamental pools, and other site water features and even for cooling towers if filtered properly. Carrying active water management a little further, clean wastewater from sinks and interior water fountains can also be stored and used for irrigation; when a small, onsite water-treatment plant is installed, other forms of wastewater can be collected, treated, and reused.

Heat builds up on a building site and the surrounding areas during the summer months in some climates and more or less all year round in other areas. Buildings, pavements, and sidewalks act as heat sinks, absorbing and storing heat during daylight hours and releasing that heat as the sun goes down. A study by the U.S. Environmental Protection Agency revealed that the mean air temperature in a city with a population of 1 million can be 1.8° to 5.4°F higher than its surroundings, and in the evening the difference can be as high as 22°F due to the slow release of absorbed heat. These high temperatures place more demands on air-conditioning and create more air pollution and greenhouse-gas emissions, not to mention more heat-related illness in some city dwellers without air conditioning.

There are numerous design options to reduce the heat load on a building. A new structure can have a portion designed to be set partially underground in an area that has no groundwater. Once below the frost line in northern climates, the temperature is fairly constant all year long. For example, in the mid–Hudson Valley area of New York State, the below-grade temperature is about 54°, which means that in the winter the heating equipment turns on above 54° instead of near 0°. The same analysis is true in the summer, when ambient air will reach the high 80s or low 90s while the underground portion remains cool.

Geothermal heating and cooling

Geothermal heating and cooling are other options to pursue when seeking energy efficiency. This technology relies on the fact below a certain level, the earth remains at a fairly constant temperature throughout the year. A geothermal system uses that constant temperature to concentrate naturally existing heat instead of producing heat derived by combustion from a fuel source. A geothermal system consists of three basic components:

1. Geothermal earth connection subsystem, a series of pipes, called a "loop," that is buried in the ground and circulates a liquid, generally water with antifreeze. The liquid absorbs heat from the surrounding soil or relinquishes heat back into the soil when cooling is required.

2. Geothermal heat pump system, which removes heat from the fluid in the loop, concentrates it, and transfers that heat to the building. The cooling cycle is the reverse: the heat pump transfers the liquid back to the soil for heat dissipation.

3. Geothermal heat and cooling distribution systems, which distribute the heating or cooling from the geothermal pump throughout the building.

When used in a residential or small commercial building's heating and cooling system, this geothermal equipment can also provide hot water.

Fenestration

Fenestration, the design and placement of exterior windows, can play a large part in providing an energy-efficient building envelope. The location, size, and construction of a building's windows can allow outside lighting to reduce some lighting loads during daylight hours, contain reflective coatings to keep out the sun's rays during hot weather, and conserve and transfer heat and cooling from the exterior to the interior of a building and vice versa.

HOW COST EFFECTIVE ARE GREEN BUILDINGS?

Capital costs, operating costs, and life-cycle costs are all important to owners, and at first blush, the cost to "go green" may appear to extract a premium in both design and construction. Nonetheless, both activities have gained considerable cost effectiveness as more companies are offering materials and equipment at competitive prices to capture market share.

In August 2000, California's then governor Gray Davis issued Executive Order D-16-00, funding research in the green building/sustainable movement, and in October 2003, *A Report to California's Sustainable Building Task Force* was issued (and is available online at *www.usgbc.org/docs/new/news477.pdf*). It revealed that green buildings may cost more than conventionally designed buildings but are 25 to 30 percent more energy efficient, which means significantly lower operating costs over the lifespan of the building.

The costs premiums associated with the various levels of LEED certification ranged from 0.66 to 6.5 percent, per Table 9-2. This California study looked at various building components and came to these conclusions:

Table 9-2 Average Green Cost Premium per Level of USGBC Certification

Level of Certification	Average Green Cost Premium
Level 1 - Certified	0.66%
Level 2 - Silver	2.11%
Level 3 - Gold	1.82%
Level 4 - Platinum	6.50%
Average of 33 buildings – 1.84%	

Source: USGB – Capital E Analysis and State of California Sustainable Building Task Force study (October 2003)

- More efficient lights, task lights, and sensors to turn off unnecessary light fixtures and more daylight transmission would reduce power requirements and also reduce heat loads, thereby reducing cooling costs.

- Increased ventilation effectiveness would help decrease cooling loads during peak periods through improved system utilization.

- Using underfloor air distribution and plenums to deliver space conditioning typically cuts fan and cooling loads.

- Using heat island reduction measures, such as roof reflectivity, will lower building temperatures and reduce cooling loads.

- Generating some power onsite via photovoltaics, in some climates, can generate 20 percent of total power consumption.

- Potable water can be used more efficiently through better design and new technologies.

- Gray water from bathroom sinks, tubs, showers, washing machines, and drinking fountains can be captured and used for lawn and landscape irrigation.

- Onsite storm water collection can be used to recharge groundwater tables.

Reduced energy use in green buildings of varying LEED certification levels is shown in Table 9-3.

The 2003 California study produced some interesting statistics regarding energy cost savings, which were calculated over a 20-year period using the net-present-value (NPV) cost-analysis method. Here are their findings:

- The consumption of electricity was reduced by 30 percent, or $5.48 per square foot.

- The additional value of peak-demand reduction from green buildings was estimated at $0.31 per square foot.

Table 9-3 Reduced Energy Use in Green Buildings as Compared with Conventional Buildings

	Certified	Silver	Gold	Average
Energy efficiency (above standard code)	18%	30%	37%	28%
On-site renewable energy	0%	0%	4%	2%
Green Power	10%	0%	7%	6%
Total	28%	30%	48%	36%

Source: USGBC Capital RE analysis and State of California-Sustainable Building Task force (October 2003)

- Together, the total 20-year present value of energy savings from a typical green building was $5.79 per square foot.

A study conducted by Gregory H. Kats in 2003, sponsored by the Barr Foundation, Environmental Business Council of New England, Equity Office Properties, and the Massachusetts Technology Collaborative, entitled *Green Building Cost and Financial Benefits*, looked at green building cost benefits after studying 33 such buildings across the country. According to the study, paved streets accounted for about 40 to 50 percent of all impervious areas in a residential development. Reducing the width of a roadway in those developments from 26 feet to 20 feet—a still acceptable area for traffic management—would reduce the overall development's impervious surfaces from 5 to 20 percent. And since these paved areas generate most of the runoff pollution from vehicles, including deicing materials, according to a study by the University of Connecticut, ground pollution would also be reduced.

GREEN BUILDING PRODUCTS AND PROCESSES

Just about every manufacturer of construction materials and equipment has jumped on the green bandwagon, with products ranging from carbon-fiber exterior wall panels that reduce cement content while increasing insulation quality to banana plant wall coverings to waterless urinals to solar hot-water heaters. *Green Source*, a McGraw-Hill quarterly publication, lists page after page of advertisers announcing new green and sustainable products. Major manufacturers such as USG, makers of drywall and other construction ceiling and wall materials, provide technical services to architects searching for sustainable building materials.

SUSTAINABILITY

Sustainability is the process applied to our quest to sustain economic growth while maintaining our long-term environmental health. Sustainability means designing structures that take advantage of technological advancements to create eco-friendly products. Inert-gas-filled insulated windows, engineered-wood products made from scrap wood shavings, sawdust and assorted wood fibers, and thermal break window frames that keep the cold and hot air out are all examples of sustainable products. These products provide the owner and occupants with the following benefits:

- Reduced maintenance and replacement costs over the life of the building
- Energy conservation
- Improved occupant health
- Productive working environment

Sustainable products incorporated into the building should follow these selection guidelines:

- Recycled content
- Natural, plentiful, or renewable materials
- Products manufactured by a resource-efficient process
- Locally available products
- Salvaged, refurbished, or remanufactured products
- Reusable or recyclable products
- Durable products

Using sustainable materials can also improve the indoor working environment and save money. Consider these advantages:

- Materials that emit few or no carcinogens or irritants, as demonstrated by the manufacturer's long-term testing results

- Minimal chemical emissions from volatile organic compounds (VOC) that out-gas (continue to emit chemical vapors after installation)

- Moisture-resistant materials that are not easily susceptible to mold growth

- Materials that are easily maintained and require simple nontoxic cleaners

- Equipment systems that promote healthy indoor air quality (IAQ) by identifying indoor air pollutants

- Products and systems that help reduce water consumption

The sustainable approach to design would include requirements to do the following:

- Simplify construction details.

- Utilize repetitive details and components.

- Standardize design components.

- Incorporate accurate dimensions in the design, as some product and material sizes many have been reduced.

- Simplify building systems so future expansion projects can take advantage of simplified designs or components.

- Consider occupant safety and worker productivity gains in the new design.

- Investigate more efficient and environmentally sensitive ways to bring underground utilities into the site with the least disruption to the existing terrain.

- Consider other ways of disposing of site drainage onsite rather than offsite.

- Adjust new site contours to provide for a balanced site where no offsite fill or off-site disposal of surplus soils is required.

- Optimize dimensions to utilize a standard product size.

- Minimize plumbing pipe and HVAC ductwork bends to reduce liquid and air friction.

- Select fittings and fasteners that permit quick assembly.

- Select sealants with the least environmental impact and longest life.

- Investigate ways to accumulate salvaged and waste materials for recycling.

- Consider donating surplus materials to a nonprofit organization such as Habitat for Humanity.

- Deconstruct all existing structures with substantial recoverable materials and dispose of them to recyclers.

When designing a new green structure, a number of goals must be set. The site must meet or exceed standards for sedimentation control and erosion:

- Prevent the loss of soil during excavation and construction due to surface water drainage; keep dust down and cover large stored piles of earth to prevent wind erosion.

- Prevent the silting up of existing storm drains in the immediate area by constructing erosion and silt fence enclosures around areas to be excavated.

- Prevent the siltation of existing nearby streams or waterways by installing erosion and silt fencing around those streams adjacent to areas to be excavated.

- Protect topsoil piles for reuse. (Topsoil piles are generated early in the construction process as soil is stripped during rough grading operations; respreading is one of the last operations, commencing as landscaping is put in place.)

The site utilities should reduce soil erosion during excavation of trenches and contain storm water runoff:

- Plan infiltration swales and basins during trenching operations to contain surface water.

- Retain or recharge existing water tables by minimizing site disturbances; leave as many trees as possible; use existing vegetation and retain natural contours.

- Consider a design to store roof runoff when the building has been completed; it can be used as gray water or reclaimed wastewater.

- Investigate a small onsite, state-of-the-art treatment plant to recycle reclaimed water.

An open-space and landscaping program can accommodate the following:

- Protect trees during construction; they enhance property values and lower cooling loads.

- Consider indigenous landscaping; it supports natural wildlife and plantings and lowers the level of irrigation as well as the need to fertilize and apply chemical treatment.

- Minimize pesticide use by installing weed cloth; use mulches and planting species that create dense planting beds.

Site-circulation and transportation programs should meet these objectives:

- Encourage carpooling.
- Provide areas for people to store bicycles during working hours.
- Encourage the use of public transportation by instituting a program of incentives.

THE RISKS AND PITFALLS IN GREEN BUILDING DESIGN

The green and sustainable movement is relatively young, having its origins in the 1970s. The United States Green Building Council (USGBC) was founded more than 20 years later in 1993, and the LEED program was introduced 1998 and revised in 2000 and 2003. All of these programs are relatively young, and life-cycle costs may not have been firmly established.

As with all new technologies, products offered for sale often do not fully meet their intended function, and architects and engineers, eager to embrace this new movement, may not have a full understanding of these new designs. They may be relying on product and equipment manufacturer's claims that do not reflect long-term data simply because the product has not been on the market for a long period of time. The contractor who is implementing green and sustainable designs may not have had much or any experience with the LEED certification program.

The building's commissioning process, always one of the most important tasks as the mechanical systems are turned over to the owner after the TAB (test adjust, balance) process and certified that they meet the contract requirements, takes on a more important role as that equipment and other components of the building must undergo review to a new standard—one that meets the green parameters.

At the Schinnerer Annual Meeting of Invited Attorneys in 2005, Thomas F. Waggoner presented the report "Green Buildings: Potential Pitfalls for Design Professionals." Mr. Waggoner, whose extensive practice in construction litigation represents architect, engineers, and surveyors, said that "unwary design professionals can find themselves in a difficult situation if they have not fully explained to the owner that the initial costs in both design and construction will be higher and that there is no guarantee that the property developer will see a return on investment in the 'short term.'" He emphasized the need for owners to engage design professionals who are experienced in this type of design and engineering. Items such as waterless urinals save on water consumption but have a higher cost; the purported increase in employee productivity within a green building may be difficult to quantify. Warranties offered by many manufacturers on "green" equipment that limit liability may not have been fully vetted because they have not been in operation over an extended period of time.

This green movement is expanding rapidly in the United States, as many more public and private building owners are appreciating the social and economic benefits. Numerous studies point to the potential savings that green and sustainable buildings can provide and the positive impact they have on our environment.

Owners who are considering green and sustainable construction should seek out design consultants and building contractors who are experienced in these types of projects and explore the capital versus operating costs of their design as well as their desire to take positive steps to improve the environment.

Disputes and claims

Try as we may, a dispute or two or more may arise on a construction project and, left unresolved, will become a formal claim. The goal of owner, architect, and contractor alike is to avoid any claims even though, given the complexity of the design and construction process, it may lead to a dispute over contract interpretation, compliance with acceptable work standards, and a multitude of other misunderstandings that frequently occur. When it appears that a problem is in the making, there are several important rules to remember:

- Document the potential problem in case it does expand into a dispute. Too often a contractor will state, "Gee, I didn't know that you thought that was a problem"—two months *after* it really became a problem.

- Don't think that by ignoring a potential problem it will go away; it usually gets worse. It could have been resolved quickly and at a lesser cost when first discovered.

- As an owner, be involved in an early discovery and resolution of the potential dispute; although advice from others may be involved, you as the owner will be in charge of the final resolution.

- As in all other dealings, be fair and reasonable in the resolution.

DOCUMENTATION IN THE BIDDING PROCESS

Although the anticipation of a dispute or claim is the thought farthest from an owner's mind when their project is being advertised to bid, careful bid preparation, follow-through documents, and preservation and retention of appropriate records are important for an endeavor as complex as a construction project. An owner and his or her design consultants must prepare succinct, detailed documents all along the

© 2010 by Elsevier, Inc. All rights reserved.
Doi_No = 10.1016/B978-1-85617-548-7.00010-0

path from soliciting bids through bid acceptance and on to preconstruction, construction, and postconstruction, not anticipating any serious problems along the way but prepared to deal with them if and when they occur. And it all starts with the invitation to bid.

All such documents, either formal or handwritten, should be kept on file. More than one owner has said, in the midst of a dispute, "Boy, I wish I had not thrown away those papers." In a previous chapter we discussed the prebid process—the method by which the bid is prepared and executed. Our concern now is retention of those instructions, forms, responses by bidders, and other data distributed or received during this process. These are the prebid records that should be retained:

- The invitation to bid and the responses to the invitation—those contractors who accepted the offer and those who did not.

- A list of those contractors who either accompanied the architect on a site tour or walked the site on their own; this will come in handy if there are any claims later on that a condition was unknown even though it would have been apparent during a site visit.

- Minutes of any prebid meetings, and questions raised and by whom and responses by the architect or engineer.

- Any proposed schedules issued by the architect and any responses or comments registered by the contractors.

- Photographs taken by the design consultants or the owner of the site at the time of the prebid conference.

- Logs of any telephone calls or copies of any e-mails relating to questions, comments, or responses to or from any contractors, architects, engineers, or the owner relating to the prebid activity.

As the bids are received, all responses should be retained, whether a bid was unacceptable or under consideration for award. The process by which bids were analyzed—questions asked of bidders, responses, related e-mails—should all be retained. If a few contractor interviews were held, detailed comments arising from the interviews should be attached to the appropriate contractor file. Any notes, scraps of paper bearing any relation to the review of bids, calculations, or estimates are to be retained.

A record of any suggestions, changes, or comments regarding the content of the contract should be kept. Was the contractor concerned about the schedule? What were his comments, and what was the owner's or architect's response? Did the contractor seem hesitant about accepting a specific provision in the contract? What were your thoughts about why he was hesitant? A year and a half or two years later, when memories become vague, these types of notes may prove beneficial.

In the public sector, some government agencies require the contractor to review and sign each plan in the set of contract drawings and to initial every page in the specification manual to defuse any claim that the builder did not notice some detail or other. This set of initialed documents is retained by the owner along with the executed contract. Any comments pertaining to the plans, specifications, contract, or any of its attachments should be recorded. A simple question such as "Do you have any questions about the project's plans, specifications, or contract provisions?" will suffice The contractor's answer is to be noted along with the question.

DOCUMENTATION DURING CONSTRUCTION

When the architect/engineer furnishes construction services, he or she should be the repository for all documentation; it may be wise for an owner to review this document accumulation with the architect from time to time. The job meeting minutes, if sufficiently detailed, will adequately record the important events taking place as the building gets built, and the following supporting documents should augment that passage:

- The project meeting minutes along with any comments regarding their content that may have been expressed orally or in writing/e-mail

- Shop drawings approved and rejected along with the log that records the review, approval, and rejection time frame

- The request for information (RFI) or clarification (RFC) log, along with detailed questions and responses and timelines from submission to conclusion

- All correspondence passing through the owner to architect/engineer to contractor/subcontractor and back again

- All correspondence generated by the general contractor and the subcontractors directed to the appropriate design consultant and owner and copies of the responses and resolution of any open items

- Change-order files, including all backup documentation for approved and rejected proposed change orders

- Copies of the contractor's daily reports; they can be produced by the contractor at the weekly progress meeting, and upon review, the owner and A/E can submit written comment, if the reports seem incomplete, erroneous, or lacking in detail.

- Copies of field reports issued by the contractor, the architect/engineer, or any subcontractors

- Copies of all site inspections and their results, whether performed by the owner or the contractor's testing services; copies of all inspections by government authorities or written notification by the contractor that these building officials have performed inspections and have either approved or rejected work

- Copies of all accident reports and OSHA inspections provided to the owner; safety violations, when reported, and written responses regarding corrective action

- During periods of extreme weather conditions, the architect's or owner's record of conditions, to be compared with the contractor's documentation

- Copies of any notification of delays, labor disputes, strikes, delivery problems, and other disruptive events reported by the general contractor or the subcontractors and vendors

- When in doubt, place any other documentation in the file!

PRINCIPAL CAUSES OF DISPUTES AND CLAIMS

The following areas most often lead to disputes and claims:

1. Plans and specifications that contain errors, omissions, and ambiguities
2. Plans that have not been properly coordinated
3. Incomplete or inaccurate responses or nonresponses to questions or problems presented by one party in the contract to another party
4. Inadequate administration of an owner's, architect's, or contractor's responsibilities
5. Unwillingness to comply with the intent of the drawings by the owner, architect, or contractor
6. Site conditions that differ materially from those represented in the contract documents
7. Unforeseen subsurface conditions
8. A change in conditions
9. Discrepancies in the plans and/or specifications
10. Breaches of contract by any party
11. Disruptions to the normal pace of construction by disputed change orders, acceleration of the work, or lack of decisiveness by the architect, owner, or contractor
12. Delay of the work caused by any party to the contract
13. Inadequate financial strength on the part of the owner, the contractor, or any of the subcontractors or vendors

Contract issues

Before we investigate some of the circumstances that can possibly trigger or lead to a dispute or claim, we should review some contract provisions dealing with disputes and claims. The AIA A201 document General Conditions to the Contract for Construction, or a derivative thereof, contains the rules and regulations that govern the relationship of the owner and architect to the contractor. The document includes an entire section on claims and disputes and is a must-read for any owner whose contract includes this document as an appendix.

Some of the more important claims-related provisions of Article 4.3 Claims and Disputes can be summarized as follows:

- Time limits on submission of claims. Claims by either party must be submitted within 21 days after the occurrence of the event. (AIA defines "day" as "calendar day" unless specified to the contrary elsewhere.) Claims must be submitted, in writing to the architect to commence the process.

- The contractor must diligently pursue contract work while the claim is being reviewed and adjudicated.

- Guidelines for preparing and presenting claims regarding concealed or unknown conditions are provided.

- Claims for additional time are also discussed, along with the documentation required for a submission of this type.

- Weather-related claims are also addressed; they require substantiating data. If the weather conditions were abnormal for the period of time, and unseasonably cold or hot temperatures could not have been reasonably anticipated, local historical weather patterns will substantiate or contradict these types of claims.

Article 4.4 Resolution of Claims and Disputes lists the steps to take to resolve these disagreements. Litigation is expensive and time-consuming, and in many cases sets up such an adversarial environment that good sense goes out the window and attempts at an equitable reconciliation go along with it.

The 1997 edition of the AIA A201 document is helpful in resolving disputes and is really fair to all parties. The first step outlined in Article 4.4 requires mediation, a process that is nonbinding but allows a disinterested party to resolve a dispute by shuttle diplomacy: moving back and forth between the aggrieved parties, who generally are physically separated. In pointing out the strong and weak points of each party to the dispute, rational heads often prevail, and the dispute is resolved.

If mediation doesn't work, however, the next step is arbitration, where each party presents its case to a professional arbitrator, often supplied by the American Arbitration Association (AAA). The AIA provision calls for binding arbitration, which means that once the arbitrator renders a decision, referred to as an "award," no further appeal can be made. If the arbitrator's award is not carried out, a judgment can be filed in the appropriate court to foreclose on the responsible party.

Contractors welcomed these provisions because more than one contractor with a legitimate claim against an owner for an amount under $25,000 has been told, "If you don't like my refusal to pay, go ahead and sue me," knowing full well that the cost of litigation would take a huge bite out of the dispute amount. Owners also welcome this mediate-arbitrate process because it is less costly, can usually be resolved more quickly by avoiding the long backlog of court cases, and somewhat reduces the antagonisms that often accompany long, drawn-out legal battles.

We have previously discussed the flow-through provisions of the owner's contract with the general contractor into the agreements with subcontractors and vendors. A look at a typical general contractor's subcontract agreement will be helpful to illustrate this flow-through process. GCs are just as wary of disputes and claims from their subcontractors and vendors as owners are of those with their contractors. A typical subcontract agreement will include a section on claims:

Claims for extra or later work changes, modifications, changed conditions, delays, equitable adjustments, and damages resulting from requirements, acts, or omissions of the OWNER shall be governed by the provisions of the General Contract. If the General Contract authorizes an appeal from the act or decision of the ARCHITECT and/or OWNER and the time for such appeal has not expired, the CONTRACTOR, upon written request of the SUBCONTRACTOR shall, at its option, either prosecute an appeal on behalf of the SUBCONTRACTOR under said General Contract provisions or permit the SUBCONTRACTOR to prosecute an appeal in the name of the CONTRACTOR, either to be at the SUBCONTRACTOR'S sole expense.

What this provision allows is that claims submitted to a general contractor by a subcontractor are governed by the same terms as those in the general contract between owner and general contractor. This makes a claims and disputes statement in the owner/contractor agreement doubly important:

Claims by the SUBCONTRACTOR against the CONTRACTOR for extra work or damages of any kind shall be set forth in detail in the next succeeding monthly requisition after the occurrence of said claim or damage; otherwise said claim or damage shall be deemed waived. SUBCONTRACTOR expressly waives all claims of rights for multiple damages

This provision places a time limit on the submission of a subcontractor claim and precludes the possibility that a claim will be announced long after the events leading up to it have long past.

Plans and specifications containing errors or omissions

The quest for the perfect set of plans and specifications is probably as elusive as the search for the Golden Fleece. Although a provision in the general conditions requires the contractor to review the plans and specifications for errors, omissions, and inconsistencies and to notify the architect of same, on the other hand, the contractor has a right to expect a set of more or less complete documents. Minor plan/specification problems, if they arise, are the contractor's responsibility to point out. The owner, after all, has hired a professional, a contractor experienced in the type of work being undertaken, who has surely experienced such minor problems previously and should be expected to question or resolve them at no cost to the owner.

Quite often a contractor will advise the owner/architect of some small errors, stating that he or she can resolve them if the A/E will allow a minor change in the specifications or details to make them simpler. And, if approved, this mutual agreement may prove useful in other more complicated design inconsistencies. Major drawing deficiencies are another matter and should be addressed and resolved among all parties rather quickly by negotiating in good faith.

An order of precedence, often included in the construction contract, is also in place when conflicts occur between one drawing and another and between the plans and specifications. This terminology is as follows:

In case of conflict between drawings and specifications as to extent of work or location of materials and/or work, the following order of precedence will govern:

1. *Large-scale drawings*
2. *Small-scale drawings*
3. *Schedules on the drawings (i.e., door schedules, window schedules, finish schedules)*
4. *Technical specifications*

In case of conflict as to the type or quality of materials, the specifications will govern.

Even if this provision has not been included in the contract for construction, this order of precedence will historically favor the party adhering to it.

Lack of proper drawing coordination

Many architects "subcontract" portions of their work to other design specialists: a civil engineer for site work, a structural engineer for foundations and superstructure, and HVAC, plumbing, and electrical work to appropriate engineering firms. All of these consultants to the architect have a responsibility to coordinate their work so everything fits together properly and within the space allotted to it. But this doesn't always happen because the team leader or captain—usually the architect—may not have paid enough attention to this process or have been requested by an owner to complete the drawings before he or she has had a chance to properly review them. Improper coordination of the plans can prove catastrophic in a multistoried project.

As an extreme example, I was working as an owner's representative on a 14-story residential apartment building, and the architectural drawings had not been properly coordinated. The structural drawings of the posttensioned concrete slabs on each floor (cast-in-place concrete floors with a series of cables inside that would be tensioned up after they were poured), which a structural engineer had prepared, contained overall floor dimensions (out-to-out) that differed from the architectural drawings. The edge of the contract slabs on each of the floors extended beyond the façade of the building, clearly an example of a total lack of coordination between architect and engineer. These errors spilled over to the mechanical and electrical trades, since all of their vertical shaft openings, designed to carry electrical conduits, air-conditioning and heating ducts, water lines, and fire-protection lines, were designed according to the architectural drawings; the shaft locations were several feet away from the openings that were indicated on the structural drawings.

The drawings were being coordinated, a time-consuming process, as construction work was in progress in unaffected areas and either ceased or, at best, slowed down to a walk in other areas. When the proper, corrected drawings were issued, the owner directed the contractor to adhere to the original schedule, although recognizing that the problem originated with the design team.

Unlimited work hours, seven days a week, were authorized, resulting in low productivity from tired workers. Equipment deliveries that had been put on hold now had to be expedited; additional subcontractors had to be employed to augment existing ones; and a multimillion-dollar claim was averted only because the owner and contractor approached the problem as businesspeople and settled with the builder for $7 million, certainly not an inconsequential sum.

On any project, but particularly on multistory buildings, proper coordination of all design disciplines must be stressed and possibly reinforced via a provision in the agreement between owner and architect. When conflicts do occur as the drawings are being prepared that involve the location and layout of one or more components, the following priorities are frequently used by the design consultants to resolve them:

1. The structure and partitions
2. Equipment location and adequate access to that equipment
3. Ceiling heights with recessed lighting fixtures
4. Gravity drain lines, storm water lines, waste lines
5. High-pressure ductwork and associated devices
6. Large pipe mains, valves, and devices and access to those valve and devices
7. Low-pressure ductwork, diffusers, registers, grills, HVAC equipment
8. Fire-protection piping and sprinkler heads
9. Small piping, tubing, electrical conduits
10. Sleeves through rated partitions

Most construction-contract specification manuals contain a provision requiring the general contractor to prepare a set of "coordination drawings" for submission and approval by the architect. To ensure proper coordination, the general contractor will meet with the subcontractors who have an interest in the coordination process and distribute floor plans among those subcontractors, who will either accept the space allotted to them diagrammatically on the contract drawings or request a change in that space. As these coordination drawings pass through this review and comment process, a final set of drawings with any requested space changes is then submitted to the design consultant for review and comment.

If minor changes are needed to make everything fit, the GC will request little or no additional money. Sometimes lowering a ceiling height two to three inches or enlarging a chase wall by six inches will accommodate the required piping or equipment. Only when time-consuming design changes are needed will extra costs be involved.

Incomplete or inaccurate responses

During the entire construction process, many participants will ask many questions, some more urgent than others, and these questions must be answered. Some of these questions may require investigation by the design consultant, such as changes in one component that may trigger changes in several other components, prior to issuing a response. But time is ticking away, and the process of building the facility goes on. Requests for complex answers to complex problems need to have a due date agreeable to owner, architect, and contractor to keep the flow of work from being seriously jeopardized, resulting in loss of time and money and a request for compensation.

These time-sensitive requests require all responsible parties to agree to a time frame and abide by it. Too often the urgency is not given the attention it deserves, and when delays occur, an owner is shocked when he or she receives a claim for a substantial sum. An owner's representative must be the watchdog in these matters; if delays go unheeded, he or she must advise the negligent party that he will be responsible. Although others are hired to supervise, monitor, and oversee the construction process, an owner must stay close to the action, because if he doesn't, there will be a lot of finger pointing—and guess who pays the price? The owner.

Inadequate administration of the project

Inadequate administration can go up or down the line, and when it is spotted, it must be quickly changed. The owner has a responsibility to respond to questions from the architect, the design team, and the contractor. A clear chain of command should be established so certain decisions can be made in the field while other require home-office review and approval. Contractors get very upset when their progress payments are overdue. Prompt payment is important to contractors, who in turn must pay their subcontractors promptly so they can demand service when required. If the owner's funds are not available, the contractor may have to extend a credit line that may already be extended too far. An important owner responsibility is to review, approve, and process contractor requests for payment within a reasonable period of time.

The project team proposed by the general contractor, whose experiences look good on paper, may turn out to be not so good. Answers to questions never seem to come when promised, and when they do, they are incomplete or even erroneous. Documents and events are delivered late or fail to meet deadlines, and the project manager or super seems to be either overworked or working well below expected levels. If the owner, architect, and engineer arrive at the same conclusion that the managers are not performing, the GC should be notified in writing and replacements—qualified ones—requested.

The same scenario can involve a design consultant or an owner's representative. If other team members have real concerns about the performance of the supervising architect, engineer, or owner's representative, changes must be made promptly.

A watchful eye must be kept on architect/engineer processing of important documents, change-order requests, requests for information, payment requests, and so forth, and the owner must ensure that the design consultants are performing in a timely and professional manner, much the same as they expect of the contractor.

Unwillingness to comply with the intent of the drawings

An architect battling with a contractor over the intent of the plans can be a disruptive force. An owner should not stand idly by while architect and contractor argue over such matters; he or she must intercede, mediate, and attempt to resolve the matter. There again, the general conditions document comes into play. Article 1.2.1 says that "the intent of the contract documents is to include all items necessary for the proper execution and completion of the Work." Article 4.2.11 states that the architect will interpret and decide matters concerning performance under the requirements of the contract documents, and Article 4.2.12 says that the interpretations and decisions of the architect must be consistent with the intent of and reasonably inferable from the contract documents. The contractor can be reacquainted with these provisions if he or she failed to note them, and that may solve the "intent" problem. It becomes the owner's responsibility to become engaged, stress reasonableness and fairness, and resolve issues of this nature.

Site conditions that differ materially from the contract documents

Site matters are some of the most frequent and complex disputes that can surface on a construction project. And they require that all parties view the matter not only from their own perspective but from that of others.

Contractors base their site work estimates on information provided by the owner's civil engineer, a geotechnical technician, referred to as the geotech. The report that is compiled by the geotech includes a host of information. Soil samples taken from test borings not only provide the structural engineer with soil-bearing capacity, information upon which to design the project's foundation, but they also provide the contractor with reasonable expectations of what he will find when he excavates for those foundations and other site work.

The geotechnical report is in narrative form, and it also diagrammatically provides a contractor with a view of the soil sample, its composition, an indication of underground springs and water levels, the location of rock formations and the type of rock encountered, and the elevation (depth) where these soils, rocks, and water were found.

As you recall, in the bid documents the contractor is invited to inspect the site to become familiar with the condition. This is a critical prebid requirement that has many manifestations if and when a site problem is encountered. The test borings are often qualified with a statement such as "The conditions found in the test borings are indicative of conditions found solely at that location and may vary widely in adjacent areas of the site," Once construction begins, the contractor may find that the composition of the soil may differ "materially" from that represented in the contract documents and the accompanying geotech report and test borings because he or she has excavated in a location where borings were not taken. This not only applies to soil composition—gravel, clay, humus, decomposed vegetation, sand, and silt—but also to rock and the presence of water. Not infrequently, what was anticipated by the soil-boring analysis and what was actually uncovered vary considerably.

Government contracting issues and resolutions are probably the standard for determining the meaning of "differing materially." Several government agencies use the "15 percent rule." If the actual quantity of material encountered exceeds 15 percent of what was a reasonable interpretation of the geotechnical report, a differing or changed condition has been established. On that basis, the contractor may issue a request for a cost proposal to cover the added costs for the work. But for the contractor to make that claim, he must present the following documentation:

1. The time, date, and condition when these differing conditions were observed
2. The project superintendent's entry of this event in the daily log or daily report
3. Photos, still or video
4. A statement from the field personnel involved in the discovery, such as the excavating contractor, the contractor's foreman, or equipment operator
5. A statement explaining the operation that was taking place at the time—for example, excavating for a water line, building foundations—and the exact location

The contractor should alert the owner's representative or architect immediately so he or she can observe these conditions and comment on them as they are occurring rather than after the removal of the differing materials has taken place.

The two types of differing-conditions claims, established in the public sector and acknowledged by the private sector, are Type I and Type II claims. For a Type I claim, the contractor must prove the following:

- The contract documents include the subsurface or latent conditions that form the basis of the claim.

- The contractor's interpretation of the contract documents is reasonable.

- The contractor relied on these interpretations when they prepared the estimate of the work.

- The subsurface or latent conditions actually encountered were materially different from those represented in the contract documents.

- The actual conditions discovered were reasonably unforeseen.

- The costs included in the claim are solely representative of the costs to correct the materially differing conditions.

For a Type II claim the contractor must prove the following:

- What would have been the usual conditions the contractor would have encountered based on the information included in the contract documents?

- What were the actual conditions encountered?

- The physical conditions encountered differed materially from the known or usual conditions.

- The encountered conditions created an increase in the cost of the work.

To counter these claims, owner and architect will present the following as their case:

- The conditions encountered by the contractor were not really different from those included in the contract documents.

- The encountered conditions should have reasonably been anticipated by the contractor.

- The project was not managed properly, and the claim was an attempt to recoup unrelated costs due to the mismanagement.

- The contractor should have conducted a more thorough site investigation, even requesting, during the bid process, permission to dig some pits to discover the real nature of the site (a rather weak owner argument but one that can be presented).

- The excavating equipment was the wrong type or wrong size and could not cope with the conditions uncovered. (This would mainly apply to loose or deteriorated rock that could have been removed with a large excavator and did not require blasting.)

- The contractor's operators or supervisors were inexperienced.

Various court cases have somewhat quantified what is meant by a "changed" condition or a "materially differing" condition, and varying court interpretations as to what constitutes "materially differing conditions" have arisen. We have already discussed the 15 percent rule, but there are others.

In the late 1990s, the Federal Highway Administration responded to a query from U.S. Senator Frank Lautenberg of New Jersey concerning what constituted "differing" conditions. This FHWA interpretation may be the clearest way to quantify this difficult topic:

We recognize that after a contract gets underway, conditions may change or circumstances may exist that were not anticipated during preparation of the plans, specifications, and estimates. Our governing legislation and our implementing regulations allow for change orders within the scope of work covered by the contract. In awarding contracts for federal-aid highway projects, the State transportation department must include a standardized clause for changed conditions to provide for an adjustment of contract terms if the altered character of the work differs materially from the original contract or if a major item of work is increased or decreased by more than 25 percent of the original contract value.

Unforeseen subsurface conditions

The scenario for unforeseen subsurface conditions would generally follow that for materially differing conditions with certain exceptions:

- The discovery of significant underground springs where none were indicated in the contract documents. The explanation could be as simple as test borings being taken during a season when underground springs recede or disappear, only to reappear at other times of the year.

- Discovery of several large tree stumps buried previously but not located on the drawings and now uncovered where the new building's foundations are to rest. (Buried tree stumps and large tree limbs must be removed because, as they decay, voids are formed, endangering the structural integrity of the foundation if built over those materials.)

- Discovery of other buried debris that was not indicated by the contract documents but must be removed.

- Abandoned wells where the well head was not indicated in the contract documents and where discovery could not be made by a contractor's visit to the site as required in the bid documents.

The geotechnical report accompanying the site work plans and specifications contains lots of disclaimers and rightfully so because it represents a limited investigation of what is sometimes a very large site. One hundred percent coverage of underground investigations is not only impractical but would be vastly expensive. Often one six-inch (15.24 cm) boring spaced every 50 (15.24 meters) to 150 feet (45.72 meters) is meant to be representative of the soil composition for multistory projects. More extensive subsurface exploration can be accomplished by digging test pits to a minimum of 3 feet (.9 meters) by 3 feet by 6 feet (1.8 meters) deep or greater.

The geotech will base the report on observations and the subsurface conditions after analyzing the sampling of test borings around the site. The civil engineer includes several disclaimers in the report that can often be applied by an owner when less-than-strong claims for site extras are proposed by the contractor. One such disclaimer typically reads as follows:

The analysis and recommendations submitted in this report are based upon information revealed by this exploration. This report does not reflect any variations which may occur beyond the locations of the test borings and test pits. Since the nature and extent of variations may not become evident until the course of construction, an allowance should be established to account for possible additional costs that may be required to construct the foundations as recommended herein.

This clause ostensibly puts the contractor on notice that he or she should include some contingency in the site work estimate for some as yet unforeseen conditions, and most contractors know that. It is only when conditions discovered during site work operations exceed what could reasonably be expected that some contractors may consider grounds for submitting a claim.

Specific terminology can be inserted in the bid and contract documents that can offer a great deal of protection from these kinds of differing conditions and unforeseen subsurface conditions—but at a price. By deeming the site an "unclassified site," the contractor is put on notice that he or she "owns" all encountered underground conditions. If unsuitable soils, rock, debris, underground structures, or tree stumps are encountered during excavation, the contractor must deal with them at his own expense. Although this protects the owner quite a bit, the contractor's estimate for site work will include a contingency large enough to cover any unanticipated site conditions. In the event that little or no unsatisfactory underground conditions are encountered, the contractor will reap the rewards. Before considering an unclassified site condition, would you rather play it safe or throw the dice and pay for only those legitimate costs if unforeseen subsurface conditions do occur? If you were a contractor submitting a bid, which approach would you prefer? Isn't it better for an owner to assume actual costs for unforeseen conditions than gambling on what may or may not be there?

A typical unclassified site provision would look like this:

Excavation shall be unclassified and shall comprise and include the satisfactory removal and disposal of all materials encountered regardless of the nature of those materials and shall be understood to include rock, shale, earth, hardpan, fill, foundations, pavements, curbs,

A change in conditions

This concept is similar to materially differing conditions but is more all-encompassing. A change-in-conditions situation can occur beyond the site work stage and well into the actual construction stage. When the nature of the work involved in building the superstructure dramatically changes, this could be cause for the contractor to claim a change in conditions.

A contractor who is preparing the estimate for construction will assume responsibility for performance associated with the requirement of the bid or contract documents and any changes affected by approved change orders. He or she will assume that managing the construction project will impose an additional workload in the field and also in the central office—accounting, project management, estimating, and purchasing—and will plan for that in the estimate.

The contractor's general conditions costs are time related, and any extension of that time, reflected in the contract schedule and as adjusted by change order(s), will cause the contractor to incur additional costs if the schedule is extended. And when those costs increase because the nature of the project has changed dramatically, a potential claim for change in conditions may be floated by the general contractor. This was the situation that occurred in the poor drawing coordination example that cost the owner $7 million; that claim was based on acceleration and a change in conditions.

This could also occur when the contract documents are vague or insufficiently detailed, resulting in the submission of hundreds of requests for information from the contractor to the design team. By its very nature, this flow of documents to and from parties will slow the project and, in some instances, so overwhelm the contractor's support staff that additional staff may be required.

An inordinate amount of change orders generated either by the owner or the contractor will also affect the pace and extent of supervision. For example, it is rather easy to see how a 25 percent addition in work via scope changes would impact every administrative function of the contractor. This major change in scope of work could easily be considered to present an entirely new project. When these types of changes occur, the contractor will simply state that the initial obligation to perform the work has been radically changed and that she will incur to meet these new conditions. In fact, a change in conditions to those originally conceived has occurred. The contractor will present additional costs that include the following:

- Increased administrative and management costs, both field and general office related

- Costs associated with the delays to fold this new work into the existing schedule, which may include hiring additional subcontractors

- Changes in the original baseline schedule, and the resultant costs associated with the revised schedule

- A requirement to expedite material and equipment deliveries and assign these responsibilities to an added staff member

- Increasing various trade crews or directing existing ones to work extended hours at premium-time rates, often leading to a considerable drop in worker productivity

Owners and design consultants should keep this change-in-conditions concept in mind whenever the flow of RFIs or RFQs increases significantly or when changes to the initial scope of work are so extensive as to create an assumption of a much different project than originally conceived.

Discrepancies in the plans and/or specifications

Some of the disagreements related to document discrepancies occur because of what appear to be conflicting requirements—for example, one drawing includes a detail that differs from a similar detail on another drawing. A door schedule on one drawing may indicate that Door 101A, for example, is a 3070 door, but the floor plan may reveal that this same Door 101A is a 2868. Which door is the contractor obligated to supply?

Over the years, court rulings have provided some answers to these questions, and when conflicts do arise, an order of precedence is often employed to hopefully resolve some of these disagreements. This is a priority standard that has come to be accepted primarily through past court decisions:

- Contract terms
- Drawings are modified by addenda
- Supplemental conditions issued as bid documents
- Drawings with schedules
- Large-scale details or drawings

When drawings and specifications are not in concurrence in quality or quantity, the contract usually requires an interpretation from the architect.

If legal action is contemplated, the courts may look at the purpose for which the disputed item is intended and rule accordingly. These are only guidelines; there are no hard-and-fast rules for many of these types of disputes, making it even more important for all parties to negotiate their grievances rather than pursue court action.

In fact, borrowing from a landmark 1918 case, *Spearin v. United States Navy*, the court decision known as the Spearin Doctrine is still being used today to resolve some of these contract-shortcoming disputes. The contractor, Spearin Construction, contested the U.S. Navy's demand that it be held responsible for what appeared to be a design defect. This was the court's ruling:

> *If the contractor is bound to build according to the plans and specifications prepared by the owner, the contractor will not be responsible for the consequences of the defects in the plans and specifications.*
>
> *The responsibility of the owner is not overcome by the usual clauses requiring bidders to visit the site, to check the plans, and to inform themselves of the requirements of the work.*

Disruptions to the normal pace of construction

Disruptions to the normal pace of construction can occur due to actions or inactions of an owner, architect, engineer, or contractor. Often these disruptions are created by indecision on the part of one party to the contract, who may be unaware of the lack of decision making and its impact on the orderly flow of work that is expected.

Construction schedules often include extra days, designated as float, that essentially involves several days in the schedule set aside for the minor delays that so often occur in this complex building process. At times the contract will stipulate that the owner "owns" all or a portion of the float and the contractor "owns" the balance. These extra days can be used to prolong a decision or await materials or equipment on the job site. But everyone must be sensitive to prompt response and resolution of decisions that can impact the day-to-day pace of construction.

Inadequate financial strength of any party

One matter that will cause a contractor's blood to boil is a substantial delay in receiving payment from an owner after the request for payment has been approved by the architect. This can set off a whole series of events that can ultimately affect the entire project. Many contractor contracts with subcontractors and vendors contain a "pay when paid clause," requiring the general contractor to pay the subcontractors only when payment is received from the owner. Subcontractors who are not paid promptly payment still have to meet payrolls on the project, and without an indication that payment is forthcoming, they may tell the GC that they are removing workers from the project and placing them on another job where there are assured of receiving payment promptly. Some states have banned the "pay when paid" clause on state contracts, so perhaps a check with your attorney on the status in your area may be in order.

This type of problem can cascade to subcontractors of subcontractors, creating a disruptive situation. Prompt payment by an owner when the architect has certified the payment request is important. If there is a delay, advise the contractor, so he can plan accordingly.

Just as important is assurance from the general contractor that he or she is not taking the funds from one project to pay subcontractors and suppliers on another project. This is referred to as "comingling funds" and is frowned upon (though it happens frequently). Lien releases from subcontractors will be one check on a general contractor's assurance of payment, since they ostensibly state that the subcontractor has been paid for work performed for the previous work period.

It's usually not necessary to worry about whether the general contractor is paying his or her workers, because the employees will simply quit if not paid. Subcontractors can present another problem when they are short of funds and begin to run out of day-to-day supplies because they may have exceeded their credit limits or are on a COD basis. An owner has as much interest in subcontractor performance as the general contractor because subcontractor default affects both parties. If there are any doubts, an owner must determine whether the general contractor's payment to the subcontractor is current or whether the subcontractor is experiencing cash-flow problems. These questions need to be posed to the general contractor and a response required within a short time frame, or the architect and owner can request a meeting with the general contractor and the subcontractor(s) to respond to any payment concerns.

If there are doubts as to whether payments to a subcontractor are being made in accordance with the requests reflected in the general contractor's application for payment, or upon learning from a subcontractor that he is not being paid the full amount he requested (and reflected in the GC's request for payment), the use of a joint check can be raised with the general contractor. A joint check is one that contains two payees: the general contractor and the subcontractor or vendor. The joint check is made out in the amount owed to that subcontractor or vendor as agreed upon by the general contractor. The check is first passed to the general contractor for signature and then on to the subcontractor or vendor. The general contractor's permission to issue a joint check is required and should be obtained in writing. In most cases he will agree to this process. If not, that should raise another red flag.

DELAYS AND THE PROBLEMS THEY CAUSE

Many of the events that result in disputes and claims create situations where work is delayed and cannot proceed along the path included in the project's baseline or mutually adjusted timeline. Delays cost money for both owner and contractor, and unless dealt with promptly, they tend to escalate in cost and tempers.

The owner has a number of reasons for wanting the project completed on time: the higher cost of construction financing versus less expensive permanent financing, plans to move into the new structure at a specific date, and delivery dates for new furniture and equipment, among others. The contractor's general conditions are mostly time related: costs of salaries for project and site supervision, field-office rentals, and equipment. Contractors with a backlog of work anticipate moving their project managers and superintendents from the current project to a new one, anticipating that they will be able to do so if that project closes out as scheduled. Along with actual additional costs being incurred, this timely shifting of supervision is of prime importance to the general contractor.

An owner must be alert to the need for prompt review of contractor-generated documents requiring either architect/engineer or owner response so as to avoid the potential for delays. As the requests for information (RFI), requests for clarification (RFC), and proposed change orders (PCO) are issued by the contractor, the owner must monitor the passage of these contractor- or architect-generated queries to avoid any potential delay claims. The owner's representative should be responsible for reviewing these logs at each project meeting and should act as a prod when one party or the other is lax in processing the necessary information to close out those documents.

The owner must always be aware of events that may create the potential for a contractor delay claim. If the architect/engineer requires a slight push to resolve any contractor-related matters, the owner, after the project meeting and in private, should have a word with the design consultants to determine why documents are not being processed quickly and how they plan to rectify the situation.

To explore the construction-delay process, we should start with some contractor-related quasi-legal terms. An *excusable delay* is a delay whereby a contractor or owner is allowed a time extension but no monetary compensation. This type of delay can occur when there are acts of God, fires, or transportation delays, over which the contractor or owner has no control; labor strikes, also beyond the owner/contractor's control; and unusually severe weather. When there are liquidated damages in the construction contract, the builder will need to extend the project's completion date even though no costs for extension can be considered. When excusable delays occur and are approved, the architect will issue a no-cost change order with an extension of the completion time agreed upon by contractor and owner.

A *concurrent delay* occurs when two or more delays are created within the same time frame that impact the completion date and are caused by events created by the contractor and the owner. They cancel each other out, and neither party can recover damages. A *compensable delay* is one where damages are created and compensation requested. It is a delay caused by one party and within the control of that party, whether it be owner, architect, or contractor. The contractor's costs may include increased crew size, increased costs of materials and equipment, extended general conditions, or loss of productivity because of the interruption to the orderly sequence of work, to name a few. The owner's costs are also varied: a landlord guaranteeing occupancy of commercial rental space, the opening of a retail store, or extending financing costs.

The project meeting and the resultant minutes may be the appropriate arena for unresolved matters to be discussed, ball-in-court determinations made, and dates set for resolution. This meeting can act as a tracking device to document the events that have the potential for a delay. The word *delay* can be incorporated into the minutes to let everyone know that the warning has been made—for example, like the following:

> *Contractor has not responded to the architect's Architect Field Instruction #23 issued on September 15, 2009, even though first and second requests have been issued previously. If no response is received by architect on September 30, 2009, any delays and associated costs will be borne by the Contractor.*

Acceleration is another one of those construction terms that owners become familiar with. When delays are attributed to an owner's or architect's actions or inactions, are recognized as such, and the owner requires the contractor to maintain the original completion schedule, acceleration of the project will be required to meet that contract completion date. The following is a more formal definition of acceleration:

> *When a project owner recognizes that there have been delays in the construction project that would ordinarily extend the completion date, but directs the contractor to complete the project as originally scheduled, "acceleration" has been created.*

An owner may take this position for many reasons, but he or she also must be aware of the potential costs this directive can create.

A requirement to maintain the initial completion date can arise when the owner's new factory must be on-stream as planned to fulfill a large influx of orders. When the owner of a baseball or football sports complex must have that facility ready by Opening Day, the reason to accelerate is obvious.

The added costs to accelerate can be staggering, and few contractors will provide any form of guarantee. These are some of the problems that can occur when a contractor has been directed to accelerate:

- *Out-of-sequence work:* Instead of working in a linear fashion, moving from activity A to activity B, subcontractors may have to shift crews to more immediate tasks, performing activity G and then going back to normal sequencing. This process of working out of the normal pattern of work as indicated on the baseline or adjusted baseline schedule will increase the total cost of that work.

- *Loss of productivity:* One of the consequences of out-of- sequence work is the loss of productivity among those workers who had to shift from their normal schedule and take up a task in another part of the building or drop what they were doing and proceed with new instructions. Extended premium-time work has been proven to reduce worker productivity. A seminal study performed by then Business Roundtable in 1993 produced the document "Scheduled Overtime Effect on Construction Projects," which graphically demonstrated the amount of worker productivity lost when construction workers worked several periods of 50- and 60-hour weeks.

- *Trade stacking:* This occurs when several trades are working in a somewhat confined space to complete a specific component of work. For example, carpenters may be installing a ceiling grid in one or two rooms, and electricians and HVAC workers are also in those rooms installing ceiling light fixtures and air-conditioning and heating ducts. These trades are stacked and working together in these rooms, whereas one trade would normally follow the preceding one. Trade stacking results in loss of productivity.

Acceleration creates other real costs such as expedited deliveries of materials and equipment and premium-time work, but many of these costs are "guesstimates," and an owner will find himself hard-pressed to analyze and agree on the actual costs. So a word of caution: Avoid requesting a contractor to accelerate, but if it is necessary, request a daily accounting, and investigate all of these costs while they are fresh in the minds of everyone who generated them.

When delays occur on a construction site and the contractor's equipment remains idle during that period of time, she will often claim loss of revenue because that equipment could have generated revenue were the delays not incurred and the equipment was performing "contract work." An idle-equipment claim involves many issues. One is whether the equipment is actually owned by the contractor or rented from an equipment rental company. Using a Caterpillar 350 excavator as an example, if this equipment is rented, rates on an hourly basis would be about $175 per hour. This excludes the subcontractor's operator cost and the cost of fuel, both of which are not provided by the rental company. If the Cat excavator was rented and the contractor estimated 10 hours to complete a certain portion of work, the estimated cost of that work would be $175 per hour plus the cost of the operator at $40 per hour and possibly $25 for fuel, for a total of $240 per hour. So the total cost of the operation would have been $2,400.

Suppose the contractor had to stop work for three hours because of a delay caused by an owner or because of unusual site conditions, such as the presence of rock in the area of the excavation, The rental of the equipment to the contractor continues to accrue even though the equipment is not working, and, all other things being equal, the contractor may be entitled to reimbursement for downtime on the idled equipment. But since no operator or fuel costs were involved, the loss of productive equipment would be the base rental rate, unless the contractor could not provide the operator with work on another actively operated piece of equipment.

If the preceding scenario involved the same Caterpillar excavator, but instead of being leased, it was owned by the contractor, then even the bare equipment cost of $175 per hour would not be justified. Although when renting equipment, downtime costs continue to accrue, when the equipment is owned, no such "fixed" costs apply. The idle or downtime causes no additional "wear and tear" on the equipment. It is not depreciating because it is not operating; no maintenance is required, and its operation life has not been shortened. The rate for downtime or idle time on a piece of owner-operated equipment is much less than its full operational cost. Some experts state that this downtime cost should be anywhere from 30 to 50 percent of the equipment's operating cost.

A court decision seems to lean toward the 50 percent figure. In the *Appeal of Dillon Construction, Inc.* (ENGBCA No. PCC-101, November 21, 1995), the court said that when contractor-owned equipment is idled, the contractor could claim 50 percent of the equipment cost as posted in an equipment cost manual. The generally accepted equipment cost manual is the *AED Green Book* of equipment rental rates. This book lists all types of equipment for excavators and contractors and average rental costs daily, weekly, and monthly, with regional adjustments.

To further guide an owner when claims for lost revenue involve equipment downtime for extended periods of time—a few days, a week, or longer—the contractor owning the downtime equipment has the option of taking it offsite and using it on another, more productive project. If no other revenue-producing source for this piece of equipment was available, the contractor may have a difficult time claiming any costs other that the discounted idle-equipment cost. The responsibility is on the contractor to prove that the equipment could have been productively used elsewhere. In *C.L. Fairly Construction Company* (ASBCA No. 32,581 90-2 BAC (CCH), par. 22,665, 1990), the court said that the contractor must establish that the equipment in question could have otherwise been productively employed before claiming downtime costs.

The following are some general rules to consider when dealing with disputes:

- Establish an environment of trust between owner and contractor, which also involves the owner's design consultants if they appear to be too aggressive in their relationship with the contractor.

- Establish an environment of reasonableness and fairness, not only with the general contractor but with their subcontractors. As the saying goes, walk a mile in the other person's shoes.

- Avoid the blame game, and recognize that some portion of the problem may lie with the owner and the design consultants.

- Resolve disputes as quickly and as equitably as possible. The longer a dispute remains open, the more hardened positions form barriers to resolution.

- And remember the old saying that in a successful negotiation neither party is entirely happy with the results.

LEGAL PRECEDENTS RELATING TO CONSTRUCTION CLAIMS

A legal precedent has been defined as an act or instance that may be used as an example in dealing with subsequent similar events; another definition is that it is a judicial decision that may be used as a standard in other similar cases. Many legal precedents pertaining to construction claims have been set that may be of interest to owners involved in construction-related disputes.

Differing site conditions

In *Randa/Madison Joint Venture III v. Dahlberg* (239 F.3d 1264, 2000 U.S. App/ U.S. Fed. Cir. February 7, 2001), Randa/Madison, the contractor, was under contract to dewater a pumphouse foundation excavation and found out that the amount of dewatering required had been grossly underestimated. Although soil test results and soil and rock samples were available for inspection, Randa/Madison did not avail itself of these reports and after submitting a claim for extra work, indicated that it was under no obligation to review that information and that the owner had an obligation to disclose that information, not just make it available. The owner's contract contained two clauses: the contractor must satisfy itself with the character and quantity of work and the bid documents must address the physical data that were available. The court ruled that the owner placed the contractor on notice that additional information existed and that the contractor was presumed to have reviewed it. The claim was denied.

Complete set of drawings

In *John McShain v. United States* (412 F.2d 1218, 1969), a contractor sued the project owner, stating that the true condition of the drawings were not known at bidding, that some illegible drawings were replaced with legible ones, and that the addenda issued did not correct many of the coordination errors in the bid documents. The court stated that although the plans and specifications need not be perfect, they must be adequate for the purpose for which they were intended and that the contractor had no legal or contractual obligation to inspect the drawings to determine their adequacy for construction. The contractor's claim was upheld.

Damages for breach of contract

Some contractors will claim that they lost potential jobs because of the problems they incurred in a breach of contract with the owner on a current project. A general rule is that lost profits are not recoverable because it is often difficult to show that an owner's actions were in fact the direct cause of losing that other job This ruling came about in *BEGL Construction Co., Inc. v. Los Angeles Unified School District* (Cal.App. 4th 154, 2007).

Electronic records

Company records are being electronically stored with more and more frequency, and although courts usually accepted computer printouts as evidence, when Rule 34 of the Federal Rules of Civil Procedure Act was enacted in 1985, more legal rulings have reinforced this determination. The presentation of electronic data and, more particularly, e-mails that were apparently erased but reconstituted has been accepted as evidence by several courts. Owners must be diligent in creating, managing, and storing all electronic data pertaining to the construction project in the event that this information may be required in a potential lawsuit.

CONTRACTOR'S GUARANTEE OF DESIGN

If an architect designs a specific component, the contractor builds the component per that design, and the installation results in poor quality or performance, is the contractor responsible? In *Teuful v. Weiner* (68 Wash. 2d 31,411 P, 2d 151, 1966), the contractor installed a curtain wall exactly as designed, but the system leaked on several occasions. The contractor had previously notified the design consultant of changes that should be considered to improve the design, but the advice went unheeded. When the contractor denied responsibility for the ensuing water leaks, the owner sued—and lost. The court ruled that the leaks were not caused by faulty materials or substandard workmanship but were the results of a design defect.

Withholding payment due to defective or incomplete work

A contractor who was hired to build 15 condominium units in Montana failed to meet the completion date on 14 of the 15 units, and the owner's contract included a schedule with penalty and bonus clauses. The architect certified a payment of $83,141, but the owner withheld a portion, claiming that the late delivery and defective work justified those actions. The contractor filed a lien, and the owner countersued. The lower court ruled that the contractor was contractually bound to perform punch list and warranty work and awarded the owner costs to complete the contractor's obligations. However, the Supreme Court of Montana reversed the decision and allowed the contractor to terminate its performance because the owner failed to make a payment that was certified by the architect.

In this case, an AIA contract was in effect: Article 9.5.1 stipulates that the architect may decide not to certify payment or withhold an amount necessary to protect the owner. The architect, however, certified the entire payment. Article 9.7.1 in the owner-contractor agreement stipulates that if the owner does not pay the contractor the amount certified by the architect within seven days, the contractor, by sending written notice to the owner within an additional seven-day period, may stop work until payment is made. The court would have most likely ruled differently if the architect had certified a lesser amount, citing poor performance and defective work as the rationale.

Claim for lost productivity

Lost productivity claims by a contractor are just as difficult to deal with as differing or changed conditions, because these claims are so hard to quantify. In *Appeal of Clark Construction Group Inc.* (WL 37542, VANCA No. 5674.00-1 BCA para. 30,870, 2000), the Board of Contract Appeals concluded, "Quantification of loss of efficiency or impact claims is a particularly vexing and complete problem and separating inefficiency costs to be both impractical and essentially impossible."

But other decisions were not so definitive. The Mechanical Contractor's Association of America (MCAA) published the study "Change Orders, Productivity Overtime—A Primer for the Construction Industry," which includes lost productivity factors that have been supported by the courts. Three court cases support the MCAA manual used to prepare labor productivity issues. The *Appeal of P.J. Dick Inc.* (WL 1219552, VABC No. 5597, 01-2 BCA, para. 31,647, 2001) found that the contractor's calculation of loss based on this method was acceptable.

Hensel Phillips Construction Company v. General Services Administration (WL 43961, GSBCA 01-1 BCA para. 31,249, 2001) found that while the "measured mile" approach is generally the preferred method of dealing with these types of costs, the MCAA manual approach was deemed reasonable. "Measured mile" refers to the process whereby a contractor shows that the previous cost of work experienced on repeated, similar projects revealed that certain performance levels and unit costs were lower. These previously assembled "measured mile" costs, when compared to the higher unit costs incurred on the troubled project, provide justification for the higher cost claim. For example, if a concrete subcontractor can show that unit costs for placement of concrete were $100 per cubic yard plus or minus 10 percent ($90 to $110 per cubic yard) on ten previous, similar projects and the costs on the current project, where loss productivity claims of $180 per cubic yard are occurring, this represents a measured mile approach to these extra costs and can be strong evidence in a claim for, say, $80 additional cost per yard of placed concrete.

In *S. Leo Harmonay Inc v. Binks Mfg Co.* (S.D.N.Y., 597 Supp. 1014, Harmonay, 1984), the U.S. District Court ruled that the measured mile approach backed up by MCAA factors was appropriate in an award of damages to Harmonay for 30 percent less productivity caused by excessive work hours, overly crowded conditions (trade stacking?), and constant revision of contract drawings.

Prompt review of shop drawings

A contractor with Peter Kiewits Sons filed a claim that it incurred damages because the architect did not process shop drawings promptly and had no justifiable reasons for the delay in the review and response processes. This claim was upheld by the court ruling in the contractor's favor in *Peter Kiewits Sons Co. v. Iowa Utility Co.* (355 F.Supp. 376,392, S.S. Iowa, 1973).

GENERIC GUIDELINES FOR COORDINATION DRAWING

The purpose of drawing coordination is to create drawings where building components fit within the horizontal and vertical dimensions and avoid interference with the structural framing, ceilings, partitions, mechanical and electrical equipment, and other building equipment. The architectural drawings are diagrammatical and require the general contractor to review them to ensure that all of the components will fit in the space allotted for them. Some problems are rather easy to resolve, such as lowering a ceiling height by a few inches or increasing the width of a partition or two by a few inches. In other instances, major coordination problems that require shifting partition locations or lowering ceiling heights to unacceptable levels can create serious problems.

The general contractor usually has a contract obligation to perform normal coordination work, and this coordination requirement is usually included in the contract specifications, directing the general contractor to perform a number of procedures. General procedures include preparing coordination drawings for areas where close coordination is required for installation of products and materials fabricated offsite by separate entities and where limited space necessitates maximum utilization of space for efficient installation of different components. Coordination drawings include but are not necessarily limited to the following:

1. Structure
2. Partition and room layout
3. Ceiling layout and heights
4. Light fixtures (installed above the ceiling)
5. Access panels (to ensure access for maintenance)
6. Sheet metal, heating coils, boxes, grilles, diffusers, and similar items
7. All heating pipes and valves
8. Smoke and fire dampers
9. Soil, waste, and vent piping
10. Major water mains and branch lines
11. Roof drain piping
12. Major electrical conduit runs, panel boards, feeder conduits, and racks of branch conduits
13. Above-ceiling miscellaneous metal items
14. All equipment
15. Equipment located above finished ceilings that require access for maintenance and service
16. Existing conditions (when renovation work is involved)

To effect this effort, the general contractor will circulate drawings to the following subcontractors, who will show actual size and location of their respective equipment:

- Elevator subcontractor
- Plumbing subcontractor
- Fire-protection subcontractor
- Heat, ventilating, and air-conditioning subcontractors
- Electrical subcontractors
- Control systems contractors

Each subcontractor will note any apparent conflict, suggest alternative solutions, and return the drawings to the general contractor, who in turn will send them on to the architect for review, comments, or approval.

If conflicts are uncovered, the general contractor will consult with the architect and appropriate engineers to develop alternative solutions that hopefully will have no additional cost impact. When changes are required that impact costs, the respective subcontractors may have cause for extra expenses to resolve those conflicts.

Effective claim development and preparation

Construction claims are complex because they usually occur over many months and involve multiple parties. Because of this gradual ascent from minor irritation to one with significant cost implications, some of the background for the claim may have gone undocumented by the owner. The first order to business is recognize as early as possible that something is not right. Quite often a general contractor will dismiss a minor glitch as not having any consequence later down the road. Although a claim flag should not be raised for every bump in the road, making note of those events could help if more bumps are encountered.

The first line of defense is the general contractor, who, while serving two masters—her company and the owner—must be alert to the rumbling of a claim by one or more subcontractors and step into the breach to either defuse the claim or deal with it quickly and effectively. If justifiable costs are involved, the general contractor must bring it to the owner's attention for resolution.

The architect and engineers, being very familiar with the terms and conditions of the construction contract, can often dissuade potential claimants by referring to provisions in the contract that protect the owner. Compromise is essential when claims are submitted. Normally, the claim is justified, although not in the amount set forth by the claimant. The task is to determine which portions of the claim are substantive and the value of those portions. This is the basis upon which negotiations can begin to resolve the claim.

As we stated previously, parties often leave negotiation sessions unhappy with the results. The involvement of lawyers and claims consultants is a costly process, and the longer a claim lingers, positions harden and resolution becomes more difficult. Rapid and reasonable actions are the key ingredients in settling a claim or dispute.

Basic construction components

11

When owners attend design and construction meetings, quite often the contractor or the architect may use technical terms and construction components that may be confusing to nontechnical participants. This chapter provides owners with a primer in construction terminology and acquaints them with some of the basic construction components they will encounter during the course of their project. In this section we discuss the following:

- Site work and site utilities
- Building foundations
- Concrete construction, both cast-in-place and precast
- Structural and miscellaneous steel
- Masonry, both brick and concrete masonry units (CMUs)
- Glass and glazing
- Roofing
- Drywall partitions
- Doors and frames
- Finishes, both paint and vinyl wall coverings
- Plumbing
- Heating, ventilation, and air conditioning (HVAC)
- Fire protection
- Electrical

SITE WORK

Site work extends from the exploration of a planned new construction site to the work associated with preparing that site for the building to the installation of all new site utilities: gas, water, sewer, paving, and landscaping. Site exploration is required prior to the building's structural design, since the information uncovered will determine the

©2010 by Elsevier, Inc. All rights reserved.
Doi_No = 10.1016/B978-1-85617-548-7.00011-2

bearing capacity of the soils, which in turn will determine the size and type of foundation required for the new structure. These explorations, which are performed by a civil engineer, may also uncover the presence of rock strata that may need to be removed and may also indicate the presence of underground water or springs that must be controlled.

This geotechnical report consists of test borings (Figure 11-1) drilled at various points around the site to discern subsurface conditions. This geological technician's (geotech) report not only provides the structural engineer with foundation design information but is also used when bid documents are prepared so contractors can be made aware of existing conditions where those borings were taken. Test pits can be dug with an excavator if a larger portion of underground strata is required for soils analysis.

Site work includes "clearing and grubbing"—removing trees in the area where the building will be placed and clearing out all brush and unwanted vegetation overgrowth. Rough grading will then commence, and topsoil will be stripped from the building area and stockpiled for respreading later in the project. Excavation and backfilling operations will include foundation work, underground utilities, and surface preparation for concrete and asphalt paving.

FOUNDATIONS

Poured-in-place concrete foundations are easily recognizable and are applicable when soil-bearing capacities are appropriate. When soil-bearing capacities cannot adequately support the building to be placed on them, other forms of foundation design are employed, most frequently piles. The two basic types of piles are friction piles and end-bearing piles.

A friction pile depends on the friction created along the entire surface of the pile—that is, around its circumference and its length—to support the force exerted by the building to be placed upon them. An end-bearing pile depends on reaching a depth where the pile will come to bear on a surface that can support its load. There are various types of materials used for piles:

- Timber piles, generally treated with a preservative

- Precast-concrete piles—concrete piles produced under controlled conditions in a precast concrete factory

- Cast-in-place concrete piles, which usually require a form—a shell—into which concrete is poured at the job site

- Steel H piles—special steel columns with a cross section approximating an "H" shape

- Steel pipe piles—cylindrical steel piles

Engineers and Scientists	Stratford cort			Baring No ___1___
SML Geotechnical	Uncasvile,CT			Page _1_ of _1_ File No. 50512 Chkd. By: JHB

Baring Co. ABC Drilling		Casing	Sampler Split Spoon	Date	Groundwater Readings				
Foreman Torm Jones	Type		Split Spoon	Date	Time	Depth	Coating	Stab. Time	
Rep.	I.D./O.D.		1 3/8"/2"	1/8/91	1420	*	out	none	
Date Start 1/8/91 End 1/8/91	Hammer Wt.		140 Lbs.						
Location NE section of proposed building	Hammer Fall		30 in.						
GS.Elev. 143 ± Datun NGVD	Other		2X" HSA AX						

DEPTH	CBALSOHWGS	Sample Information				Field Testing (ppm)	SAMPLE DESCRIPTION & CLASSIFICATION	Stratum Description	REHKS	Equipment Installed
		No.	Pen./Rec.	Depth (Ft.)	Blown/6"					
		S-1	24/18	0-2.0	1-3-5-8		Top 6" Loose, brown TOPSOIL, Next 6" Loose, brown, fine SAND, Little Silt. Bottom 6": Medium dense, brown, fine to medium SAND, Little fine Gravel, trace Silt.	TOPSOIL		
								0.5' LOOSE SILTY SAND		
								1.0' MEDIUM DENSE SAND		
5		S-2	18/12	5.0-6.5	16-28-70		Very dense, gray and brown, fine to coarse SAND, Scot fine to Coarse Gravel, troce Silt.	5.0' VERY DENSE SAND		
10		S-3	5/5	10.0-10.4	100/5ᴴ		Very dense, gray, weathered ROCK.	10.0' VERY DENSE WEATHERED ROCK		
15		C-1	60/34	13.5-18.5	7 min/ft. 8 8 7 6		GNEISS (ROO=0)	13.5' BEDROCK		
20								18.5' E.O.8.		
25										

REHARKS	
1.	Auger refusal at 13.5'.
*	Water used for coring altered the water level.
Note:	Ground surface elevation is interpolated from topographic plan. Borings were located in the field by taping from existing site features.

Stratification Lines represent approximate boundaries between soil types, transitions may be gradual. water level readings have been made at times and under conditions stated. Fluctuations of groundwater may occur due to factors other than those present at the time measurements were made.

Boring No. 1

FIGURE 11-1

A typical soil test boring. This one shows drilling to a depth of 18.5 feet, finding weathered rock at a depth of 10 feet, bedrock at 13.5 feet, and no underground water.

- Minipiles or micropiles—small-diameter steel piles, bored into the soil and then filled with grout, a cementitious material

- Caissons—typically large-diameter drilled or augered holes filled with concrete

SITE UTILITIES

The installation of underground site utilities encompasses potable (drinking) water, incoming electrical power, sanitary sewers and manholes, storm sewers and storm inlets, gas mains, and fire-protection mains. Sanitary sewer and storm sewer lines are installed to allow for gravity flow and potable water, and fire-protection mains are under pressure and generally require pressure-reducing valves to regulate the pressure and volume of water after it enters the building.

High-voltage electrical conduits (another name for pipe), usually plastic, are often encased in concrete to prevent damage if inadvertently uncovered at some later date. A variety of conduit materials are available for each of these underground utilities:

- Potable water—copper pipe, cement-lined steel pipe, cast-iron pipe, ductile iron pipe, plastic, cross-linked polyethylene

- Electrical cable—PVC or metallic conduit

- Sanitary sewers—PVC, cast iron, ductile iron, manholes (generally precast-concrete segments that fit together, although some are constructed of concrete masonry units)

- Storm sewers—concrete pipe, PVC, corrugated metal, high-density polyethylene, manholes (similar to sanitary with storm sewer inlets placed around the site)

- Gas mains—steel, corrugated stainless steel

- Fire protection—steel, cement-lined ductile iron

CONCRETE

This seemingly universal material is really quite complex. Concrete is a combination of Portland cement, fine and coarse aggregate (sand and crushed stone), and water. Concrete is very strong in compression (density) but weak in tensile (bending) strength, which is why reinforcement is required. Steel rods, called rebars, placed in the lower portion of a concrete slab prevent that slab from cracking when subjected to that tensile bending force. When rebars are placed in concrete walls, flexural strength is provided. The strength of concrete is measured in pounds per square inch (psi) of compressive strength—that is, a crushing force. The strength of concrete depends on the following:

- The precise amount of water; too much and the strength will weaken
- The amount of Portland cement

- The type, size, and shape of aggregate
- The proper curing procedures when placed in normal, hot, or cold environments

Concrete comes in three types:

- Normal weight, which is cast in place at the job site with standard stone ingredients
- Lightweight, which is cast in place at the job site with special lightweight aggregates, often used to raise the level of an existing concrete slab while limiting the amount of weight of that overlay
- Precast, which is high-strength concrete with steel reinforcement; manufactured in an offsite plant to specific architectural and engineering dimensions and specifications and often steam-cured to attain high strength—a process known as "autoclaving." (Precast concrete is further divided into "structural" slabs, used as floor or roof panels or interior wall sections, and "architectural" concrete, cast in a variety of forms that will provide a decorative finish when used as exterior wall panels.)

Concrete has the following characteristics:

- Curing: As concrete cures, a chemical reaction takes place called hydration. If water disappears too rapidly via evaporation (in hot temperatures) or freezes when exposed to cold weather, the ultimate strength of the concrete will suffer. In hot weather, the contractor either mists the freshly placed concrete or places a protective cover over it to reduce evaporation. In the winter, an enclosure will be formed and heated to protect the concrete and keep the water from freezing.
- Concrete shrinks as it cures, so joints are created to prevent random cracks from appearing as it cures. These joints are called control joints and are placed specific distances apart at the direction of the design engineer. This differs somewhat from a construction joint, which is a wood or steel bulkhead placed after a floor slab has been poured when another pour is scheduled for the next day.
- Various chemicals and different types of aggregate can dramatically change the character of the concrete; these additives are called admixtures. Water-reducing admixtures improve workability of the concrete while increasing strength and reducing the amount of water required in the mix. High-range water-reducing admixtures, referred to as superplasticizers, allow up to a 30 percent reduction in water while creating a soupy, flowable mixture and retaining high strength.
- Accelerating admixtures speed up the "set," or curing time, of concrete, which is helpful in cold weather. Retarder admixtures slow curing time and are desirable in hot weather. Air-entrained mixtures create millions of microscopic bubbles to be formed in the concrete, which is desirable for use in exterior

concrete such as walks and curbs, since they give the water a place to expand during freezing weather. Fly ash and silica fume are two separate materials, and both, when added to the concrete, will greatly increase its strength.

Several events take place when each batch of concrete is delivered by truck: a slump test and the preparation of sample cylinders. The slump test involves taking a sample of concrete from the truck and placing it in a cone-shaped form; the form is then inverted over a flat surface, allowing the concrete to flow out into what is called a "slump," the height of the resultant pile. This height is then measured; three to four inches are considered ideal, but a shorter slump may indicate too much water in the mix, which can weaken the concrete.

Samples of fresh concrete taken from the chute of the delivery truck will be used to fill at least three cardboard or plastic cylinders, each approximately 6 inches in diameter and about 16 inches high. These three cylinders will be allowed to cure (age)—one for 7 days, one for 14 days, and one for 28 days—at a testing laboratory, where they will be subjected to a compression (crushing) test to determine if they meet design strength. It is not unusual for the 7-day test to be slightly below design strength, while the 14- and 28-day tests exceed the design strength for a well-proportioned batch of concrete.

STRUCTURAL STEEL

Iron was produced until the mid-1880s, after which Henry Bessemer, an English metallurgist, developed a process to force oxygen into the smelting process to burn away impurities and create steel, a more malleable metal. Today, several grades of steel are produced by adding other materials such as nickel, sulfur, manganese, carbon, molybdenum, vanadium, and phosphorous to enhance specific performance qualities. These various grades of steel are assigned numerical designations by the American Society of Testing and Materials (ASTM); the two most prevalent grades for building construction columns and beams are Grade 36 and higher-strength Grade 50.

Basic structural steel shapes are I-beams, channels, and angles (Figure 11-2). Wide-flange beams and columns (Figure 11-3) are the most commonly employed shapes in a building's structural steel framework. These wide-flange beams and columns are manufactured in a variety of sizes and strengths to meet the engineer's load requirements. Wide-flanged structural members are identified by the depth of the beam and its weight per lineal foot, so a W12 × 84 refers to a beam or column with a 12-inch depth weighing 84 pounds per lineal foot. When engineers refer to these steel members, they often call them "rolled sections" because of the way they are manufactured; a rough steel blank is rolled back and forth between rollers at the steel mill to exert pressure to reshape the material into the desired configuration.

The steel columns and beams are joined together with either welded connections (Figure 11-4) or bolted connections (Figure 11-5) or a combination of both. Open-web steel joists (Figure 11-6) are frequently used as intermediary support members in multistory steel frameworks to support metal decking (Figure 11-7) that is used as a form for cast-in-place concrete floors or roof support.

Steel construction has many advantages. It is competitively priced, adaptable to many design innovations, and has the ability to erect its framework in all but severe weather. Structural steel requires protection from fire, which at high temperatures causes softening and collapse. Steel is "fireproofed" by either applying a spray fireproofing material or encasing the steel members in a fire-resistant material such as gypsum drywall or masonry.

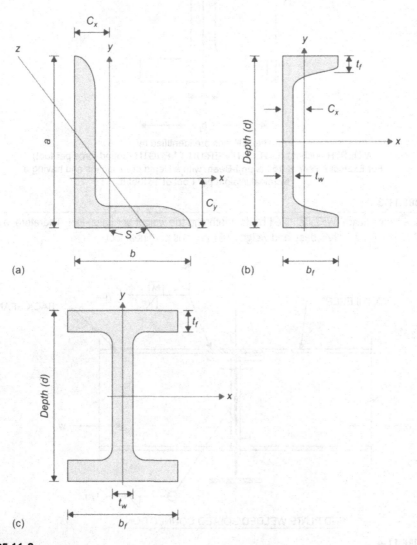

FIGURE 11-2

Basic Structural steel shapes: I-beams, channels, and angles.

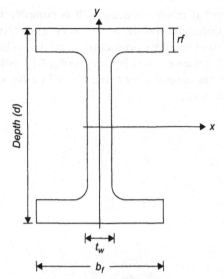

The I-Beams are identified by:
W DEPTH (inches) × WEIGHT PER UNIT LENGTH (pound force per foot)
For Example: **W27 × 161** is an I-Beam with a Depth of 27 inches and having a
Nominal Weight per Foot of 161 ibf/ft.

FIGURE 11-3

A wide flange beam (WF) identified by its depth and the weight per lineal foot. Therefore, a
W27 × 161 is 27 inches deep and weighs 161 pounds per lineal foot.

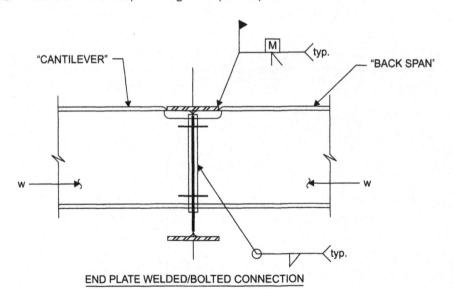

END PLATE WELDED/BOLTED CONNECTION

FIGURE 11-4

A welded beam connection.

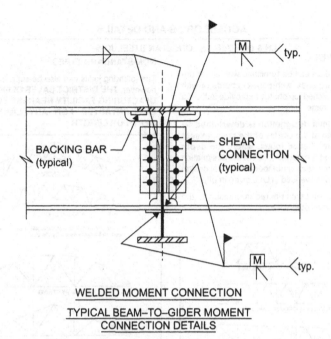

WELDED MOMENT CONNECTION

TYPICAL BEAM–TO–GIDER MOMENT
CONNECTION DETAILS

FIGURE 11-5

A bolted connection; this is a beam to girder connection called a "moment" connection.

MASONRY

The four basic components of masonry construction are brick, concrete-masonry units (CMU), mortar, and wall reinforcement. Masonry construction is often used in residential foundations and in interior and exterior walls in commercial construction. Landscape architects use decorative masonry products to build low walls, pavers, and walkways. Mortar is the glue that holds the masonry units together to achieve structural and architectural integrity, and reinforcement is added to provide greater strength and seismic stability.

Bricks

Bricks are familiar to everyone, and they come in a variety of shapes, colors, and sizes, from common brick (4 inches by $2^3/_8$ inches by 8 inches) to roman brick (4 inches by 2 inches by 12 inches) to utility brick (4 inches by 4 inches by 12 inches). Bricks can be positioned in a wall to create various patterns and wall treatments. Figure 11-8 reflects various brick positions, and Figure 11-9 shows how these positions together with a variety of colors can produce dramatic architectural effects.

ACCESSORIES AND DETAILS

LH & DLH SERIES LONGSPAN STEEL JOISTS

STANDARD TYPES

Longspan steel joists can be furnished with either underslung or square ends, with parallel chords or with single or double pitched top chords to provide sufficient slope for roof drainage.

The Longspan joist designation is determined by its nominal depth at the center of the span, except for offset double pitched joists, where the depth should be given at the ridge. A part of the designation should be either the section number or the total design load over the design live load (TL/LL given in plf).

All pitched joists will be cambered in addition to the pitch unless specified otherwise.

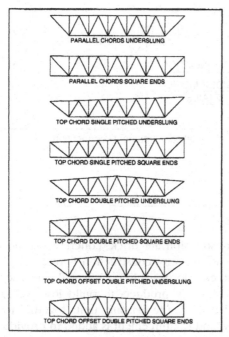

CAMBER

Non-Standard Types: The design professional shall provide on the structural drawings the amount of camber desired in inches. If camber is not specified, Vulcraft will use the camber values for LH and DLH joists based on top chord length.

Standard Types: The camber listed in the table will be fabricated into the joists unless the design professional specifically states otherwise on the structural drawings.

NON-STANDARD TYPES

The following joists can also be supplied by Vulcraft, however, **THE DISTRICT SALES OFFICE OR MANUFACTURING FACILITY NEAREST YOU SHOULD BE CONTACTED FOR ANY LIMITATIONS IN DEPTH OR LENGTH.**

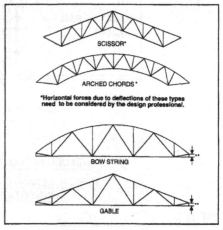

**Contact Vulcraft for minimum depth at ends.

CAMBER FOR STANDARD TYPES

LH &DLH series joists shall have camber in accordance with the following table:***

Top Chord Length		Approx. Camber	
20'-0"	(6096 mm)	1/4"	(6 mm)
30'-0"	(9144 mm)	3/8"	(10 mm)
40'-0"	(12192 mm)	5/8"	(16 mm)
50'-0"	(15240 mm)	1"	(25 mm)
60'-0"	(18288 mm)	1 1/2"	(38 mm)
70'-0"	(21336 mm)	2"	(51 mm)
80'-0"	(24384 mm)	2 3/4"	(70 mm)
90'-0"	(27432 mm)	3 1/2"	(89 mm)
100'-0"	(30480 mm)	4 1/4"	(108 mm)
110'-0"	(33528 mm)	5"	(127 mm)
120'-0"	(36576 mm)	6"	(152 mm)
130'-0"	(39621 mm)	7"	(178 mm)
140'-0"	(42672 mm)	8"	(203 mm)
144'-0"	(43890 mm)	8 1/2"	(216 mm)

*** NOTE: If full camber is not desired near walls or other structural members please note on the structural drawings.

FIGURE 11-6

Open-web steel joists.

_/ **VULCRAFT**

NON-COMPOSITE & COMPOSITE DECK DETAILS

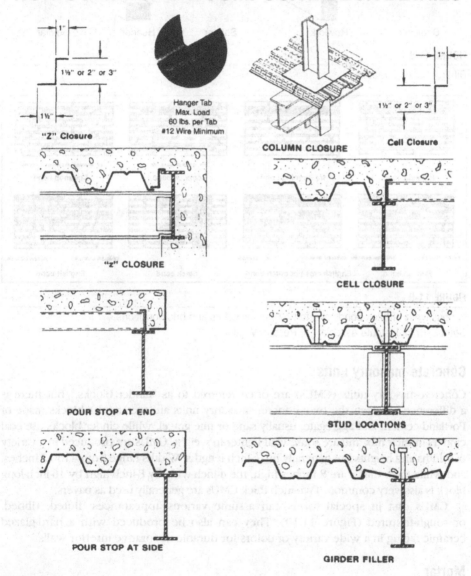

FIGURE 11-7

Profile of a metal deck that will receive a cast-in-place concrete floor.

Source: By permission of Nucor Vulcraft® © 2008, Charlotte, North Carolina.

VARIED BRICK POSITIONS

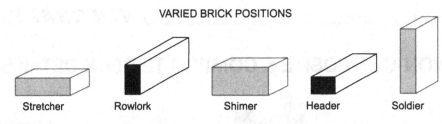

| Stretcher | Rowlork | Shimer | Header | Soldier |

FIGURE 11-8

Different brick positions.

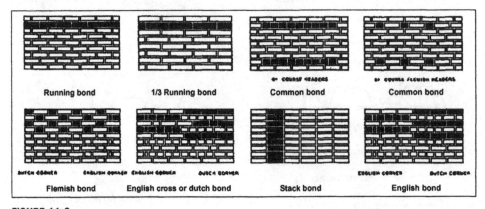

FIGURE 11-9

Various types of brick bonds created by varying colors and brick positions.
Source: Brick Institute Association, Reston, Virginia.

Concrete-masonry units

Concrete-masonry units (CMUs) are often referred to as "cinder blocks," but there is a difference between the two: concrete-masonry units are rectangular blocks made of Portland cement and aggregate, usually sand or fine gravel, while cinder blocks use coal cinders in their mix and are somewhat lighter in weight. CMUs are produced in a variety of widths and heights but generally in 16-inch lengths. Widths range from 4 to 12 inches, and while most CMUs are 8 inches high, the 4-inch-wide by 8-inch-high by 16-inch-long block is also very common. Two-inch-thick CMUs are generally used as pavers.

CMUs cast in special forms can assume various appearances: fluted, ribbed, or rough-textured (Figure 11-10). They can also be produced with a hard-glazed ceramic facing in a wide variety of colors for durable decorative interior walls.

Mortar

The correct type and proportion of ingredients and the proper application of mortar are essential for the structural integrity of a block or brick wall. Mortar serves many functions:

■ It bonds the masonry units together.

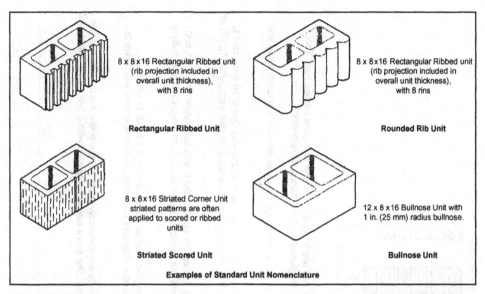

8 x 8 x 16 Rectangular Ribbed unit
(rib projection included in
overall unit thickness),
with 8 rins

Rectangular Ribbed Unit

8 x 8 x 16 Rectangular Ribbed unit
(rib projection included in
overall unit thickness),
with 8 rins

Rounded Rib Unit

8 x 8 x 16 Striated Corner Unit
striated patterns are often
applied to scored or ribbed
units

Striated Scored Unit

12 x 8 x 16 Bullnose Unit with
1 in. (25 mm) radius bullnose.

Bullnose Unit

Examples of Standard Unit Nomenclature

FIGURE 11-10

Decorative types of concrete-masonry units: ribbed, rounded rib, striated, and bullnose.
Source: Featherlite Building Products, a division of Acme Brick, Fort Worth, Texas.

- It assists in retaining a level wall, using the mortar bed as compensation for the small dimensional differences between one brick or block and another.

- It bonds installed wall reinforcement together into one structural unit.

By varying the color of the mortar or the type of tooled joints, mortar provides an additional aesthetic appeal. Figure 11-11 shows a cross section of typical mortar joints in masonry walls.

Mortar is a mixture of cement, sand, and water, mixed to the consistency of a heavy paste. This workability or plasticity of mortar provides both cohesive and adhesive properties. There are five basic type of mortar, each applicable to specific functions of the masonry wall:

- Type M mortar is a high-compressive-strength mix providing greater durability, generally used on unreinforced masonry walls below grade.

- Type S mortar is also a high-strength mortar with slightly less compressive strength than Type M. It has greater tensile strength and is the product of choice when reinforced masonry walls above grade are built.

- Type N mortar is a midrange compressive-strength product often used on interior non-load-bearing masonry walls.

- Type O mortar has lower compressive strength than Type N and is also used for interior non-load-bearing walls.

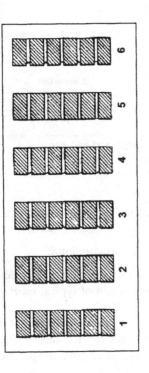

Concave Joint (1) and V-Shaped Joint (2). These joints are normally kept quite small and are formed by the use of a steel jointing tool. These joints are very effective in resisting rain penetration and are recommended for use in areas subjected to heavy rains and high winds.

Weathered Joint (3). This joint requires care as it must be worked from below. However, it is the best of the troweled joints as it is compacted and sheds water readily.

Struck Joint (4). This is a common joint in ordinary brickwork. As American mechanics often work from the inside of the wall, this is an easy joint to strike with a trowel. Some compaction occurs, but the small ledge does not shed water readily, resulting in a less watertight joint than joints (1). (2) or (3).

Rough Cut or Flush Joint (5). This is the simplest joint for the mason, since it is made by holding the edge of the trowel flat against the brick and cutting in any direction. This produces an uncompacted joint with a small hairline crack where the mortar is pulled away from the brick by the cutting action. This joint is not always watertight.

Raked Joint (6). Made by removing the surface of the mortar, while it is still soft. While the joint may be compacted, it is difficult to make weather-tight and is not recommended where heavy rain, high wind or freezing is likely to occur. This joint produces marked shadows and tends to darken the overall appearance of the wall.

Colored mortars may be successfully used to enhance the patterns in masonry. Two methods are commonly used: (1) the entire mortar joint may be colored or (2) where a tooled joint is used, tuck pointing is the best method. In this technique, the entire wall is completed with a 1-in. deep raked joint and the colored mortar is carefully filled in later.

FIGURE 11-11

Cross section of mortar joints in a masonry wall.
Source: Brick Institute Association, Reston, Virginia.

- Type K mortar has the lowest compressive strength and can be used for some non-load-bearing walls if local building codes allow.

Mortar is similar to concrete and shares two basic components: cement and water. Like concrete, the rapid dissipation of water in hot weather and the freezing of water in cold weather must be avoided to provide a high-quality masonry wall.

Wall reinforcement

Most load-bearing walls above grade require reinforcement to provide the flexural strength required, as these masonry walls expand and contract due to temperature changes, loads imposed by strong winds, and the weight of the wall itself. Reinforcement also provides stability of CMU walls during seismic events. Walls can be solid masonry or "veneer" or "cavity" walls.

Cavity wall construction often consists of a brick outer wall, an air space, and a structural inner wall of steel or wood studs faced with gypsum board. These two wall types must be joined together to provide structural integrity and steel wall reinforcement. Reinforcements can be truss and ladur types or wall ties that tie the outer wall to the inner wall.

Sections of masonry walls are known as wythes; a single wythe is one masonry wall thick, and a double-wythe wall consists of one outer and one inner wall, often of different masonry types: one brick (outer) and one CMU (inner).

GLASS AND GLAZING

Fenestration is a term associated with the design and placement of windows and other exterior wall openings. Many commercial buildings have fixed-glass windows, as opposed to operative windows, to better control the quality of the interior working environment. These fixed *lights,*—the term used for architectural glass and window glazing—can be of three basic types:

- *Annealed,* commonly used in architectural glass because it does not produce distortion, which can occur when the glass is tempered. Annealed glass has very good surface flatness but tends to break into large, sharp shards upon strong impact.

- *Heat-strengthened* glass has about twice the strength of annealed glass and is more resistant to wind loads and thermal stress, but it does produce some distortion during the heating process. Heat-strengthened glass will break just like annealed glass.

- *Fully tempered* glass is four times stronger than annealed glass, imparts some distortion, and breaks into small, slightly rounded fragments upon impact.

Specialty glasses also come in many varieties:

- *Laminated* glass is made by adhering two lights of glass together with a clear plastic interlayer, thereby preventing glass shards from being distributed when it is fractured. Laminated glass also provides some protection from ultraviolet-ray penetration and has some acoustical qualities.

- *Tinted* glass distributes color uniformly throughout its surface and both looks attractive and provides protection from ultraviolet rays.

- *Coated* glass is glass that has been coated with a reflective or low-emissivity (Low-E) coating to reduce thermal absorption from the sun's rays, thereby lowering the heat load on the surface of the building where it is installed.

- *Insulated* glass is different from double-pane glass. It incorporates a vacuum in the dead space between the inner and outer layers before they are sealed together. *Double-pane* glass, also called double glazing, is simply an outer and inner pane installed in a frame with an air space in between but no vacuum, sealed only by the frame (wood, metal, or vinyl) in which it has been installed.

ROOFING

Most commercial buildings nowadays use what is called a single-ply membrane roofing system. The most commonly used material for this membrane is EDPM—ethylene propylene diene monomer. Available in various thicknesses and colors, this single-ply membrane is installed over rigid roof insulation in two ways. In ballasted installation, the membrane is weighted down with small-diameter smooth-edged stones such as river gravel. The membrane can also be attached to the roof substrate with mechanical fasteners. The advantages of single-ply roofing are many: It can be applied quickly, it has a long life, and tears or damage to the surface can be spotted easily and repaired quickly using solvent or heat welding.

Built-up roofing surfaces require the installation of several layers of two-, three-, or four-ply roofing felt, an interlocking material with vegetable fibers held together with a binder. Prior to each layer being installed, the underlying surface is coated with a hot asphalt liquid. These built-up roofs also enjoy long life, are priced competitively, and can be easily repaired if damaged.

Roll roofing is a roll of coated felts, either smooth-faced or mineral-surfaced, that is rolled out on the roof surface in sheets about three feet wide. Edges are over-lapped and either fastened with roofing nails or mopped with a liquid asphalt. Roll roofing is an inexpensive roofing system that is often used on small utility sheds.

Metal roofing, such as standing seam roofing, is often selected for decorative pur-poses for shed-type sloped-roof configurations. High-performance painted steel or aluminum makes up the bulk of metal roofing systems. Shingles, wood, composition, metal, and ceramic are rarely used in commercial construction but can be used on sloped roofing surfaces.

DRYWALL PARTITIONS AND CEILING CONSTRUCTION

Partition framing can be constructed with either wood or metal framing studs of two-, four-, and six-inch nominal widths. When various combinations of gypsum wallboard (drywall) are attached to this framing, the resultant partition can meet a multitude of fire-resistance, moisture-resistance, and sound-resistance standards.

Fire-rated partitions

Figure 11-12 shows a wall assembly with a $3^5/_8$- or 6-inch metal stud with one layer of drywall on each side, a typical non-fire-rated partition wall. To achieve a one-hour or higher fire rating using metal studs, other types and layers of gypsum board must be installed:

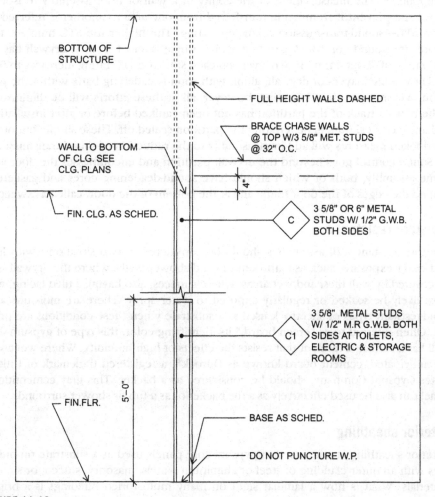

BOTTOM OF STRUCTURE

FULL HEIGHT WALLS DASHED

BRACE CHASE WALLS @ TOP W/3 5/8" MET. STUDS @ 32" O.C.

WALL TO BOTTOM OF CLG. SEE CLG. PLANS

FIN. CLG. AS SCHED.

4"

C 3 5/8" OR 6" METAL STUDS W/ 1/2" G.W.B. BOTH SIDES

C1 3 5/8" METAL STUDS W/ 1/2" M.R G.W.B. BOTH SIDES AT TOILETS, ELECTRIC & STORAGE ROOMS

5'-0"

FIN. FLR.

BASE AS SCHED.

DO NOT PUNCTURE W.P.

FIGURE 11-12

A Typical stud partition wall using either 3 5/8" wood or metal studs to just above ceiling height to extended to the structure above.

45-minute fire rating: one layer of 1/2-inch fire-rated sheetrock on each side of the stud
1-hour fire rating: one layer of 5/8-inch fire-rated drywall on each side of the stud
2-hour fire rating: two layers of 5/8-inch fire-rated drywall on each side of the stud
3-hour fire rating: three layers of 5/8-inch fire-rated drywall on one side of the
    stud and one layer of 1-inch fire-rated coreboard on the other side

Sound-rated partitions

To lessen the transmission of sound from one area or room to another, various sound-control systems can be instituted using stud, drywall, and other acoustical components. The measurement of the ability of a wall or floor assembly to isolate sound and prevent it from being transferred from one area to another is referred to as an STC—sound-transmission-coefficient rating. The higher the STC number, the greater the sound control. A partition with a single layer of $^5/_8$-inch drywall has an STC rating of about 44; with two layers on each side, the STC rating increases to 54.

These added layers of drywall, along with sound-deadening batts within the partition, will increase the STC rating; however, all of these efforts will be diminished if the bottom track of the partition has not been caulked before or after installation and any electrical outlets penetrating the partition sealed off. These simple but often overlooked measures will add immeasurably to the reduction of sound transmission.

Sound control goes beyond the drywall partition and encompasses the door and frame assembly, both of which should have sound-deadening cores and gasketing around the edges of the door frame and at the bottom of the door, called a "sweep."

Moisture resistance

Moisture-resistant wall assemblies should be considered in two situations: with limited water exposure, such as bathroom tub and shower walls, where the drywall will be covered by wall tiles, and wet areas, where surfaces, also having a tiled facing, will most likely be soaked or regularly exposed to water spray. There are moisture- and mold-resistant types of sheetrock used as a substrate when these conditions are present. Often referred to as "green board," its identifying color, this type of gypsum drywall has a surface treatment that resists the effects of high humidity. Where wet areas are anticipated, cement board known as Durock®, a registered trademark of United States Gypsum Company, should be considered as a backer. This gray cementitious panel can also be used effectively as a tile backer or as a tub or shower surround.

Exterior sheathing

Exterior sheathing is composed of gypsum-core panels used as a substrate on buildings with an outer cladding of steel or aluminum panels, masonry, stucco, or similar materials. What is now a familiar sight on many multistoried buildings is a bright yellow material called DensGlass®, manufactured by Georgia-Pacific. These gypsum-board panels have a yellow exterior fiberglass mat coating that provides a stronger resistance to wind loads and moisture.

DOORS AND FRAMES

In commercial construction, four types of doors and frames are used most often:

- Wood, used primarily for architectural and aesthetic value
- High-pressure laminate faced with engineered-wood-product cores, combining aesthetics and ease of maintenance
- Steel, used for utilitarian and security purposes
- Aluminum, combining architectural value with low maintenance

Wood doors can span the entire gamut of cost, function, and appearance from medium-density fiberboard (MDF) to exotic veneers, from non-fire-rated to two-hour fire-rated when installed in similarly rated frames, from low-sound-rating (STC) to high acoustical ratings when installed in the proper door frame.

Door components for each of these four types are much the same, consisting of stiles—the vertical side members; rails—the top and bottom components; the core—the center portion of the door, and the face—the visible panel (Figure 11-13). The door's core serves several functions, adding stability, strength, and, when mineral cores are used, fire and sound ratings. Wood doors can be either flush or contain various types of glass-panel inserts, referred to as "borrowed lights."

Hollow-metal doors and frames are used in offices, corridors, and utility and maintenance areas, and have a reputation for being both cost-effective and durable. Hollow-metal door frames are manufactured KD (knocked down), where the jambs and head must be assembled onsite, or "set up and welded," where the manufacturer welds the jambs to the head, ready for installation when shipped to the job. Figure 11-14a shows some of the stock hollow-metal door configurations available, and Figure 11-14b illustrates the various types of cores in those doors. Figure 11-14c shows sections through installed hollow-metal frames: abutting a masonry/concrete wall opening, wrapped around a steel or wood stud and drywall partition, and wrapped around a masonry/concrete door opening. Figure 11-14d illustrates the many custom configurations available for hollow-metal door and window frames. Figure 11-15 diagrammatically displays left- and right-hand door swings and what are called right-hand-reverse and left-hand-reverse swings.

FINISHES

Different types of wall finishes range from wood paneling to paint or wall coverings, generally vinyl, which are used in a commercial or institutional environment. Plywood veneer wood paneling may be specified for executive offices or conference room walls, with choices spanning wide cost and appearance levels. Interior paint products commonly used today are acrylic and alkyd.

Although the term *latex* is applied to some types of paint, the correct name would be *acrylic latex,* since the rubber-type additive is now replaced by a resin.

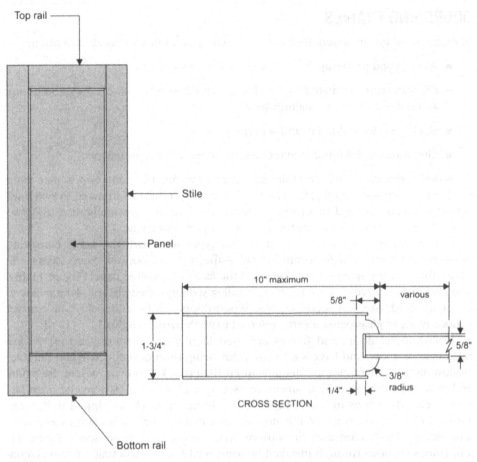

Top rail

Stile

Panel

Bottom rail

10" maximum

5/8"

various

1-3/4"

5/8"

3/8" radius

1/4"

CROSS SECTION

FIGURE 11-13

Typical wood door construction with top and bottom rails, stiles, and panel. Also, a cross section through a panel door where the panel is recessed from the stile.

This type of paint is easy to apply by brush or roller, dries rapidly, and presents a reasonably washable surface. Water-based acrylic paint is environmentally friendly and can be disposed of by washing down the drain. Alkyd interior paints are often referred to as "high-traffic" paints because they tend to resist normal wear and tear and are applied to moldings, doors, and high-touch areas.

Wiping stains, which are used over unfinished woods, can be brushed, sprayed, or wiped on, and after drying are usually sealed with a clear acrylic lacquer. "High-performance" coatings are available for long-life exposure to the elements that are applied to exterior steel or aluminum components.

Vinyl wall coverings, which are available in a wide range of colors, patterns, and textures, are divided into three groupings: Type I for light duty, Type II for medium duty, and Type III for heavy duty.

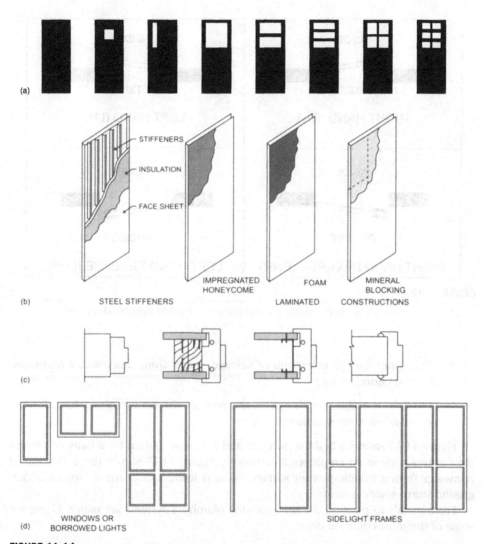

FIGURE 11-14

Hollow-metal doors. (a) Solid or with glass panels known as borrowed lights. (b) Core materials, including acoustical and fire rated. (c) Installation of hollow-metal door frame, butting a masonry opening and wraparound frames at a wood stud, a metal stud, and a masonry wall. (d) Custom hollow-metal door and glass panel (borrowed light).
Source: Steel Door Institute, Cleveland, Ohio.

PLUMBING

Building plumbing systems are comprised of potable-water supply and storm water and sanitary sewer effluent collection and discharge. Plumbing pipe layouts in a multistoried building consist of three basic components:

1. Piping for potable (drinking) water and piping for plumbing fixtures

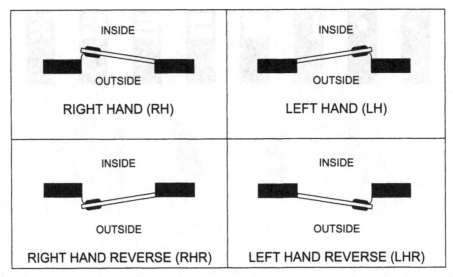

FIGURE 11-15

Determining a door swing: right and left, and left-hand-reverse and right-hand-reverse.

2. Waste lines for the collection of sewage, water from sinks, water fountains, and rain (storm) water

3. Vent lines to exhaust sewer gases and provide the necessary ambient air pressure to allow wastes to flow freely

Figure 11-16 shows a building's waste and vent line system for a bank of plumbing fixtures such as an employee's bathroom. Figure 11-17 shows the collection of rainwater from a building's roof surface via rain leaders discharged into an underground storm sewer system.

From metals to plastics, the materials for plumbing systems are many. These are some of the most common ones:

■ *Copper,* primarily used for potable-water supply lines and in some heating and cooling systems. Copper piping is available in a wide range of sizes from $1/8$ inch to 8 inches in diameter. Different types of copper piping have different wall thicknesses; Type K tube has the thickest walls, followed by Type L and Type M. Copper pipe can be joined by threading, soldering, brazing, or compression-type fittings.

■ *Cast iron* was first used in the United States in the early part of the nineteenth century. Later, cast-iron pipe was the material of choice for storm and sanitary waste pipes. Both rugged and durable, it was used for underground drain lines, and because it had soundproofing qualities, it was used in vertical storm drainpipe installation so the rush of flowing water would be somewhat silenced. Some of these applications have been replaced by plastic pipe, which is less expensive to purchase and install. Cast-iron pipe is made with hubs and hubless, each requiring a different joining method. Cast-iron pipe with a hub is joined with a compression push-on gasket, and hubless pipe and fittings are joined and connected with stainless-steel retaining clips.

■ The four most common *plastic* pipe materials are ABS (acrylonitrile-butadiene-styrene), PVC (polyvinyl chloride), CPVC (chlorinated polyvinyl chloride), and PE (polyethylene). All of these plastic pipes exhibit similar qualities, such as ease of installation, corrosion resistance, low frictional loss of liquids passing through them, longevity, and relatively low cost. These pipes, when installed above ground, generally require more support than metal pipes, and plastic pipe is not recommended for high-pressure air and gas usage. Plastic pipes can become brittle at low temperatures. Most plastic pipes are joined together by solvent welding—applying a liquid that makes the pipe surface tacky and seals the pipe after insertion into a fitting.

HEATING, VENTILATING, AND AIR CONDITIONING

Heat in buildings is accomplished by liquid (hydronic water) or steam generated from a central heating plant, such as a boiler or heat exchanger, and distributed by heating coils, radiators, or baseboard radiation. The central-heating source can also provide hot air to be distributed via an air-handling unit (AHU) and distribution ductwork.

Ventilation is required to control indoor air quality by exchanging the tempered air within the building with outside air, in the process filtering out air pollutants both inside and outside the building. This process is accomplished by air-handling units. These are some of the most common components in a commercial air-conditioning system:

■ Air handler (AHU)—either a roof- or exterior-mounted concrete pad, this device moves air through the building's ductwork system, exiting through terminal devices—grills and diffusers. AHUs also exhaust and recirculate the air through return-air ductwork and grills.

■ Chiller—this device contains an evaporator, condenser, compressor, and expansion valve. The evaporator, with a liquid refrigerant in its tube bundle, absorbs heat, and the refrigerant vapor is taken into the compressor, which pumps that vapor to the condenser, raising its pressure and temperature. The high-pressure liquid refrigerant then passes through the expansion device, which reduces the refrigerant's temperature and pressure as it flows over the chilled water coils, absorbing heat from the water in those coils.

■ Condensers—heat exchangers that condense a substance, usually a refrigerant such as Puron®, from a gaseous to a liquid state and in doing so give off latent heat, which will be transferred to the condenser coolant.

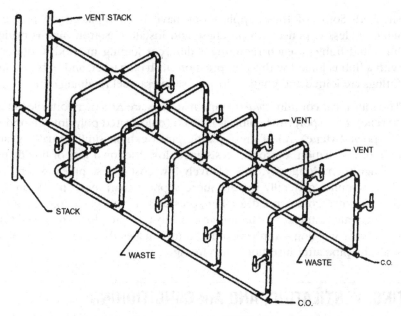

VENT STACK

VENT

VENT

STACK

WASTE

WASTE

C.O.

C.O.

Drainage for a Battery of Fixtures with a Wide Pipe Space Available.

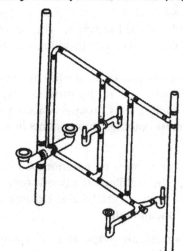

**PIPING FOR TUB, LAVATORY & WATER CLOSET
EACH FIXTURE VENTED**

**Typical Piping Arrangement for a Water Closet, Lavatory and Tub.
Piping May Be Either Hubless or Hub and Spigot.**

FIGURE 11-16

Typical building plumbing layout showing a bank of bathroom fixtures with waste and vent lines. Also, detail of tub, lavatory, and water closet waste and vent layout.

Source: By permission of the Cast Iron Soil Pipe Institute, Chattanooga, Tennessee.

TYPICAL LAYOUTS

Cast Iron Soil Pipe Underground to Point of Disposal

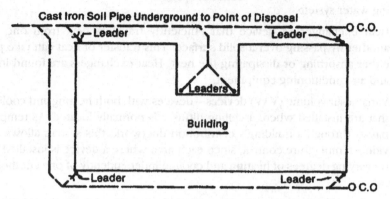

Roof Leaders and Drains Outside Building.

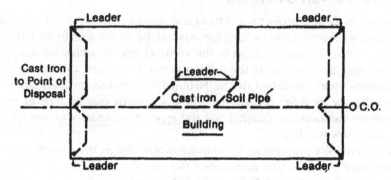

Roof Leaders and Drains Inside Building.

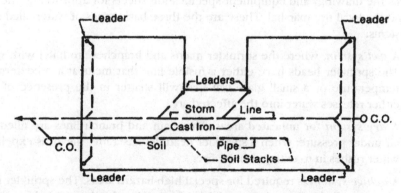

Combination Sewer (Sanitary and Storm) Where Permitted by Code.

FIGURE 11-17

Collection of a building's rain—storm water—from the roof rain leaders to below-grade cast iron piping with cleanouts.

Source: By permission of the Cast Iron Soil Pipe Institute, Chattanooga, Tennessee.

- Cooling tower—equipment that removes heat absorbed in a circulating cooling water system.

- Heat exchanger—a device that efficiently transfers heat from one fluid to another by passing over a solid surface. This transfer of heat can take place by either absorbing or dissipating the heat. Heat exchangers are found in boilers and air-conditioning equipment.

- Variable-air-volume (VAV) devices—devices with both heating and cooling coils that are installed where a ceiling diffuser is normally located. As tempered air passes through a building's distribution ductwork, this system allows for individual temperature control, since each area where a device is installed can call on varying degrees of heating and cooling independently of one another.

FIRE-PROTECTION SYSTEMS

In most instances, the project's mechanical engineer does not furnish a design for the building's fire-protection or sprinkler system; he or she usually includes what is called a performance specification in the contract specifications manual. This fire-protection specification directs the subcontractor to prepare a design to meet certain standards, one established by the National Fire Protection Association (NFPA) and the other by local fire regulations. These standards vary by type of protection mandated by the local fire marshal and the type of occupancy or storage requirements of the new building.

The selected fire-protection subcontractor will design the system with a computer software program that determines the coverage required to meet the performance specifications (and NFPA and local regulations). The subcontractor then prepares the drawings and equipment specification sheets for approval by the engineer and the local fire marshal. These are the three basic types of water-filled sprinkler systems:

1. A *wet system,* where the sprinkler mains and branches are filled with water. The sprinkler heads have either a fusible link that melts at a predetermined temperature or a small glass vial that will shatter in the presence of heat; either releases water into the affected area.

2. A *dry system* for unheated areas. The main and branch lines are filled with air under pressure; when a sprinkler head detects a fire, the air is expelled as water rushes in to extinguish the fire.

3. A *deluge system* is required for special high-hazard areas. The sprinkler heads are open all the time, and a special deluge valve that is actuated by its own control panel allows a rapid flow of water through the open heads to quickly extinguish the fire.

Foam suppression systems are also available, and commercial kitchens have an Ansul® fire-protection system installed in the hood over the cooking areas.

A typical fire-protection system will consist of the following:

- An incoming underground main, generally separate from the incoming potable-water main

- A fire pump to supply water at the proper pressure and volume to all areas in the building

- In multistoried buildings, a sprinkler riser emanating from ground floor to all upper floors; main and branch piping will extend from this riser

- A fire-detection alarm and control panel

- Fire hose stations

- Sprinkler heads that will be activated by either smoke or heat

Some typical sprinkler heads for both a dry and wet system are shown in Figure 11-18. Concealed pendants are installed with their head flush to the ceiling; extended pendants extend below the ceiling surface.

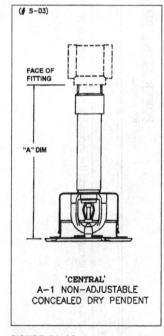

(# S-D3)

FACE OF FITTING

"A" DIM

'CENTRAL'
A-1 NON-ADJUSTABLE
CONCEALED DRY PENDENT

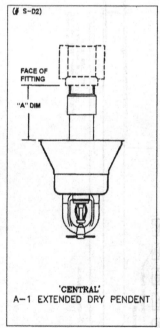

(# S-D2)

FACE OF FITTING

"A" DIM

'CENTRAL'
A-1 EXTENDED DRY PENDENT

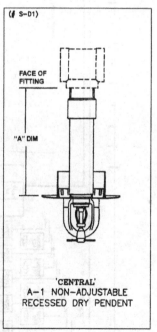

(# S-D1)

FACE OF FITTING

"A" DIM

'CENTRAL'
A-1 NON-ADJUSTABLE
RECESSED DRY PENDENT

FIGURE 11-18

Wet and dry system sprinkler heads.

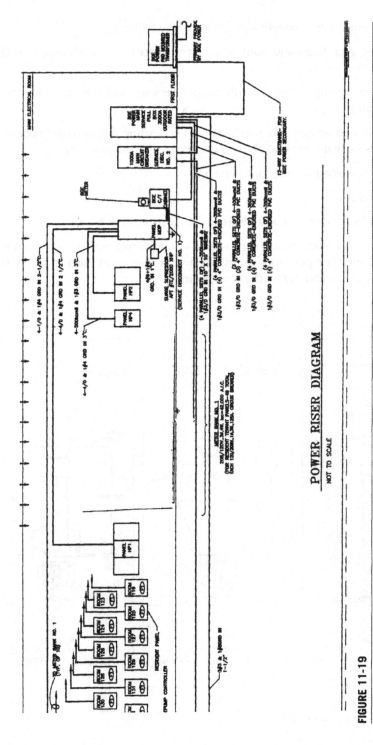

POWER RISER DIAGRAM

NOT TO SCALE

FIGURE 11-19

An electrical one-line drawing showing the incoming primary service on the right and the MDP in the building's electrical room.

ELECTRICAL SYSTEMS

A building's electrical system commences at the connection to the existing local utility company service, known as the primary service. Upon entering the construction site, this underground primary service connects to a transformer that is mounted on a concrete pad or installed in a transformed vault onsite. The electrical cables on the building side of the transformer are known as the secondary service; these are the cables that enter into the building. The secondary service connects to a main distribution panel (MDP), or switchgear, which then divides into other service panels, some for lighting and some for power panels. The lighting panels, designated LP, distribute electrical circuitry to the building's lighting system, and the power panels, designated PP, distribute electrical circuitry to the building's HVAC system and various motors and power-driven devices in the building.

Most buildings have two types of electrical systems: line voltage of 277 volts or 240/120 volts for lighting and equipment, and a low-voltage system for voice and data communication. Figure 11-19 shows a diagrammatical representation of a building's electrical-distribution system. This drawing presents a diagram of the pad-mounted transformer with primary service and secondary service entering the structure via a duct bank (series of electrical conduits bundled together) terminating at the main service switch in the electrical room. From there, the electrical cables are fed to disconnect switches with circuit breakers and then to a series of panels for power (PP) and lighting (LP). The project's electrical plans usually consist of one or more one-line drawings: one for power and lighting and one for low-voltage systems. On occasion, another one-line drawing may be included for a building's security system if it is not included in the voice/data scheme.

Glossary of architectural and construction terms

ABS plastic Acrylonitrile-butadiene-styrene pipe used for plumbing.

Access door A removable panel, usually small in size, installed in a ceiling or wall to allow access to a piece of mechanical equipment or plumbing pipes, or valves for inspection, maintenance, or repair.

ACI American Concrete Institute.

ACT Acoustical tile.

Acoustical duct lining A fiberglass blanket material installed as a lining in a sheet-metal duct to reduce noise transmission.

Actual total price The sum of the cost of the work plus the contractor's fee.

Addendum A supplement to the documents, issued prior to receipt of bids for the purpose of clarifying, correcting, or changing the bid documents.

Admixture A material other than water, lime, or cement used in either mortar or concrete to change its properties.

Aerator fitting A device that allows air into an existing stream of water.

Aggregate An inert granular material such as sand, stone, gravel, or ceramic particles, bound together in a mass by a matrix, forming mortar or concrete.

Air change A measure of the volume of air supplied to or exhausted from a building or room, usually expressed as cubic feet per minute (cfm).

Air content The amount of air voids in cement paste, mortar, or concrete, expressed as a percentage of total volume of the mixture.

Air curtain A stream of moderately high-velocity air directed downward across an opening to prevent the transfer of heat across the opening.

Air diffuser An air-outlet device located in the ceiling, wall, or floor containing deflecting devices to direct air supply and air flow in a space or room.

Air entrainment A chemical used in a concrete mixture to infuse million of microscopic bubbles in the mix to increase its resistance to damage during exposure to freezing weather.

Air leakage The volume of air that flows through a closed window or door or through the joints of ductwork.

AISC American Institute for Steel Construction.

Alligatoring Cracking or wrinkling of a painted or asphalt paving surface in a pattern similar to the hide of an alligator.

Allowable bearing pressure The maximum allowable pressure on a building's foundation that provides adequate support against rupture of the soil mass on which it rests.

Allowance An agreed-upon amount included in the schedule of values comprising the lump-sum or GMP contract total encompassing the cost of labor, materials, and equipment for a specific aspect of the work.

Alternate An alternate or substitute item or method of completing some aspect of work, which the owner may incorporate into the work at a predetermined cost identified in the contract.

Ambient noise An all-encompassing average of background noise associated with a specific environment.

Ambient temperature The temperature of the surrounding air.

Amperage The flow of electric current in a circuit.

Anchor bolt A steel bolt fixed in a building's foundation or superstructure with its threaded portion exposed to secure framework or equipment bases.

Anodize A process that provides a hard, durable surface on aluminum, either clear or colored, created by electrolytic action.

Appurtenance A built-in, nonstructural portion of a building, such as partitions, doors, or windows.

Apron That portion of a concrete slab that extends beyond the building's face; a flat piece of wood trim affixed to the base of a cabinet; the wood trim placed directly under a wood windowsill.

Architect's scale A scale or ruler having various divisions along its edge that, when applied to the lengths depicted on the project plans, can provide actual feet or meters from the reduced-size lines.

Areaway The open space between a row house and the sidewalk.

As-built drawings Construction plans (drawings) that present the work as it is actually installed rather than as presented on the plans (drawings).

Ashlar A range of quarry-faced stone or rectangular masonry units scarfed to resemble hacked stone and arranged in a wall in a random fashion.

ASHRAE American Society of Heating, Refrigerating, and Air-Conditioning Engineers.

ASME American Society of Mechanical Engineers.

Asphalt concrete A mixture of asphalt and aggregate used in paved parking areas.

Asphalt-seal coat A bituminous coating, generally without aggregate, applied to the source of asphalt paving to preserve the surface.

ASTM American Society for Testing Materials.

Astragal A plain or bead molding applied to one leaf of a double-leaf door assembly to hide the joint where the two meet in the center.

At-risk construction management A project-delivery system where the construction manager commits to deliver a project at a guaranteed maximum price.

Attic stock The extra materials the contractor is obligated to provide to the owner as defined in the contract documents. These materials would be stored in the "attic" until required and include extra acoustical ceiling pads, extra boxes of ceramic tile, cans of paint, spare hardware parts, and so forth.

Autoclave curing Curing of precast concrete by high-pressure steam at the manufacturer's plant.

AWG American Wire Gauge, a method of determining and identifying electrical conductors by size; the larger the number, the small the diameter of the conductor.

Awning window A window with a top-hinged sash, the bottom edges of which swing outward.

Backer board A sheet of plywood, oriented-strand board (OSB), or medium-density fiberboard (MDF) to which another surface is applied.

Backflow preventer A device fitted into a pipe that prevents water or other liquids from siphoning back into the system.

Backplate A metal plate that serves as a backing for a structural member.

Backset The horizontal distance from the edge of the door to the center of the hole bored for the doorknob.

Ballast The stone that weighs down the roofing membrane, keeping it from lifting off the surface of the roof under negative air-pressure flow over the roof's surface.

Balluster or baluster Any of the short vertical members in a stair rail that join the top rail to the bottom rail.

Bank-run gravel Material consisting of aggregate (stones) of various sizes and fines taken directly from its natural deposit.

Barrel bolt A door bolt where the cylindrical horizontal steel member slides within an outer metal casing to lock into a receptor at the door jamb.

Batt A section of fiberglass insulation.

Batten A narrow strip of wood or metal that spans two or more parallel panels or structural members.

Bay A regularly defined space repeating a division of a façade or structural column.

Beam A structural member made of steel, wood, or concrete that supports a load from above.

Bearing wall A wall that also acts as a support for a load imposed from above.

Bedding A layer of sand or crushed stone placed on top of an excavated ditch to serve as support or cushion for a buried utility line.

Benchmark A marked reference established by a survey crew to create a point from which all other height (elevation) measurements can be taken.

Beneficial occupancy Uhe use of a constructed facility by the owner when it reaches the stage of completion to allow for occupancy or use as designed.

Berm A mound of earth formed into a sloped rise higher than the corresponding elevation on each side of that mound.

Bevel siding Tapered wood siding where the joints overlap for weather protection.

Bid bond Assurance from surety (bond company) to pay the bond amount if the bidder defaults on a commitment to enter into a contract for the bid price.

Binder course As applied to asphalt paving, the course with a larger aggregate overtop to which the finish course of paving will be applied. Also known as the **base course**.

Blocking Pieces of wood or metal used to secure a face panel to a partition, ceiling, or wall; pieces of wood or metal placed in between two vertical wood or metal studs to provide stability.

Board and batten Wood siding consisting of vertical placed boards with narrow vertical strips (battens) installed at the board joints.

Board foot A unit of measure that takes into account the thickness and width of the board and how that measurement compares to a 1-inch board that is 12 inches wide.

Bond strength A measure of the resistance of mortar to allow separation of two units of masonry installed in a masonry wall.

Borrow When applied to earthwork, material taken or "borrowed" from one source for transport to another source, whether the source is on the site or brought from a source off- site.

Bottom rail The bottom horizontal member of a door or window that joins the other parts together.

Box beam A fabricated hollow beam, either wood or steel, rectangular or square in shape.

Branch piping Any part of a plumbing piping system other than a main riser or stack.

Breeching The duct or pipe that connects a boiler to an exhaust stack.

Brick anchor A metal device that secures a brick outer wall (veneer) to a structural backup.

Bridging Wood or metal cross-bracing installed between wood or steel joists to stiffen them.

Building paper A black asphalt–impregnated paper used as a moisture barrier behind a brick-cavity wall or placed under residential roof shingles.

Bulletin A supplement to the documents, issued after a construction contract has been executed, for the purpose of clarifying, correcting, or changing the contract documents.

Bus bar A rigid copper electrical conductor that provides a connection between electrical circuitry.

Butt joint A joint created when two pieces of material with square edges are joined to each other.

Caisson A round, watertight chamber, filled with concrete, used to create a foundation below existing water tables either on land or in a waterway.

Camber A slight curvature to a wood or steel member that is used to compensate for any loads imposed upon it, thereby returning to a true horizontal plane.

Cantilever A projecting beam, girder, or other structural member that is supported only on one end.

Cap flashing A waterproof, flexible metal sheet that seals the top of a cornice or a wall.

Carrier A steel framework installed behind the surface of a wall panel that provides structural integrity to a plumbing fixture attached to that wall.

Casement window A window with one or more vertical sashes that swings outward.

Cavity wall An exterior wall where the outer portion, generally masonry, is separated from the inner wall, creating an air space in between.

Certificate of occupancy A document issued by the local building official signaling that the structure complies with all applicable building codes, rules, and regulations and is fit for occupancy.

Chalk line A line made on a floor or wall by snapping a taut cord dusted with colored chalk against the wall or floor surface.

Chamfer A bevel usually cut at a 45-degree angle to create a corner of a masonry wall, a wood-paneled wall, or decorative wood trim.

Changed conditions Conditions or events that alter the circumstances on which the contract for construction was based.

Chase A recess in a wall or enclosure that allows vertical mechanical ductwork, plumbing pipes, or electrical conduits to traverse from floor to floor in a multistoried building.

Check A small crack that runs parallel to the grain in a wood stud or trim piece.

Cleanout A pipe fitting that allows access for cleaning.

Clerestory A glazed wall section at the top of an exterior partition or wall.

Closed-cell polystyrene or **polyethylene** A strong, rigid material with great resistance to air and water-vapor infiltration.

Coefficient of expansion The change in the dimensions of a material due to a change in temperature, expressed as a unit of dimension per degree change.

Cofferdam A temporary watertight structure installed around an area of water or water-bearing soil to allow construction to take place within this enclosure.

Cohesion The quality of some soil particles to attract and stick together.

Coin or quoin The stones or bricks that form the corner of a building.

Cold joint A joint formed when a concrete surface hardens before the next concrete pour takes place, thereby producing a poor bond.

Commissioning The start-up, testing, and adjustment process to certify to the design functioning of contractor-installed equipment.

Common bond (brickwork) Every fifth or sixth course is a header, and other courses are stretchers.

Compaction The process of closely packing soil particles by rolling or tamping to reduce voids and increase bearing capacity.

Completion percentage Percentage of work actually completed by the contractor as of the period covered by the application for payment.

Concealed suspension system A method of installing ceiling tiles where the suspension system is not visible.

Concrete-masonry unit (CMU) A brick or block cast of Portland cement and an aggregate, commonly referred to as a "cinder block."

Concrete plank A precast structural concrete panel primarily used as floor or roof decking.

Condenser a heat-exchange piece of equipment in which a refrigerant is liquefied by the removal of heat.

Conduit An empty pipe that will house electrical cables.

Consent of surety Written consent of the surety (bonding company) that all conditions relating to the contractor's bond (payment or performance) have been achieved.

Constructability The ease in which the design documents allow a project to be constructed and achieve its overall objectives.

Continuous beam A structural beam that extends over two or more supports.

Controlled fill Fill designated to become a bearing surface for the structure above, placed in layers, each of which is compacted to the engineer's specifications.

Contour line A line on a site plan that represents a point(s) of elevation on the ground. Proposed contours are those to be created; existing contours are those present prior to any grading operations.

Convection heating Heating provided by the movement of air, gas, or liquid through a device that transfers heat from a hotter surface to a cooler environment.

Corbel A bracket produced by a course of wood or masonry that extends outward in successive stages from the vertical wall surface.

Cornice The uppermost projecting portion of a vertical wall, frequently expressed with decorative brackets.

Cost of the work The actual direct costs of work, including general conditions, permits, and insurance but excluding the contractor's fee.

Counterflashing Formed metal or heavy coated fabric secured onto or into a wall to cover and protect the upper edge of metal or fabric flashing installed over the base flashing.

Coursing Masonry that is laid in place horizontally in a wall to form a pattern in that wall; a continuous row or layer of brick, concrete-masonry units, or stone.

Crawl space An interior space of limited height sufficient for workers to enter to perform work but of insufficient height to occupy, often used to conceal utilities as they enter or exit a building.

Cricket A small, valley-shaped area on a roof that channels water away from roof-mounted equipment or a chimney.

Critical-path method Referred to as CPM, a management technique used to plan and construct a project, combing all relevant information into a single plan. The critical path is the longest sequence of activities in a planning network that establishes the minimum length of time required to complete those activities.

Crook The warp of a board's edge, determined by striking a line between the two ends.

Crown the center elevation of a road surface that encourages drainage.

Cup The deviation in the face of a board from a horizontal or vertical plane.

Cut and cover A method of installing an underground utility or structure such as a tunnel by excavating a deep trench, laying or installing the utility or structure, and then backfilling it.

Dado A rectangular groove cut across the width of a board to receive the end of another board.

Datum Any level surface used as a plane of reference to measure elevation (height).

Deck The floor or floor forms of a building.

Deficient work Work that is deemed incomplete, insufficient, or exhibiting poor workmanship.

Deformed reinforcing bar A steel bar inserted into a concrete pour to add tensile strength. Ridges along the surface of the bar give the appearance of being deformed.

Density of soil The mass of solid particles in a sample of soil or rock.

Dentil One of a row of small blocks in a row used as a decorative frieze or cornice; it looks like a row of widely spaced teeth.

Design strength The load-bearing capacity of a structural member as determined by the engineer.

Dewater Pumping water from an excavation or substructure to maintain a dry environment.

Differential settlement The movement of different parts of a structure due to the uneven sinking or consolidation of the soil below.

Direct costs The costs incurred in the field directly attributable to the construction of the project, consisting of labor, materials, and equipment.

Direct digital control (DDC) A series of equipment controls in which all logic is performed by computers that send signals to the building's HVAC system to regulate the building's internal working environment.

Ditch witch A machine that creates a continuous narrow ditch for the installation of small-diameter underground pipe or conduits.

Door jamb The vertical members on each side of the door opening.

Door rail A horizontal door member that joins the two vertical side members (the stiles) together.

Dormer A structural extension of a building roof whose purpose is to provide light and added headroom in a half-story structure with a window in the vertical face.

Double-hung Refers to a window where the upper and lower sashes are movable; a single-hung window is where one sash is fixed and the other movable.

Dovetail A triangular tendon cut into the corners of a cabinet's side and rear panel that joins both pieces together, similar to the shape of a dove's tail.

Dressed lumber Lumber planed smooth on one or more surfaces.

Drip edge A metal flashing or plastic strip with an outward projecting lower edge, used to control water dripping from a roof edge or exterior windowsill.

Drip line An area around a tree that approximates the diameter of a circle encompassing the tips of its branches.

Dry-pack Forcibly ramming mortar with a stiff, claylike consistency into a confined masonry or concrete recess.

Dry-pipe sprinklers A sprinkler system installed in an area subject to freezing that releases water on command from a source outside of that area.

Drywall The common name for a gypsum-filled, paper-backed panel attached to a backer stud to build partitions.

Dutch door A door consisting of two separate leaves, upper and lower, operating independently.

Dynamic load Any load that is not static, such as a load imposed by wind.

Eave The lower portion of a sloped roof that extends beyond the exterior wall.

Efflorescence The white, saltlike appearance on the exterior of masonry walls caused by alkalis in the mortar mix leaching out onto the surface of the wall.

Effluent Sewage, treated or partially treated, flowing from sewage-treatment equipment.

Elastic limit The greatest stress that a material can stand without permanent deformation after release of that stress.

Elastic modulus Also referred to as **modulus of elasticity**; the limit to which an elastic material can be stretched and return to its initial shape.

Elephant trunk A cylindrical, flexible chute attached to a concrete transit-mix truck or to a concrete pump that allows dispersion of the concrete while preventing freefall that could cause aggregate separation.

Elevation An exterior face of a building as depicted on a plan or drawing.

EMT Electrical metallic tubing, also called "thinwall," lighter and less rigid than heavier metal conduit.

End-bearing pile A pile used as a structural support that rests on rock or other structurally sound strata.

Engineered-wood products A series of scrap-wood components formed under pressure with suitable binders to create framing lumber, panels, and trim pieces.

English bond Brick wall patterns with alternate courses of headers and stretchers.

Entrained air Microscopic voids created in exterior placed concrete by the use of a chemical additive that allows for room for expansion during freezing temperatures without damaging the concrete.

Estimated cost to complete A current assessment of remaining costs to be incurred to complete a construction project up to some point in time.

Evaporative cooling Cooling accomplished by evaporating water using a fine spray; the concept employed by cooling towers.

Expansion bolt A fastening device that fits into a predrilled hole and expands after insertion and tightening, thereby creating sufficient friction to provide holding power.

Expansion joint A joint or gap between adjacent parts of a building or building component that allows each part or component to expand and contract without damaging the adjacent part or component.

Expansion-joint cover A prefabricated plate that spans the expansion joint and offers protection from damage.

Façade The architectural face of the building.

Face brick Brick made especially for structural durability and maximum appearance quality.

Falsework Temporary bracing or cribbing to support work in progress under construction.

Fan coil unit A unit used in an air-conditioning system containing a heating and/or cooling component, an air filter, and a fan to distribute the heat or the cool air.

Fascia A flat, horizontal exterior or interior member that closes off the top of the section of wall where it is applied.

Fast track The process of dividing a project into segments or phases to permit one segment or phase to start before the entire design has been completed and to overlap construction of purchasing activities to proceed with the design phase.

Feeder An electrical conductor(s) originating at the switchgear that distributes power to panels located elsewhere in the building.

Fenestration The design and arrangement of exterior windows in a building.

Field order A written directive from an architect or engineer directing a contractor to proceed with work not generally involving an increase in cost or contract time.

Final completion When the work has been fully completed and approved by the architect in accordance with the plans, specifications, and other contract documents, including all punch list items.

Fine grading The grading required to provide finished contours as required by the contract; performed after initial rough-grading operations.

Fines The smallest soil particles (less than 0.002 mm).

Finial A crowning ornament of a pointed element such as a spire.

Fire rating The results of a laboratory fire-endurance test to determine the length of time it takes for a building component–wall, door, ceiling–to lose its integrity when surrounded by fire.

Fire resistance The ability of a material or component of construction to provide protection against fire.

Fishplate A wood or steel plate that adds to the structural strength when the two pieces are joined together.

Fissured soil Soil material having a tendency to break off along definite planes of fracture.

Flame-spread rating A numerical rating assigned to a construction material, representing its ability to resist flaming when exposed to fire.

Flange A projecting collar or rib along the length of a structural-steel beam or column.

Flanged union A circular disk on the end of a plumbing pipe, allowing two sections to be joined together by bolting through holes in each companion flange.

Flashing A thin sheet of metal or impervious fabric placed in mortar joints and around exterior doors, windows, and roof appurtenances to prevent water from infiltrating into the building.

Flemish bond A brick wall pattern consisting of headers and stretchers laid alternately in each course.

Float Contingency time included in a schedule of activities to be used in the event that minor unforeseen events occur that might otherwise impact the construction schedule.

Floor-to-floor dimensions The vertical dimension from one floor surface to the floor surface of the floor above.

Flue A noncombustible, heat-resistant chimney liner or a metal double-walled fire-rated assembly to carry away combustion from a furnace or boiler to the atmosphere.

Footcandle A unit of illumination that equals one lumen per square foot.

Force account Directed work performed by the contractor that exceeds the scope of the original contract.

Fret An architectural ornament composed of raised, incised continuous lines arranged in a rectilinear pattern.

Full bond A masonry wall where all bricks are installed as headers. *See* **Header**.

Furring strips Spacers, either wood or metal, affixed to a wall or ceiling that serve as points of attachment for wall or ceiling panels.

Gable The triangular section of an end wall of a gabled roof, also called the gable end.

Gambrel roof A roof with two slopes on opposite sides of the ridge.

Gasket A linear or circular strip of resilient material that provides a tight seal between the joined pieces.

Gel coat A thin outer layer of resin applied to a fiberglass component as its finish coat.

Geodetic survey A land survey where the curvature of the earth is taken into consideration when establishing a precise location.

Girder A principal structural member, either wood, concrete, or steel, used to support heavy loads.

Glue lam A manufactured-wood product consisting of several thin layers of wood, jointed together with adhesive under pressure, used as a floor or ceiling joist.

Gneiss A metamorphic rock.

Gooseneck A section of pipe or duct curved like the profile of a goose's neck.

Gothic sash A window sash composed of mullions that cross and form a pointed arch.

Gradient The degree of incline of a road, pipe, or other surface expressed as a percent.

Granular soil Gravel, sand, or silt with little or no clay content, with no cohesive strength.

Gravel stop A metal strip attached horizontally to the edge of a roof to prevent loose gravel from washing off the edge of the roof; also provides a decorative band around the roof that diverts water away from the edge.

Grease interceptor A tank constructed in such a manner that grease and other waste water can be separated, allowing the grease to remain in the tank but allowing the waste water to pass through.

Grease trap A tank installed to collect liquid grease from a restaurant operation so it can be emptied as required.

Ground-fault interrupter An electric-shock protector, installed in a location where moisture or a water source is nearby.,

Grout A mortar-type mixture with a consistency that allows it to be poured or placed in a space or crack to provide structural strength to that space.

Grouted door frame A hollow metal door frame with the empty space filled with grout to provide extra rigidity to the frame.

Grouted masonry wall Grout placed in the hollow cores of concrete masonry units (block walls) to provide added structural stability, particularly when combined with other seismic components.

Grub Clearing a site of existing plants, shrubs, roots, and stumps.

Gypsum flooring A liquefied, pastelike, self-leveling gypsum material applied to a floor deck; it hardens to provide a smooth finish for other flooring materials.

Hairline crack A very fine crack that appears randomly in cast-in-place concrete that may present more of an aesthetic than a structural concern.

Half-mortise hinge A hinge where one plate is surface-applied to a door or door frame and the other plate is mortised (recessed) in that frame or door surface.

Hardboard A generic name for panels composed of medium-density wood fibers or other types of wood fibers compressed into a sheet.

Hardpan Soil that has become rocklike because of the accumulation of cementing materials such as calcium carbonate.

Header A brick that is laid in a wall with its end exposed as opposed to its long side (stretcher); a wood or metal member installed across of top of a door or window opening.

Heartwood The center portion of a log.

Heat exchanger A device that efficiently transfers heat between one medium (generally a liquid) and another.

Heat pump A device where low-level heat is transferred from one location to another and in the process raises the temperature to a higher level; when cooling is required, the process is reversed.

Heat sink Heat transferred to another medium after it has been removed from its source.

Hiding power The ability of a paint film to cover an existing color or wall decoration.

High-early concrete Concrete batched with high-early cement or a concrete additive that permits the concrete, when placed, to reach its design strength more quickly.

Hip roof A roof structure that slopes downward from all four sides of a building.

Holiday When referring to the application of paint, an area where a full coat has not been applied and where the prior coat or surface shows through.

Honeycomb When concrete is placed in a form and a vibrator has not be used to consolidate the mixture, upon the removal of the forms small voids can be seen, much like a section through a honeycomb.

Hose bib An outside water faucet, also referred to as a **sill cock**.

House trap A configuration of plumbing pipe and fittings similar in shape to the letter P with the vertical member omitted, preventing sewer gases from passing below the trap into the room.

Hydraulic cement A cementitous mixture that allows hardening under moist or trickling water conditions.

Impervious soil Fine-grained soil such as clay that slows or prevents the passage of water.

Indirect-waste pipe A pipe that is not attached directly to a building's waste system but discharges through a plumbing fixture's trap.

Induced draft Forced movement of air or gas created by a fan's suction.

Inertia block A mass of concrete onto which is mounted a piece of mechanical equipment that generates vibration, reducing the transmission of that vibration to other parts of the building.

Initial set The first sign of curing or strength of poured concrete or of mastic applied to a surface when a film appears prior to installation of floor, wall coverings, or tile.

Interceptor A device to trap or remove hazardous materials prior to discharging into a public system.

Interstitial The space above the ceiling between floors.

Invert A line that runs lengthwise along the base of a channel or pipe at the lowest point of its wetted perimeter; the bottom portion of pipe or channel.

Invisible hinge A door or cabinet front hinge that is not visible when that door or cabinet front is closed.

Jacked pile A pile forced into suitable soil strata that will provide structural integrity for the load superimposed above it.

Jalousie window A window consisting of a series of overlapping glass louvers that open and close together.

Jamb A vertical wood or metal member on each side of a window or door that provides a place for anchoring.

Jetting Sinking of piles by driving in via the use of a high-pressure water source.

Joinery Cabinetry or woodworking involving the production and installation of wood doors, paneling, and cabinets to distinguish it from other carpentry work, such as framing.

Joist A parallel structural member made of wood or steel used to support floor and ceiling loads; steel joists are referred to as open-web joists; some fabricated-wood joists are referred to as truss joists.

Joist girder A primary steel framing member whose design is a simple span that will support concentrated loads from open-web steel joists.

Junction box A metallic or plastic electrical box where multiple wiring connections can be terminated; typical boxes are round or octagonal; both have secured covers.

Kalamein door An obsolete term referring to a door with a core of wood and a face of metal.

KD frame Knocked-down metal or wood frame, requiring a carpenter to fasten the jambs and head together.

Kerfing Making a series of closely spaced, small cuts in wood, primarily trim pieces, to enable the wood to bend without snapping.

Keystone The top masonry unit in an arch that provides structural integrity to the arch.

Kickplate A plastic or metal protective plate fastened to the lower portion of a door to prevent damage to the face of the door.

Kiln-dried lumber Lumber that has been seasoned by artificial means to remove excess moisture.

Labeled door A designation affixed to a door identifying the fire resistance of that door in terms of "hour ratings."

Labeled frame A labeled door must be installed in a corresponding labeled frame to retain the unit's fire rating.

Labeled hardware Special type of hardware required for a labeled opening such as a door and frame.

Labeled window Same concept but applied to the fire rating of a window to be installed in a fire-rated partition or wall.

Laminated beam A structural and/or decorative beam composed of a number of built-up layers bonded together with adhesive under pressure.

Land survey A survey of the property lines, referred to as "metes and bounds," that define the area of ownership of the property.

Lamp The bulb used in a lighting fixture.

Lath Wood or metal backing to support a wall finish such as plaster or stucco.

Latticework A panel composed of narrow strips of wood or metal interlaced usually in a diagonal pattern.

Lean mix A concrete mix with low cement content, used as a nonstructural thin slab over exposed earth under a crawl space.

Ledger board A horizontal board to which vertical members are attached.

Leveling coat A thin layer of material spread over an existing wall or floor to provide a level surface prior to the application of a wall or floor covering, such as tile or vinyl fabric.

Life-cycle cost All costs incident to the planning, design, construction, operation, and maintenance of a facility or system for a specified life expectancy in terms of present value.

Lintel A horizontal structural member, made of wood, steel, or stone, to support a door, wall, or window opening.

Lintel block A U-shaped concrete-masonry unit (CMU) serving as the top course in a wall, where a reinforcing bar can be placed longitudinally and continuously and then filled with mortar to create a strong integral structure.

Liquidated damages A cost, generally applied on a daily basis, that a contractor agrees to incur in the owner's favor for delaying the completion of the construction project beyond its original or adjusted contract time.

Load A force placed on a structure. Dead load is the dead weight of the building; live load is a load imposed by the building's furnishings, equipment, and occupants.

Loess Silty material with high cohesive ability.

Loose lintel A lintel not attached to any structural member but placed across an opening.

Low boy A trailer for hauling excavating equipment to and from the site; built low to the ground to allow the equipment to easily ramp up and board.

Low-voltage electrical work Systems requiring less than 120 volts of power to operate effectively, used for such work as data communications and decorative lighting.

Luminaire A complete lighting fixture including the lamp (bulb).

Luminous ceiling A ceiling with a continuous surface of light-transmitting material.

Magnetic switch A switch whose on-off cycle is controlled by an electromagnet.

Main In HVAC or plumbing work, a major duct or pipe carrying air or liquid to a series of distribution ducts or pipes of smaller size.

Manhole A covered opening providing access to an underground utility; a method to provide a 90-degree turn in an underground storm or sanitary line to allow for a noninterrupted and smooth flow.

Mansard roof A roof with a double slope on all four sides, with the lower slope quite steep.

Masonry anchors Metal straps affixed to metal door frames that allow them to be securely fastened to their masonry opening.

Master schedule A summary schedule identifying the major components of a construction project, including sequence and durations.

Master switch The main switch in the electrical switchgear that controls the power to the building.

Mechanic's lien A lien on the owner's property in favor of a vendor or contractor who supplied labor, material, or equipment that was incorporated into the owner's project but for which they had not been paid.

Medium-density fiberboard A material used in panels, dark brown in color, made of compressed-wood fibers including as masonite®.

Meeting rail The horizontal member in a double-hung window where both upper and lower sash meet.

Metal stair pan A metal form in a flight of stairs requiring concrete or mortar infill to create the treads.

Milestone schedule A schedule reflecting important key events in a construction schedule rather than detailed components.

Mill finish The finish or appearance of a metal sheet, bar, or extrusion after manufacturing but before decoration.

Miter (or mitre) An angled cut to join two pieces of material together, usually in a 90-degree angle.

Miter box A tool with precut slots at varying angles, allowing a hand saw to follow any angle with precision.

Mockup A model of an actual detail in terms of both construction technique and appearance, to be reviewed and accepted by an architect as the standard.

Modular construction A manufacturing system where repetitive units are built under factory-controlled conditions, shipped to the construction site, and assembled.

Modulus of elasticity A material that has been strained to a point below its limit.

Moment connection A steel beam-to-steel column connection, either welded or bolted.

Monolithic concrete pour An uninterrupted placement of concrete to create a single component.

Mortar A plastic mixture of cement, sand, and lime to bond brick, block, or stone together.

Mortise A cavity into which another member or material can fit.

Mud jacking A process of pumping concrete below an existing slab to fill a void beneath that slab.

Mullion A vertical piece that divides a window sash.

Muntin Pieces of wood that make up the divisions in a multipaned window.

Neat cement Mortar made without sand.

Needling A process of inserting steel or wood beams over an area in a masonry wall where a new or enlarged opening will be cut. These "needle" beams are supported by columns to take the weight off the area where the opening will be made.

Newel A central post or column that supports a staircase.

Nonbearing wall A wall that supports no load other than its own.

Noncohesive soil A soil such as gravel or sand that lacks cohesiveness.

Nonconforming work Work that does not meet the requirements of the plans and/or specifications.

Norman brick A brick whose dimensions are four inches longer on one side than regular modular brick.

Notice to proceed A formal document authorizing the contractor to commence work under the contract, officially marking the start of the project.

Obscure glass Frosted or stippled glass that provides privacy.

Oculus A round opening in an exterior wall.

Open-cell foam Foam whose structure is made of soft interconnecting cells with lots of airfill and void space, used in packing materials.

Orange peel The finish on a painted surface that looks like the texture of an orange.

Outdoor makeup air An air-handling unit that draws in outside air to make up for the air lost to the exterior.

Overflow pipe Pipe used to remove excess water from a storage tank due to expansion.

Packaged boiler A boiler unit having all components attached and shipped as one unit.

Palisade A series of strong poles, pointed on top and driven into the earth to form a fence.

Panelboard An electrical component that permits the distribution of smaller circuits for lighting and power.

Parabolic reflector A light reflector in a lighting fixture that concentrates the beam of light.

Parapet A low wall often marking the edge of a significant change in grade.

Parapet wall The portion of the exterior wall that rises above the roof line.

Parge To apply a thin coat of plaster or cement over a masonry, concrete, or plaster surface.

Partial occupancy Occupancy of a portion of a new building while the other parts are still under construction.

Party wall A wall dividing two or more adjoining buildings.

Pass A single progression of a welding operation; the first in a series.

Pavement base The stone or compacted area under a concrete slab on grade of concrete or asphalt paving.

Paver A block of stone, masonry, or precast concrete used in a sidewalk or area paving.

Payment bond A bond issued by surety to provide an owner with protection in case the general contractor fails to pay for labor, materials, or equipment.

Peat A soft, light swamp soil consisting of mostly decayed material.

Pediment The triangular face of a roof gable used as a decoration over windows, doors, or dormers.

Pendant A fixture suspended by means of a flexible cord or rigid tube through which electrical wiring is placed.

Penny A unit denoting the size of a nail.

Pent or **shed roof** A roof with a slope in one direction.

Percolation The seepage of water through soil, measured by rate of seepage within a specific time frame.

Perforated brick Brick with a series of holes in its core not to exceed 25 percent of its volume.

Performance bond A commitment from surety to pay the owner (oblige) in the event of a default in performance of the contractor's contract obligations.

Permeability The property of a porous material that permits water vapor to pass through it.

PERT schedule Project evaluation and review technique, a schedule used on many government projects.

Phased construction When each sequential phase of defined work is considered a separate project.

Pier A column designed to support a load.

Piezometer A measuring device for liquid pressure.

Pigtail An electrical conductor attached to one electrical component that provides a means for it to be connected to another electrical component.

Pilaster A flat, plain, or fluted pier, with little projection, that is attached to the surface of a wall.

Pile A structural support, either wood, steel, or concrete, driven or bored into the ground and meant to carry a vertical load.

Pile cap A concrete cap placed on a pile or spanning several piles to act as a single support for the load to be imposed on it.

Pile driver A machine that delivers repeated blows to a pile that is driven into the ground via a drop hammer.

Pile, friction A pile that attains its structural strength from the frictional forces imposed on its circumference.

Pipe hanger An individual support along one length of pipe or a tray in which several runs of horizontal pipes are supported.

Pipe, soil Any pipe that carries a building sewer to a discharge point outside the building.

Pipe, waste A pipe that carries liquid or liquid-borne waste that is free of fecal matter.

Pitch The slope of a roof as a ratio of vertical rise to horizontal dimensions; the slope of a pipe carrying a gravity flow of liquid.

Pitch pocket A small metal enclosure around a pipe penetrating a roof that is filled with pitch to seal the opening.

Plain sliced A wood veneer sliced parallel to the pith (soft central core of a log) and approximately tangent to the growth rings.

Plasticizer A chemical additive to mortar or concrete to increase its plasticity and make it more workable.

Plastic limit of soil The lowest water content of a soil sample, at which point the soil begins to crumble when rolled into a cylinder approximately $1/_8$ inch in diameter.

Plate girder A steel beam comprised of sheets, cut and welded together to form a beam.

Plate vibrator A motorized, walk-along vibrating machine used to consolidate and compact soil.

Plenum A separate space between the structural ceiling and the dropped ceiling, used for heating and air-conditioning distribution.

Plumb A term meaning exact vertical, attained by using a weighted bronze tear-shaped instrument attached to sturdy cord line.

Pneumatic controls An HVAC system using air pressure to operate control devices regulating the building's working environment.

Pocket A recess in a masonry wall that will act as a structural support for an inserted beam.

Pockmarking Blemishes in a painted surface.

Pointing up Removal of old, failed, or deteriorating mortar joints and replacement with new joint material, sometimes referred to as **tuck pointing**.

Ponding Accumulation of surface water on a roof, parking lot, or grassy that indicates there is inadequate pitch to drain properly.

Portland cement A cementitious binder for normal use in mortar and concrete; special types such as high-early (quick setting) or sulfate-resistant cement are also available.

Posttensioned concrete Concrete beams and floor slabs containing steel braided strands that are placed into tension by tightening after the beam or slab has been formed.

Precast concrete Concrete components—walls, beams, columns, slabs—produced under quality-controlled factory conditions.

Prestressed concrete Concrete beams, columns, and floor slabs that contain internal steel braids embedded under tension to overcome weakness in concrete-tensile strength.

Prime contract Direct contract with the owner.

Proctor test A test to determine the density-moisture relationship in soil and thus the degree of compaction.

P.S.I. Pounds per square inch, used in defining soil compaction or compressive strength of steel or concrete.

Puddle weld A process of filling a hole burned in the upper sheet of steel with a puddle of material to fuse the upper sheet to the lower.

Punch list A list prepared by the architect/engineer as the project nears completion, indicating that certain items of work are either unfinished or do not meet contract requirements and must either be completed or corrected prior to completing the terms of the contract.

Purge To evacuate air or gas from a duct or pipe.

Quartered Wood veneer produced by cutting in a radial direction to the pith.

Quirk An indentation separating one molding from another; a V groove in a finish coat of plaster to provide a controlled crack.

Quit-claim deed Conveyance by a seller of only the interest he or she has in the property, with no representation as to the nature of that interest.

Quoin A unit of masonry in the corner of a building, often decoratively different from adjacent masonry units.

Rabbet A longitudinal groove cut into the face of a wood-trim piece to receive another member.

Raceway A channel provided to enclose electrical conductors.

Rafter An inclined member to which roof sheathing is applied.

Rafter plate A bottom wood member to which rafters are attached to fix them in place.

Rail The cross or horizontal pieces in the assembly of a wood door.

Rake A slope on a wall or roof.

Raked joint A mortar joint depressed while it is still soft to produce a shadow line on the entire wall.

Rat slab A thin concrete slab with low cement content whose purpose is to seal off the surface over which it is placed.

Ready mix Concrete that is prepared at a concrete batch plant and delivered to the construction site via truck.

Recovery schedule A schedule that depicts special efforts required to recover lost time in the master or contract schedule.

Refractory A nonmetallic material that withstands heat.

Refractory brick Brick that can withstand high temperatures.

Refrigerant A medium by which heat is absorbed by evaporation at low temperature and pressure and given up when condensed at high temperature and pressure.

Reglet A sheet-metal receiver inset into the joint of a masonry wall for the attachment of flashing.

Reheat coil A coil in an air-distribution system that heats the air, generally for a specific area.

Reinforced concrete Concrete with reinforcing bars (rebars) that provide tensile strength to a material high in compressive strength but low in tensile (bending) strength.

Reinforced masonry A masonry wall with vertical rebars installed and grouted in their core or with steel mesh incorporated horizontally in alternate coursing.

Release of lien A legal document that releases a mechanic's lien against the property where labor, materials, and equipment had been incorporated.

Rendering A detailed artistic drawing prepared by an architect to reflect a potential building design.

Reproducible A drawing or sketch or sections from a specification manual that will be reproduced for distribution to a general contractor, subcontractor, or vendor.

Resilient flooring A manufactured flooring material that is resilient in nature, usually sheet vinyl or individual vinyl tiles.

Retainage The portion of the actual total price for which the owner temporarily withholds payment.

Retarder A chemical added to mortar or concrete to delay the set of that mix.

Return air Air returned from an air-conditioned or heated area for reheating or recooling.

Return wall A horizontal short wall perpendicular to a longer wall.

Reveal The space on the side of a window or door frame where a slight depression in the adjacent wall creates a shadow line.

Rift cut A wood veneer produced by cutting at a slight angle to the radius.

Rigid-metal conduit A conduit to house electrical wiring made of rigid steel coated with a rust-resistant finish inside and out.

Riprap Large irregularly broken stone laid randomly on a sloped area or in a swale to prevent erosion.

Riser The vertical part of a stair tread.

Rodding The consolidation of grout poured in vertical masonry cores or concrete placed in a confined area by repeated up-and-down movements with a small-diameter steel rod.

Roll roofing A smooth or mineral-coated surface roofing material that comes in a roll and is installed by rolling over the roof sheathing secured by nailing.

Rotary cut A method of producing a wood veneer by centering the entire log and turning it against a broad cutting knife.

Router A machine with a high-speed revolving spindle that is used to cut a groove to create a decorative edge on a piece of wood.

Rowlock The sill of a window created in a brick wall with all the bricks laid on edge in the same direction.

Rubble stone Irregularly shaped stone of a rough texture laid up in an irregular pattern.

S4S, S2S, S1S A finish-carpentry term denoting a board that is planed smooth on four, two, or one side.

Sanitary cove A vinyl or ceramic piece curved like the letter J that makes a smooth transition from a vertical to a horizontal plane, providing a curved corner surface for easy cleaning.

Sash The window's framework that supports the glass.

Scab A short, flat piece of wood bolted or nailed to adjacent pieces to splice them together.

Schedule of values A listing of specific contract cost segments including related labor, materials, and equipment, a brief explanation of the work task, and the sum, equaling the contract sum.

Sconce A light fixture designed to mount on a wall.

Screed A grade strip designed to level off newly poured concrete slabs or platforms prior to troweling.

Seal coat A coating on asphalt paving, concrete, or masonry to deter weathering.

Seasoning check A longitudinal crack in lumber that develops during the drying-out process.

Sewage ejector A pump designed to lift sewage to a higher elevation, thereby allowing for gravity flow.

Shaft A vertical segment of a column or pilaster between the base and the capital; the empty vertical space provided for the installation of plumbing pipes, mechanical ductwork, and electrical conduits.

Shake A thick-cut wood shingle with a rough-hewn appearance.

Sheathing The covering of studs or other structural members in an exterior wall with various types of panels, usually with a gypsum core.

Sheeting Vertical and horizontal wood, steel, or concrete members used to hold up the face of an excavated area.

Sheet piles A series of interconnected steel plates forming a barrier to retain the vertical face of an excavated area.

Shoe A metal or wood baseplate.

Shop drawings Drawings or manufacturer's detailed diagrams in sufficient detail to allow the reviewer (architect/engineer) to determine that what is proposed is consistent with the design and intent of the contract requirements.

Shoring A lumber or steel framework to temporarily support a wall or suspended floor.

Shotcrete Concrete or mortar forced through a nozzle under pressure to provide a stabilized surface for a wall or other surface.

Shrinkage cracks Small cracks that appear as concrete cures as the reduction of water takes place in the mix, usually aesthetic in nature and not structural.

Sidelight A glass panel to either side of a door.

Sill The horizontal structure of a window, door, or wall opening.

Skin The hardwood or composition panel that provides the architectural face of a flush wood door; the outer shell or exterior of a building.

Slip form A form designed to move along as concrete is poured into it; often used to produce concrete curbing.

Slump A measure of consistency in fresh concrete; concrete is poured into an inverted cone, and when the cone is removed, the concrete's diminishing height (slump) is measured.

Small tools Tools used by the contractor to complete the work, each of which cost $500 or less.

Soffit A flat wood or composite material used to finish the undersurface of any overhead exposed part of the building.

Soffit vent A premanufactured air inlet located on the back side of a soffit to provide ventilation to the enclosed space above.

Soldier course Bricks laid up by standing them on end like a row of soldiers.

Solid masonry wall A masonry wall laid up contiguously with all joints and spaces filled with mortar.

Sound-transmission coefficient A rating of the transmission or lack of transmission of sound through a wall, ceiling, or floor assembly.

Spalling A condition of brick, stone, or exterior concrete where layers break or peel off due to the presence of internal pressure caused by crystallizing water or chemicals within the mixture.

Spandrel beam An exterior beam extending from column to column that carries an imposed load.

Splash block A small masonry or composition block laid on the ground under a downspout to help prevent soil erosion.

Spot elevation The elevation datum relating to a specific reference point on a site plan.

Standpipe The vertical main that carries fire-protection sprinkler water to each floor for distribution via branch piping to the sprinkler heads.

Starter course The first layer of roofing, applied along a line adjacent to the downslope perimeter of the roof.

Static pressure In an air-distribution system, the air pressure required to overcome the resistance formed as air flows through the ductwork system.

Step flashing Individual pieces of flashing installed around chimneys and dormers, each one overlapped up the vertical surface.

Stile The upright or vertical pieces of a wood door.

S-trap A plumbing fitting shaped like an S that provides a seal against sewer gas.

Stirrup A U-shaped bent rebar installed in a concrete form as part of its reinforcing.

Stretcher course A brick pattern in which all units are laid up lengthwise.

Striker or **striker plate** A recessed receptacle or notch in the jamb of a door frame that receives the door latch or bolt.

Subbase The area under the base course of asphalt or concrete paving.

Subframe A secondary frame for a window in an exterior wall that accepts the primary window frame.

Submittal The transmittal of contract-required information such as shop drawings, samples, and test and inspection reports.

Substantial completion The date, as certified by the architect, that the contractor has attained the stage of the project's completion at which the facility meets the purpose for which it is intended; not final completion.

Superplasticizer A chemical additive in a concrete batch that permits easy flow without reducing the concrete's compressive strength.

Supplementary general conditions Additions, changes, or modifications to the general conditions included in the bid documents and part of the contract for construction.

Switchgear or **switchboard** A large electric control panel that receives the incoming secondary electrical service for distribution to a series of electrical panels located throughout the building.

Tensile strength The resistance of a material when subjected to a bending movement.

Terminal unit In an air-distribution heating and cooling system, the unit at the end of the duct through which the heated or cooled air enters the space.

Thermal expansion The movement of a building material as it changes volume when exposed to heat, such as materials in an exterior wall when exposed to intense sunlight.

Throat The width of a door frame measured along the width of the door jamb.

Tongue and groove Planks with a convex cut on one side and a concave cut on the other, allowing each piece to be firmly joined to an adjacent piece.

Top rail The top horizontal framing member of a window or door.

Total float The amount of contingency time included in a construction schedule.

Transformer An electrical device that transforms a supply of one electrical voltage to another electrical voltage.

Transom A glass panel placed over a door or window to provide additional natural light or ventilation; the transom can either be fixed or operable.

Tread The flat horizontal surface of a stair.

Tremie concrete A method of placing concrete underwater through a tube.

Trim Woodwork moldings used as baseboard, door, and window finishing materials to cover joints and provide decoration.

Trade contractors Specialty contractors, also referred to as **subcontractors**.

Tuck-pointing Replacing deteriorated mortar from a masonry wall by troweling or tooling new mortar into the joint.

Two-week look-ahead A schedule of intended work to be performed in the next two-week work period.

UL label Underwriters Laboratory identification label signifying that the product has achieved its prescribed rating.

Undercut door Additional clearance provided between the flooring and bottom edge of the door to allow for more air movement in and out.

Underlayment A sheet material such as plywood or medium-density fiberboard placed on a sub-floor to provide a more even surface for the finish flooring material.

Unit heater A small heating element with a fan in an enclosure to provide heat to an isolated area or room.

Unit prices Prices charged by the contractor for specific services or materials.

UPS Uninterrupted power supply, a device to provide temporary power in case the primary power source is interrupted.

Upstanding beam or **upturned beam** A structural beam that is built on the upside of the floor slab instead of below it.

Valley An internal angle formed by the intersection of two sloping roof planes.

Value engineering A cost-control process utilizing a systematic analysis of a particular operation or component of construction to determine if costs can be reduced while retaining the intent, function, operation, reliability, and life-cycle properties of that operation or component.

Vapor barrier A waterproof material such as a polyethylene sheet, used to prevent moisture from migrating from the soil into the concrete floor slab.

Vent pipe A pipe attached to a waste line that allows outside air to enter the system and prevents the water seal in a plumbing-fixture trap from being broken by siphoning action.

Vibration isolation A mechanical device on which a vibrating piece of equipment, such as an air handler is mounted to decrease or eliminate the transfer of vibration from the equipment to the building's structure.

Vitrified-clay pipe A glazed clay pipe with resistance to corrosion, used years ago in underground sanitary or storm-sewer systems.

Vomitory An entry or opening through a section of seats that provides special access in a sports stadium or theater.

Waiver of lien A legal document whereby a contractor waives his or her right to place a lien on the owner's property in exchange for payment for labor, materials, and equipment installed in the owner's building.

Warp Deviation from a true or plane surface.

Weep hole A small opening in an exterior concrete or masonry wall through which trapped condensation or moisture can escape.

Weighted criteria An owner-established point system to judge a contractor's qualitative rating and price to arrive at "best qualified."

Well point A hollow rod or tube with perforations that, when driven into the ground and attached to a suction pump, lowers the water level in that area.

Wicking The process of water moving uphill due to capillary action.

Wind clip An attachment device that fits over the roofing tiles or shingles to assist in securing the individual units from wind uplift.

Wind load The load placed on a structure or a building component by the force of the wind.

Window stool A horizontal wood or metal trim piece fitted under the windowsill.

Wythe or with A single wall width, of masonry.

Wye or Y fitting A plumbing fitting shaped like a Y that joins two lines into one.

Construction management owner-contractor contract

CMAA Document A-1 (2005 Edition)

THE CONSTRUCTION MANAGEMENT ASSOCIATION OF AMERICA, INC.

CMAA Document A-1 (2005 Edition)

Standard Form of Agreement Between
OWNER AND CONSTRUCTION MANAGER
(Construction Manager as Owner's Agent)

This document is to be used in connection with the Standard Form of Contract Between Owner and Contractor (CMAA Document A-2), the General Conditions of the Construction Contract (CMAA Document A-3), and the Standard Form of Agreement Between Owner and Designer (CMAA Document A-4), all being 2005 editions.

CONSULTATION WITH AN ATTORNEY IS RECOMMENDED WHENEVER THIS DOCUMENT IS USED.

AGREEMENT
Made this _____ day of _____ in the year of Two Thousand and
_____

BETWEEN The Owner:

and the Construction Manager, (hereinafter, referred to as the "CM"):

For services in connection with the Project known as:

hereinafter called the "Project," as further described in Article 2:

The Owner and CM, in consideration of their mutual covenants herein agree as set forth below:

Copyright Construction Management Association of America, Inc., 2005. All rights reserved. Reproduction or translation of any part of this Document without the permission of the copyright owner is unlawful.

By permission of the Construction Management Association of America, 7926 Jones Beach Drive, McLean, Virginia

CMAA Document A-1 (2005 Edition)

Copyright Construction Management Association of America, Inc., 2005. All rights reserved. Reproduction or translation of any part of this Document without the permission of the copyright owner is unlawful.

CMAA Document A-1 (2005 Edition)

TABLE OF CONTENTS

Article:

Copyright Construction Management Association of America, Inc., 2005. All rights reserved. Reproduction or translation of any part of this Document without the permission of the copyright owner is unlawful.

CMAA Document A-1 (2005 Edition)

ARTICLE 1
RELATIONSHIP OF THE PARTIES

1.1 Owner and Construction Manager

1.1.1 Relationship: The CM shall be the Owner's principal agent in providing the CM's services described in this Agreement. The CM and the Owner shall perform as stated in this Agreement. Nothing in this Agreement shall be construed to mean that the CM is a fiduciary of the Owner.

1.1.2 Standard of Care: The CM covenants with the Owner to furnish its services hereunder properly, in accordance with the standards of its profession, and in accordance with federal, state and local laws and regulations specifically applicable to the performance of the services hereunder which are in effect on the date of this Agreement first written above.

1.2 Owner and Designer

1.2.1 Owner-Designer Agreement: The Owner shall enter into a separate agreement, the "Owner-Designer Agreement", with one or more Designers to provide for the design of the Project and certain design-related services during the Construction Phase of the Project. The Project is defined in Article 2 of this Agreement.

1.2.2 Changes: The Owner shall not modify the Agreement between the Owner and Designer in any way that is prejudicial to the CM. If the Owner terminates the Designer's services, a substitute acceptable to the CM shall be appointed.

1.3 Owner and Contractors

1.3.1 Construction Contract: The Owner shall enter into a separate contract with one or more Contractors for the construction of the Project (hereinafter referred to as the "Contract"). The Contractor shall perform the Work, which shall consist of furnishing all labor, materials, tools, equipment, supplies, services, supervision, and perform all operations as required by the Contract Documents.

1.3.2 Form of Contract: Unless otherwise specified, the form of Contract between the Owner and Contractor shall be the CMAA Standard Form of Contract Between Owner and Contractor, CMAA Document A-2 (2005 Edition). The General Conditions for the Project shall be the CMAA General Conditions of the Construction Contract Between Owner and Contractor, CMAA Document A-3 (2005 Edition).

1.4 Relationship of the CM to Other Project Participants

1.4.1 Working Relationship: In providing the CM's services described in this Agreement, the CM shall endeavor to maintain, on behalf of the Owner, a working relationship with the Contractor and Designer.

1.4.2 Limitations: Nothing in this Agreement shall be construed to mean that the CM assumes any of the responsibilities or duties of the Contractor or the Designer. The Contractor will be solely responsible for construction means, methods, techniques, sequences and procedures used in the construction of the Project and for the safety of its personnel, property, and its operations and for performing in accordance with the contract between the Owner and Contractor. The Designer is solely responsible for the design requirements and design criteria of the Project and shall perform in accordance with the Agreement between the Designer and the Owner. The CM's services shall be rendered compatibly and in cooperation with the services provided by the Designer under the Agreement between the Owner and Designer. It is not intended that the services of the Designer and the CM be competitive or duplicative, but rather complementary. The CM will be entitled to rely upon the Designer for the proper performance of services undertaken by the Designer pursuant to the Agreement between Owner and the Designer.

ARTICLE 2
PROJECT DEFINITION

2.1 The term "Project", when used in this Agreement, shall be defined as all work to be furnished or provided in accordance with the Contract Documents prepared by the Designer.

Copyright Construction Management Association of America, Inc., 2005. All rights reserved. Reproduction or translation of any part of this Document without the permission of the copyright owner is unlawful.

CMAA Document A-1 (2005 Edition)

2.2 The Project name and location is as follows:

_____
_____
_____
_____
_____
_____

2.3 The Project is intended for use as:

_____
_____
_____
_____
_____
_____
_____
_____
_____

2.4 The term "Contract Documents" means the Instruction to Bidders, the Contract, the General Conditions and any Supplemental Conditions furnished to the Contractor, the drawings and specifications furnished to the Contractor and all exhibits thereto, addenda, bulletins and change orders issued in accordance with the General Conditions to any of the above, and all other documents specified in Exhibit B of the Standard Form of Contract Between Owner and Contractor, CMAA Document A-2, 2005 edition.

ARTICLE 3
BASIC SERVICES

3.1 CM's Basic Services

3.1.1 Basic Services: The CM shall perform the Basic Services described in this Article. It is not required that the services be performed in the order in which they are described.

3.2 Pre-Design Phase

3.2.1 Project Management

3.2.1.1 Construction Management Plan: The CM shall prepare a Construction Management Plan for the Project and shall make recommendations to the plan throughout the duration of the Project, as may be appropriate. In preparing the Construction Management Plan, the CM shall consider the Owner's schedule, budget and general design requirements for the Project. The CM shall then develop various alternatives for the scheduling and management of the Project and shall make recommendations to the Owner. The Construction Management Plan shall be presented to the Owner for acceptance.

3.2.1.2 Designer Selection: The CM shall assist the Owner in the selection of a Designer by developing lists of potential firms, developing criteria for selection, preparing and transmitting the requests for proposal, assisting in conducting interviews, evaluating candidates and making recommendations.

3.2.1.3 Designer Contract Preparation: The CM shall assist the Owner in review and preparation of the Agreement between the Owner and Designer.

3.2.1.4 Designer Orientation: The CM shall conduct, or assist the Owner in conducting, a Designer orientation session during which the Designer shall receive information regarding the Project scope, schedule, budget, and administrative requirements.

3.2.2 Time Management

3.2.2.1 Master Schedule: In accordance with the Construction Management Plan, the CM shall prepare a Master Schedule for the Project. The Master Schedule shall specify the proposed starting and finishing dates for each major project activity. The CM shall submit the Master Schedule to the Owner for acceptance.

3.2.2.2 Design Phase Milestone Schedule: After the Owner accepts the Master Schedule the CM shall prepare the Milestone Schedule for the Design Phase, which shall be used for judging progress during the Design Phase.

Copyright Construction Management Association of America, Inc., 2005. All rights reserved. Reproduction or translation of any part of this Document without the permission of the copyright owner is unlawful.

CMAA Document A-1 (2005 Edition)

3.2.3 Cost Management

3.2.3.1 Construction Market Survey: The CM shall conduct a Construction Market Survey to provide current information regarding the general availability of local construction services, labor, material and equipment costs and the economic factors related to the construction of the Project. A report of the Construction Market Survey shall be provided to the Owner and Designer.

3.2.3.2 Project and Construction Budget: Based on the Construction Management Plan and the Construction Market Survey, the CM shall prepare a Project and Construction Budget based on the separate divisions of the Work required for the Project and shall identify contingencies for design and construction. The CM shall review the budget with the Owner and Designer and the CM shall submit the Project and Construction Budget to the Owner for acceptance. The Project and Construction Budget shall be revised by the CM as directed by the Owner.

3.2.3.3 Preliminary Estimate and Budget Analysis: The CM shall analyze and report to the Owner and the Designer the estimated cost of various design and construction alternatives, including CM's assumptions in preparing its analysis, a variance analysis between budget and preliminary estimate, and recommendations for any adjustments to the budget. As a part of the cost analysis, the CM shall consider costs related to efficiency, usable life, maintenance, energy and operation.

3.2.4 Management Information System (MIS)

3.2.4.1 Establishing the Project MIS: The CM shall develop a MIS in order to establish communication between the Owner, CM, Designer, Contractor and other parties involved with the Project. In developing the MIS, the CM shall interview the Owner's key personnel, the Designer and others in order to determine the type of information for reporting, the reporting format and the desired frequency for distribution of the various reports.

3.2.4.2 Design Phase Procedure: The MIS shall include procedures for reporting, communications and administration during the Design Phase.

3.3 Design Phase

3.3.1 Project Management

3.3.1.1 Revisions to the Construction Management Plan: During the Design Phase the CM shall make recommendations to the Owner regarding revisions to the Construction Management Plan. The Construction Management Plan shall include a description of the various bid packages recommended for the Project. Revisions approved by the Owner shall be incorporated into the Construction Management Plan.

3.3.1.2 Project Conference: At the start of the Design Phase, the CM shall conduct a Project Conference attended by the Designer, the Owner and others as necessary. During the Project Conference the CM shall review the Construction Management Plan, the Master Schedule, Design Phase Milestone Schedule, the Project and Construction Budget and the MIS.

3.3.1.3 Design Phase Information: The CM shall monitor the Designer's compliance with the Construction Management Plan and the MIS, and the CM shall coordinate and expedite the flow of information between the Owner, Designer and others as necessary.

3.3.1.4 Progress Meetings: The CM shall conduct periodic progress meetings attended by the Owner, Designer and others. Such meetings shall serve as a forum for the exchange of information concerning the Project and the review of design progress. The CM shall prepare and distribute minutes of these meetings to the Owner, Designer and others as necessary.

3.3.1.5 Review of Design Documents: The CM shall review the design documents and make recommendations to the Owner and Designer as to constructibility, scheduling, and time of construction; as to clarity, consistency, and coordination of documentation among Contractors; and as to the separation of the Project into contracts for various categories of the Work. In addition, the CM shall give to the Designer all data of which it or the Owner is aware concerning patents or copyrights for inclusion in Contract Documents. The recommendations resulting from such review shall be provided to the Owner and Designer in writing or as notations

Copyright Construction Management Association of America, Inc., 2005. All rights reserved. Reproduction or translation of any part of this Document without the permission of the copyright owner is unlawful.

CMAA Document A-1 (2005 Edition)

on the design documents. In making reviews and recommendations as to design documentation or design matters the CM shall not be responsible for providing nor will the CM have control over the Project design, design requirements, design criteria or the substance of contents of the design documents. By performing the reviews and making recommendations described herein, the CM shall not be deemed to be acting in a manner so as to assume responsibility or liability, in whole or in part, for any aspect of the project design, design requirements, design criteria or the substance or contents of the design documents. The CM's actions in making such reviews and recommendations as provided herein are to be advisory only to the Owner and to the Designer.

3.3.1.6 Owner's Design Reviews: The CM shall expedite the Owner's design reviews by compiling and conveying the Owner's review comments to the Designer.

3.3.1.7 Approvals by Regulatory Agencies: The CM shall coordinate transmittal of documents to regulatory agencies for review and shall advise the Owner of potential problems resulting from such reviews and suggested solutions regarding completion of such reviews.

3.3.1.8 Other Contract Conditions: The CM shall assist the Owner to prepare the Supplemental Conditions of the Construction Contract and separate General Conditions for materials or equipment procurement contracts to meet the specific requirements of the Project, and shall provide these to the Designer for inclusion in the Contract Documents.

3.3.1.9 Project Funding: The CM shall assist the Owner in preparing documents concerning the Project and Construction Budget for use in obtaining or reporting on Project funding. The documents shall be prepared in a format approved the Owner.

3.3.2 Time Management

3.3.2.1 Revisions to the Master Schedule: While performing the services provided in Paragraphs 3.3.1.1, 3.3.1.2 and as necessary during the Design Phase, the CM shall recommend revisions to the Master Schedule. The Owner shall issue, as needed, change orders to the appropriate parties to implement the Master Schedule revisions.

3.3.2.2 Monitoring the Design Phase Milestone Schedule: While performing the services provided in Paragraphs 3.3.1.3 and 3.3.1.4, the CM shall monitor compliance with the Design Phase Milestone Schedule.

3.3.2.3 Pre-Bid Construction Schedules: Prior to transmitting Contract Documents to bidders, the CM shall prepare a Pre-Bid Construction Schedule for each part of the Project and make the schedule available to the bidders during the Procurement Phase.

3.3.3 Cost Management

3.3.3.1 Cost Control: The CM shall prepare an estimate of the construction cost for each submittal of design drawings and specifications from the Designer. This estimate shall include a contingency acceptable to the Owner, CM and the Designer for construction costs appropriate for the type and location of the Project and the extent to which the design has progressed. The Owner recognizes that the CM will perform in accordance with the standard of care established in this Agreement and that the CM has no control over the costs of labor, materials, equipment or services furnished by others, or over the Contractor's methods of determining prices, or over competitive bidding or market prices. Accordingly, the CM does not represent or guarantee that proposals, bids or actual construction costs will not vary from budget figures included in the Construction Management Plan as amended from time to time. If the budget figure is exceeded, the Owner will give written consent to increasing the budget, or authorize negotiations or rebidding of the Project within a reasonable time, or cooperate with the CM and Designer to revise the Project's general scope, extent or character in keeping with the Project's design requirements and sound design practices, or modify the design requirements appropriately. Instead of the foregoing, the Owner may abandon the Project and terminate this Agreement in accordance with Article 10. The estimate for each submittal shall be accompanied by a report to the Owner and Designer identifying variances from the Project and Construction Budget. The CM shall facilitate decisions by the Owner and Designer when changes to the design are required to remain within the Project and Construction Budget.

Copyright Construction Management Association of America, Inc., 2005. All rights reserved. Reproduction or translation of any part of this Document without the permission of the copyright owner is unlawful.

CMAA Document A-1 (2005 Edition)

3.3.3.2 Project and Construction Budget Revision: The CM shall make recommendations to the Owner concerning revisions to the Project and Construction Budget that may result from design changes.

3.3.3.3 Value Engineering Studies: The CM shall provide value engineering recommendations to the Owner and Designer on major construction components, including cost evaluations of alternative materials and systems.

3.3.4 Management Information Systems (MIS)

3.3.4.1 Schedule Reports: In conjunction with the services provided by Paragraph 3.3.2.2, the CM shall prepare and distribute schedule maintenance reports that shall compare actual progress with scheduled progress for the Design Phase and the overall Project and shall make recommendations to the Owner for corrective action

3.3.4.2 Project Cost Reports: The CM shall prepare and distribute Project cost reports that shall indicate actual or estimated costs compared to the Project and Construction Budget and shall make recommendations to the Owner for corrective action.

3.3.4.3 Cash Flow Report: The CM shall periodically prepare and distribute a cash flow report.

3.3.4.4 Design Phase Change Report: The CM shall prepare and distribute Design Phase change reports that shall list all Owner-approved changes as of the date of the report and shall state the effect of the changes on the Project and Construction Budget and the Master Schedule.

3.4 Procurement Phase

3.4.1 Project Management

3.4.1.1 Prequalifying Bidders: The CM shall assist the Owner in developing lists of possible bidders and in prequalifying bidders. This service shall include preparation and distribution of questionnaires; receiving and analyzing completed questionnaires; interviewing possible bidders, bonding agents and financial institutions; and preparing recommendations for the Owner. The CM shall prepare a list of bidders for each bid package and transmit to the Owner for approval.

3.4.1.2 Bidder's Interest Campaign: The CM shall conduct a telephone and correspondence campaign to attempt to increase interest among qualified bidders.

3.4.1.3 Notices and Advertisements: The CM shall assist the Owner in preparing and placing notices and advertisements to solicit bids for the Project.

3.4.1.4 Delivery of Bid Documents: The CM shall expedite the delivery of Bid Documents to the bidders. The CM shall obtain the documents from the Designer and arrange for printing, binding, wrapping and delivery to the bidders. The CM shall maintain a list of bidders receiving Bid Documents.

3.4.1.5 Pre-Bid Conference: In conjunction with the Owner and Designer, the CM shall conduct pre-bid conferences. These conferences shall be forums for the Owner, CM and Designer to explain the Project requirements to the bidders, including information concerning schedule requirements, time and cost control requirements, access requirements, contractor interfaces, the Owner's administrative requirements and technical information.

3.4.1.6 Information to Bidders: The CM shall develop and coordinate procedures to provide answers to bidder's questions. All answers shall be in the form of addenda.

3.4.1.7 Addenda: The CM shall receive from the Designer a copy of all addenda. The CM shall review addenda for constructibility, for effect on the Project and Construction Budget, scheduling and time of construction, and for consistency with the related provisions as documented in the Bid Documents. The CM shall distribute a copy of all addenda to each bidder receiving Bid Documents.

Copyright Construction Management Association of America, Inc., 2005. All rights reserved. Reproduction or translation of any part of this Document without the permission of the copyright owner is unlawful.

CMAA Document A-1 (2005 Edition)

3.4.1.8 Bid Opening and Recommendations: The CM shall assist the Owner in the bid opening and shall evaluate the bids for responsiveness and price. The CM shall make recommendations to the Owner concerning the acceptance or rejection of bids.

3.4.1.9 Post-Bid Conference: The CM shall conduct a post-bid conference to review Contract award procedures, schedules, Project staffing and other pertinent issues.

3.4.1.10 Construction Contracts: The CM shall assist the Owner in the assembly, delivery and execution of the Contract Documents. The CM shall issue to the Contractor on behalf of the Owner the Notice of Award and the Notice to Proceed.

3.4.2 Time Management

3.4.2.1 Pre-Bid Construction Schedule: The CM shall emphasize to the bidders their responsibilities regarding the Pre-Bid Construction Schedule specified in the Instructions to Bidders or the Contract Documents.

3.4.2.2 Master Schedule: The CM shall recommend to the Owner any appropriate revisions to the Master Schedule. Following acceptance by the Owner of such revisions, the CM shall provide a copy of the Master Schedule to the Designer and to the bidders.

3.4.3 Cost Management

3.4.3.1 Estimates for Addenda: The CM shall prepare an estimate of costs for all Addenda and shall submit a copy of the estimate to the Designer and to the Owner for approval.

3.4.3.2 Analyzing Bids: Upon receipt of the bids, the CM shall evaluate the bids, including alternate bid prices and unit prices, and shall make a recommendation to the Owner regarding the award of the Construction Contract.

3.4.4 Management Information System (MIS)

3.4.4.1 Schedule Maintenance Reports: The CM shall prepare and distribute schedule maintenance reports during the Procurement Phase. The reports shall compare the actual bid and award dates to scheduled bid and award dates and shall summarize the progress of the Project.

3.4.4.2 Project Cost Reports: The CM shall prepare and distribute project cost reports during the Procurement Phase. The reports shall compare actual contract award prices for the Project with those contemplated by the Project and Construction Budget.

3.4.4.3 Cash Flow Reports: The CM shall prepare and distribute cash flow reports during the Procurement Phase. The reports shall be based on actual contract award prices and estimated other construction costs for the duration of the Project.

3.5 Construction Phase

3.5.1 Project Management

3.5.1.1 Pre-Construction Conference: In consultation with the Owner and Designer, the CM shall conduct a Pre-Construction Conference during which the CM shall review the Project reporting procedures and other requirements for performance of the Work..

3.5.1.2 Permits, Bonds and Insurance: The CM shall verify that the Contractor has provided evidence that required permits, bonds, and insurance have been obtained. Such action by the CM shall not relieve the Contractor of its responsibility to comply with the provisions of the Contract Documents.

3.5.1.3 On-Site Management and Construction Phase Communication Procedures: The CM shall provide and maintain a management team on the Project site to provide contract administration as an agent of the Owner, and the CM shall establish and implement coordination and communication procedures among the CM, Owner, Designer and Contractor.

3.5.1.4 Contract Administration Procedures: The CM shall establish and implement procedures for reviewing and processing requests for clarifications and interpretations of the Contract Documents; shop drawings, samples and other submittals; contract schedule adjustments; change order proposals; written proposals for substitutions; payment applications; and the maintenance of logs. As the Owner's representative at the construction site, the CM shall be the party to whom all such information shall be submitted.

Copyright Construction Management Association of America, Inc., 2005. All rights reserved. Reproduction or translation of any part of this Document without the permission of the copyright owner is unlawful.

3.5.1.5 <u>Review of Requests for Information, Shop Drawings, Samples, and Other Submittals:</u> The CM shall examine the Contractor's requests for information, shop drawings, samples, and other submittals, and Designer's reply or other action concerning them, to determine the anticipated effect on compliance with the Project requirements, the Project and Construction Budget, and the Master Schedule. The CM shall forward to the Designer for review, approval or rejection, as appropriate, the request for clarification or interpretation, shop drawing, sample, or other submittal, along with the CM's comments. The CM's comments shall not relate to design considerations, but rather to matters of cost, scheduling and time of construction, and clarity, consistency, and coordination in documentation. The CM shall receive from the Designer and transmit to the Contractor, all information so received from the Designer.

3.5.1.6 <u>Project Site Meetings:</u> Periodically the CM shall conduct meetings at the Project site with each Contractor, and the CM shall conduct coordination meetings with the Contractor, the Owner and the Designer. The CM shall prepare and distribute minutes to all attendees, the Owner and Designer.

3.5.1.7 <u>Coordination of Other Independent Consultants:</u> Technical inspection and testing provided by others shall be coordinated by the CM. The CM shall receive a copy of all inspection and testing reports and shall provide a copy of such reports to the Designer. The CM shall not be responsible for providing, nor shall the CM control, the actual performance of technical inspection and testing. The CM is performing a coordination function only and the CM is not acting in a manner so as to assume responsibility or liability, in whole or in part, for all or any part of such inspection and testing.

3.5.1.8 <u>Minor Variations in the Work:</u> The CM may authorize minor variations in the Work from the requirements of the Contract Documents that do not involve an adjustment in the Contract price or time and which are consistent with the overall intent of the Contract Documents. The CM shall provide to the Designer copies of such authorizations.

3.5.1.9 <u>Change Orders:</u> The CM shall establish and implement a change order control system. All changes to the Contract between the Owner and Contractor shall be only by change orders executed by the Owner.

3.5.1.9.1 All proposed Owner-initiated changes shall first be described in detail by the CM in a request for a proposal issued to the Contractor. The request shall be accompanied by drawings and specifications prepared by the Designer. In response to the request for a proposal, the Contractor shall submit to the CM for evaluation detailed information concerning the price and time adjustments, if any, as may be necessary to perform the proposed change order Work. The CM shall review the Contractor's proposal, shall discuss the proposed change order with the Contractor, and endeavor to determine the Contractor's basis for the price and time proposed to perform the changed Work.

3.5.1.9.2 The CM shall review the contents of all Contractor requested changes to the Contract time or price, endeavor to determine the cause of the request, and assemble and evaluate information concerning the request. The CM shall provide to the Designer a copy of each change request, and the CM shall in its evaluations of the Contractor's request consider the Designer's comments regarding the proposed changes.

3.5.1.9.3 The CM shall make recommendations to the Owner regarding all proposed change orders. At the Owner's direction, the CM shall prepare and issue to the Contractor appropriate change order documents. The CM shall provide to the Designer copies of all approved change orders.

3.5.1.10 <u>Subsurface and Physical Conditions:</u> Whenever the Contractor notifies the CM that a surface or subsurface condition at or contiguous to the site is encountered that differs from what the Contractor is entitled to rely upon or from what is indicated or referred to in the Contract Documents, or that may require a change in the Contract Documents, the CM shall notify the Designer. The CM shall receive from the Designer and transmit to the Contractor all information necessary to specify any design changes required to be responsive to the differing or changed condition and, if necessary, shall prepare a change order as indicated in Paragraph 3.5.1.9.

Copyright Construction Management Association of America, Inc., 2005. All rights reserved. Reproduction or translation of any part of this Document without the permission of the copyright owner is unlawful.

CMAA Document A-1 (2005 Edition)

3.5.1.11 Quality Review: The CM shall establish and implement a program to monitor the quality of the Work. The purpose of the program shall be to assist in guarding the Owner against Work by the Contractor that does not conform to the requirements of the Contract Documents. The CM shall reject any portion of the Work and transmit to the Owner and Contractor a notice of nonconforming Work when it is the opinion of the CM, Owner, or Designer that such Work does not conform to the requirements of the Contract Documents. Except for minor variations as described in Paragraph 3.5.1.8, the CM is not authorized to change, revoke, alter, enlarge, relax or release any requirements of the Contract Documents or to approve or accept any portion of the Work not conforming with the requirements of the Contract Documents. Communication between the CM and Contractor with regard to quality review shall not in any way be construed as binding the CM or Owner or releasing the Contractor from performing in accordance with the terms of the Contract Documents. The CM will not be responsible for, nor does the CM control, the means, methods, techniques, sequences and procedures of construction for the Project. It is understood that the CM's action in providing quality review under this Agreement is a service of the CM for the sole benefit of the Owner and by performing as provided herein, the CM is not acting in a manner so as to assume responsibility of liability, in whole or in part, for all or any part of the construction for the Project. No action taken by the CM shall relieve the Contractor from its obligation to perform the Work in strict conformity with the requirements of the Contract Documents, and in strict conformity with all other applicable laws, rules and regulations.

3.5.1.12 Contractor's Safety Program: The CM shall require each Contractor that will perform Work at the site to prepare and submit to the CM for general review a safety program, as required by the Contract Documents. The CM shall review each safety program to determine that the programs of the various Contractors performing Work at the site, as submitted, provide for coordination among the Contractors of their respective programs. The CM shall not be responsible for any Contractor's implementation of or compliance with its safety programs, or for initiating, maintaining, monitoring or supervising the implementation of such programs or the procedures and precautions associated therewith, or for the coordination of any of the above with the other Contractors performing the Work at the site. The CM shall not be responsible for the adequacy or completeness of any Contractor's safety programs, procedures or precautions.

3.5.1.13 Disputes Between Contractor and Owner: The CM shall render to the Owner in writing within a reasonable time decisions concerning disputes between the Contractor and the Owner relating to acceptability of the Work, or the interpretation of the requirements of the Contract Documents pertaining to the furnishing and performing of the Work.

3.5.1.14 Operation and Maintenance Materials: The CM shall receive from the Contractor operation and maintenance manuals, warranties and guarantees for materials and equipment installed in the Project. The CM shall deliver this information to the Owner and shall provide a copy of the information to the Designer.

3.5.1.15 Substantial Completion: The CM shall determine when the Project and the Contractor's Work is substantially complete. In consultation with the Designer, the CM shall, prior to issuing a certificate of substantial completion, prepare a list of incomplete Work or Work which does not conform to the requirements of the Contract Documents. This list shall be attached to the certificate of substantial completion.

3.5.1.16 Final Completion: In consultation with the Designer, the CM shall determine when the Project and the Contractor's Work is finally completed, shall issue a certificate of final completion and shall provide to the Owner a written recommendation regarding payment to the Contractor.

3.5.2 Time Management

3.5.2.1 Master Schedule: The CM shall adjust and update the Master Schedule and distribute copies to the Owner and Designer. All adjustments to the Master Schedule shall be made for the benefit of the Project.

3.5.2.2 Contractor's Construction Schedule: The CM shall review the Contractor's Construction Schedule and shall verify that the schedule is prepared in accordance with the requirements of the Contract Documents and that it establishes completion dates that comply with the requirements of the Master Schedule.

Copyright Construction Management Association of America, Inc., 2005. All rights reserved. Reproduction or translation of any part of this Document without the permission of the copyright owner is unlawful.

CMAA Document A-1 (2005 Edition)

3.5.2.3 <u>Construction Schedule Report:</u> The CM shall, on a monthly basis, review the progress of construction of the Contractor, shall evaluate the percentage complete of each construction activity as indicated in the Contractor's Construction Schedule and shall review such percentages with the Contractor. This evaluation shall serve as data for input to the periodic Construction Schedule report that shall be prepared and distributed to the Contractor, Owner and Designer by the CM. The report shall indicate the actual progress compared to scheduled progress and shall serve as the basis for the progress payments to the Contractor. The CM shall advise and make recommendations to the Owner concerning the alternative courses of action that the Owner may take in its efforts to achieve Contract compliance by the Contractor.

3.5.2.4 <u>Effect of Change Orders on the Schedule:</u> Prior to the issuance of a change order, the CM shall determine and advise the Owner as to the effect on the Master Schedule of the change. The CM shall verify that activities and adjustments of time, if any, required by approved change orders have been incorporated into the Contractor's Construction Schedule.

3.5.2.5 <u>Recovery Schedules:</u> The CM may require the Contractor to prepare and submit a recovery schedule as specified in the Contract Documents.

3.5.3 Cost Management

3.5.3.1 <u>Schedule of Values (Each Contract):</u> The CM shall, in participation with the Contractor, determine a schedule of values for the construction Contract. The schedule of values shall be the basis for the allocation of the Contract price to the activities shown on the Contractor's Construction Schedule.

3.5.3.2 <u>Allocation of Cost to the Contractor's Construction Schedule:</u> The Contractor's Construction Schedule shall have the total Contract price allocated by the Contractor among the Contractor's scheduled activities so that each of the Contractor's activities shall be allocated a price and the sum of the prices of the activities shall equal the total Contract price. The CM shall review the Contract price allocations and verify that such allocations are made in accordance with the requirements of the Contract Documents. Progress payments to the Contractor shall be based on the Contractor's

percentage of completion of the scheduled activities as set out in the Construction Schedule reports and the Contractor's compliance with the requirements of the Contract Documents.

3.5.3.3 <u>Effect of Change Orders on Cost:</u> The CM shall advise the Owner as to the effect on the Project and Construction Budget of all proposed and approved change orders.

3.5.3.4 <u>Cost Records:</u> In instances when a lump sum or unit price is not determined prior to the Owner's authorization to the Contractor to perform change order Work, the CM shall request from the Contractor records of the cost of payroll, materials and equipment incurred and the amount of payments to each subcontractor by the Contractor in performing the Work.

3.5.3.5 <u>Trade-off Studies:</u> The CM shall provide trade-off studies for various minor construction components. The results of these studies shall be in report form and distributed to the Owner and Designer.

3.5.3.6 <u>Progress Payments:</u> The CM shall review the payment applications submitted by the Contractor and determine whether the amount requested reflects the progress of the Contractor's Work. The CM shall make appropriate adjustments to each payment application and shall prepare and forward to the Owner a progress payment report. The report shall state the total Contract price, payments to date, current payment requested, retainage and actual amounts owed for the current period. Included in this report shall be a Certificate of Payment that shall be signed by the CM and delivered to the Owner.

3.5.4 Management Information System (MIS)

3.5.4.1 <u>Schedule Maintenance Reports:</u> The CM shall prepare and distribute schedule maintenance reports during the Construction Phase. The reports shall compare the projected completion dates to scheduled completion dates of each separate contract and to the Master Schedule for the Project.

Copyright Construction Management Association of America, Inc., 2005. All rights reserved. Reproduction or translation of any part of this Document without the permission of the copyright owner is unlawful.

CMAA Document A-1 (2005 Edition)

3.5.4.2 Project Cost Reports: The CM shall prepare and distribute Project cost reports during the Construction Phase. The reports shall compare actual Project costs to the Project and Construction Budget.

3.5.4.3 Project and Construction Budget Revisions: The CM shall make recommendations to the Owner concerning changes that may result in revisions to the Project and Construction Budget. Copies of the recommendations shall be provided to the Designer.

3.5.4.4 Cash Flow Reports: The CM shall periodically prepare and distribute cash flow reports during the construction phase. The reports shall compare actual cash flow to planned cash flow.

3.5.4.5 Progress Payment Reports (Each Contract): The CM shall prepare and distribute the Progress Payment reports. The reports shall state the total Contract price, payment to date, current payment requested, retainage, and amounts owed for the period. A portion of this report shall be a recommendation of payment that shall be signed by the CM and delivered to the Owner for use by the Owner in making payments to the Contractor.

3.5.4.6 Change Order Reports: The CM shall periodically during the construction phase prepare and distribute change order reports. The report shall list all Owner-approved change orders by number, a brief description of the change order work, the cost established in the change order and percent of completion of the change order work. The report shall also include similar information for potential change orders of which the CM may be aware.

3.6 Post-Construction Phase

3.6.1 Project Management

3.6.1.1 Record Documents: The CM shall coordinate and expedite submittals of information from the Contractor to the Designer for preparation of record drawings and specifications, and shall coordinate and expedite the transmittal of such record documents to the Owner.

3.6.1.2 Operation and Maintenance Materials and Certificates: Prior to the final completion of the Project, the CM shall compile manufacturers' operations and maintenance manuals, warranties and guarantees, and certificates, and index and bind such documents in an organized manner. This information shall then be provided to the Owner.

3.6.1.3 Occupancy Permit: The CM shall assist the Owner in obtaining an occupancy permit by coordinating final testing, preparing and submitting documentation to governmental agencies, and accompanying governmental officials during inspections of the Project.

3.6.2 Time Management

3.6.2.1 Occupancy Plan: The CM shall prepare an occupancy plan that shall include a schedule for location for furniture, equipment and the Owner's personnel. This schedule shall be provided to the Owner.

3.6.3 Cost Management

3.6.3.1 Change Orders: The CM shall continue during the post-construction phase to provide services related to change orders as specified in Paragraph 3.5.3.3.

3.6.4 Management Information Systems (MIS)

3.6.4.1 Close Out Reports: At the conclusion of the Project, the CM shall prepare and deliver to the Owner final Project accounting and close out reports.

3.6.4.2 MIS Reports for Occupancy: The CM shall prepare and distribute reports associated with the occupancy plan.

ARTICLE 4
ADDITIONAL SERVICES

4.1 At the request of the Owner, the CM shall perform Additional Services and the CM shall be compensated for same as provided in Article 8 of this Agreement. The CM shall be obligated to perform Additional Services only after the Owner and CM have executed a written amendment to this Agreement providing for performance of such services. Additional Services may include, but are not limited to:

Copyright Construction Management Association of America, Inc., 2005. All rights reserved. Reproduction or translation of any part of this Document without the permission of the copyright owner is unlawful.

CMAA Document A-1 (2005 Edition)

4.1.1 Services during the design or construction phases related to investigation, appraisal or evaluation of surface or subsurface conditions at or contiguous to the site or other existing conditions, facilities, or equipment that differs from what is indicated in the Contract Documents, or determination of the accuracy of existing drawings or other information furnished by the Owner;

4.1.2 Services related to the procurement, storage, maintenance and installation of the Owner-furnished equipment, materials, supplies and furnishings;

4.1.3 Services related to determination of space needs;

4.1.4 Preparation of space programs;

4.1.5 Services related to building site investigations and analysis;

4.1.6 Services related to tenant or rental spaces;

4.1.7 Preparation of a Project financial feasibility study;

4.1.8 Preparation of financial, accounting or MIS reports not provided under Basic Services;

4.1.9 Performance of technical inspection and testing;

4.1.10 Preparation of an operations and maintenance manual;

4.1.11 Services related to recruiting and training of maintenance personnel;

4.1.12 Services provided in respect of a dispute between the Owner and the Contractor after the CM has rendered its decision thereon in accordance with Paragraph 3.5.1.13;

4.1.13 Performing warranty inspections during the warranty period of the Project;

4.1.14 Consultation regarding replacement of Work or property damaged by fire or other cause during construction and furnishing services in connection with the replacement of such;

4.1.15 Service made necessary by the default of the Contractor;

4.1.16 Preparation for and serving as a witness in connection with any public or private hearing or arbitration, mediation or legal proceeding;

4.1.17 Assisting the Owner in public relations activities, including preparing information for and attending public meetings; and

4.1.18 Assisting the Owner with procurement and preparation of contracts in connection with the occupancy of the Project, and providing personnel to oversee the location of furniture and equipment;

4.1.19 Services related to the initial operation of any equipment such as start-up, testing, adjusting and balancing.

4.1.20 Any other services not otherwise included in this Agreement.

ARTICLE 5
DURATION OF THE CONSTRUCTION
MANAGER'S SERVICES

5.1 The commencement date for the CM's Basic Services shall be the date of the execution of this Agreement.

5.2 The duration of the CM's Basic Services under this Agreement shall be _____ consecutive calendar days from the commencement date.

5.3 The duration of the CM's Basic Services may be changed only as specified in Article 6.

Copyright Construction Management Association of America, Inc., 2005. All rights reserved. Reproduction or translation of any part of this Document without the permission of the copyright owner is unlawful.

CMAA Document A-1 (2005 Edition)

ARTICLE 6
CHANGES IN THE CONSTRUCTION MANAGER'S BASIC SERVICES AND COMPENSATION

6.1 Owner Changes

6.1.1 The Owner, without invalidating this Agreement, may make changes in the CM's Basic Services specified in Article 3 of this Agreement. The CM shall promptly notify the Owner of changes that increase or decrease the CM's compensation or the duration of the CM's Basic Services or both.

6.1.2 If the scope or the duration of the CM's Basic Services is changed, the CM's compensation shall be adjusted equitably. A written proposal indicating the change in compensation for a change in the scope or duration of Basic Services shall be provided by the CM to the Owner within thirty (30) days of the occurrence of the event giving rise to such request. The amount of the change in compensation to be paid shall be determined on the basis of the CM's cost and a customary and reasonable adjustment in the CM's Fixed Fee, Lump Sum, or multipliers and rates consistent with the provisions of Article 8.

6.2 Authorization

6.2.1 Changes in CM's Basic Services and entitlement to additional compensation or a change in duration of this Agreement shall be made by a written amendment to this Agreement executed by the Owner and the CM. The amendment shall be executed by the Owner and CM prior to the CM performing the services required by the amendment.

6.2.2 The CM shall proceed to perform the services required by the amendment only after receiving written notice from the Owner directing the CM to proceed.

6.3 Invoices for Additional Compensation

6.3.1 The CM shall submit invoices for additional compensation with its invoice for Basic Services and payment shall be made pursuant to the provisions of Article 8 of this Agreement.

ARTICLE 7
OWNER'S RESPONSIBILITIES

7.1 The Owner shall provide to the CM complete information regarding the Owner's knowledge of and requirements for the Project. The Owner shall be responsible for the accuracy and completeness of all reports, data, and other information furnished pursuant to this Paragraph 7.1. The CM may use and rely on the information furnished by the Owner in performing services under this Agreement, and on the reports, data, and other information furnished by the Owner to the Designer.

7.2 The Owner shall be responsible for the presence at the site of any asbestos, PCB's, petroleum, hazardous materials and radioactive materials, and the consequences of such presence.

7.3 The Owner shall examine information submitted by the CM and shall render decisions pertaining thereto promptly.

7.4 The Owner shall furnish legal, accounting and insurance counseling services as may be necessary for the Project.

7.5 The Owner shall furnish insurance for the Project as specified in Article 9.

7.6 If the Owner observes or otherwise becomes aware of any fault or defect in the Project or CM's services or any Work that does not comply with the requirements of the Contract Documents, the Owner shall give prompt written notice thereof to the CM.

7.7 The Owner shall furnish required information and approvals and perform its responsibilities and activities in a timely manner to facilitate orderly progress of the Work in cooperation with the CM consistent with this Agreement and in accordance with the planning and scheduling requirements and budgetary restraints of the Project as determined by the CM.

7.8 The Owner shall retain a Designer whose services, duties and responsibilities shall be described in a written agreement between the Owner and Designer. The services, duties, and responsibilities of the Designer set out in the agreement

Copyright Construction Management Association of America, Inc., 2005. All rights reserved. Reproduction or translation of any part of this Document without the permission of the copyright owner is unlawful.

CMAA Document A-1 (2005 Edition)

between the Owner and Designer shall be compatible and consistent with this Agreement and the Contract Documents. The Owner shall, in its agreement with the Designer, require that the Designer perform its services in cooperation with the CM, consistent with this Agreement and in accordance with the planning, scheduling and budgetary requirements of the Project as determined by the Owner and documented by the CM. The terms and conditions of the agreement between the Owner and the Designer shall not be changed or waived without written consent of the CM, whose consent shall not be unreasonably withheld.

7.9 The Owner shall approve the Project and construction budget and any subsequent revisions as provided in Paragraph 3.2.3.2 of this Agreement.

7.10 The Owner shall cause any and all agreements between the Owner and others to be compatible and consistent with this Agreement. Each of the agreements shall include waiver of subrogation and shall expressly recognize the CM as the Owner's agent in providing the CM's Basic and Additional Services specified in this Agreement.

7.11 At the request of the CM, sufficient copies of the Contract Documents shall be furnished by the Owner at the Owner's expense.

7.12 The Owner shall in a timely manner secure, submit and pay for necessary approvals, easements, assessments, permits and charges required for the construction, use or occupancy of permanent structures or for permanent changes in existing facilities.

7.13 The Owner shall furnish evidence satisfactory to the CM that sufficient funds are available and committed for the entire cost of the Project. Unless such reasonable evidence is furnished, the CM is not required to commence the CM's services and may, if such evidence is not presented within a reasonable time, suspend the services specified in this Agreement upon fifteen (15) days written notice to the Owner. In such event, the CM shall be compensated in the manner provided in Paragraph 10.2.

7.14 The Owner, its representatives and consultants shall communicate with the Contractor only through the CM.

7.15 The Owner shall send to the CM and shall require the Designer to send to the CM copies of all notices and communications sent to or received by the Owner or the Designer relating to the Project. During the construction phase of the Project, the Owner shall require that the Contractor submits all notices and communications relating to the Project directly to the CM.

7.16 The Owner shall designate, in writing, an officer, employee or other authorized representatives to act in the Owner's behalf with respect to the project. This representative shall have the authority to approve changes in the scope of the Project and shall be available during working hours and as often as may be required to render decisions and to furnish information in a timely manner.

7.17 The Owner shall make payments to the Contractor as recommended by the CM on the basis of the Contractor's applications for payment.

7.18 In the case of the termination of the Designer's services, the Owner shall appoint a new Designer who shall be acceptable to the CM and whose responsibilities with respect to the Project and status under the new Agreement with the Owner shall be similar to that of the Designer under the Owner-Designer Agreement and the Contract Documents.

ARTICLE 8
COMPENSATION FOR CM SERVICES
AND PAYMENT

8.1 Compensation Basis

8.1.1 The CM shall receive compensation for its services in accordance with Paragraph 8.2 (Cost Plus Fixed Fee), Paragraph 8.3 (Lump Sum) or 8.4 (Fixed Billable Rates).

8.2 Cost Plus Fixed Fee

8.2.1 Compensation for Basis Services: The Owner shall compensate the CM for performing the Basic Services described in Article 3 on the basis of the CM's cost plus a fixed fee in accordance with the terms and conditions of this Agreement and specifically as follows:

Copyright Construction Management Association of America, Inc., 2005. All rights reserved. Reproduction or translation of any part of this Document without the permission of the copyright owner is unlawful.

CMAA Document A-1 (2005 Edition)

8.2.1.1 A Fixed Fee of_____
_____Dollars ($_____);

8.2.1.2 The cost of employees working on the Project, other than principals, in an amount which equals the multipliers established in Paragraphs 8.2.1.2.1 and 8.2.1.2.2, multiplied by the personnel expense for each such employee. Personnel expense for an employee shall be _____ times the base hourly wage. Personnel expense includes the base hourly wage, payroll taxes, employee benefits and Workers' Compensation insurance. The cost of the CM's principals shall be paid at the rate specified in Paragraph 8.2.1.3. The specified multipliers and rates shall remain constant for a twelve (12) month period following the date of this Agreement. Thereafter, the multipliers established in the referenced paragraphs shall be adjusted by the CM if the CM's personnel cost or expense changes;

8.2.1.2.1 Employees assigned to the Project and working at the construction site of employees for which the Owner provides all office facilities and services, excluding the project manager and assistant project managers, a multiplier of
_____
(_____);

8.2.1.2.2 Employees assigned to the Project and working in the CM's administrative office, including the project manager and assistant project managers, a multiplier of
_____
(_____);

8.2.1.3 Principals of the CM who participate in the Project, a fixed rate of _____ dollars ($_____) per hour. The principals to be compensated according to these terms are:_____
_____
_____
_____
_____
_____
_____
_____
_____
and;

8.2.1.4 Independent engineers, architects and other consultants employed by the CM and performing services related to the Project, a multiplier of (   ) times the amount of the invoice for such services.

8.2.2 Direct Expenses: In addition to the compensation for Basic and Additional Services stated herein, the CM shall be reimbursed for its direct expenses for Basic and Additional Services. Direct expenses are those actual expenditures made by the CM, its principals, employees, independent engineers, architects and other consultants in the interest of the Project, including, without limitation:

8.2.2.1 Long distance telephone calls, telegrams and fees paid for securing approval of authorities having jurisdiction over the Project;

8.2.2.2 Handling, shipping, mailing and reproduction of materials and documents;

8.2.2.3 Transportation and living expenses when traveling in connection with the Project;

8.2.2.4 Computer equipment rental or service fees;

8.2.2.5 Computer software purchased;

8.2.2.6 Electronic data processing service and rental of electronic data processing equipment;

8.2.2.7 Word processing equipment rental;

8.2.2.8 Premiums for insurance beyond the limits normally carried by the CM that are required by the terms of this Agreement;

8.2.2.9 Relocation of employees and their families.

8.2.2.10 Temporary living expenses of employees who are not relocated, but assigned to the Project;

8.2.2.11 Gross receipts taxes, sales or use taxes, service taxes and other similar taxes required to be paid as a result of this Agreement;

Copyright Construction Management Association of America, Inc., 2005. All rights reserved. Reproduction or translation of any part of this Document without the permission of the copyright owner is unlawful.

CMAA Document A-1 (2005 Edition)

8.2.2.12 Field office expenses including the cost of office rentals, telephones, utilities, furniture, equipment and supplies; and

8.2.2.13 The cost of premium time.

8.2.2.14 Legal cost reasonably and properly incurred by the CM in connection with the performance of its duties under this Agreement.

8.2.3 CM's Accounting Records: Records of the CM's personnel expense, independent engineers', architects' and other consultants' fees and direct expenses pertaining to the Project shall be maintained on the basis of generally accepted accounting practices and shall be available for inspection by the Owner or the Owner's representative at mutually convenient times for a period of two years after completion of the CM's Basic Services.

8.2.4 Payment: Payments to the CM shall be made monthly, not later than fifteen (15) days after presentation of the CM's invoice to the Owner, as follows:

8.2.4.1 Payment of the Fixed Fee as indicated in Paragraph 8.2.1.1 shall be in amounts prorated equally over the duration of the CM's Basic Services. The duration shall be as set out in Article 5;

8.2.4.2 Payment of personnel expense and the fixed hourly rate for principals shall be in amounts equal to the actual hours spent during the billing period on the Project multiplied by the rates and multipliers stated in Paragraphs 8.2.1.2, 8.2.1.2.1, 8.2.1.2.2 and 8.2.1.3;

8.2.4.3 Payment of independent engineer, architect and other consultant services shall be in amounts equal to the invoice in receipt by the CM for the billing period times the multiplier stated in Paragraph 8.2.1.4;

8.2.4.4 Reimbursement for direct expenses shall be in amount equal to expenditures made during the billing period and during previous billing periods not yet invoiced;

8.2.4.5 No deductions shall be made from the CM's compensation due to any claim by the Owner, Contractor or others not a party to this Agreement or due to any liquidated

damages, retainage or other sums withheld from payments to the Contractor or others not a party to this Agreement; and

8.2.4.6 Payments due the CM that are unpaid for more than thirty (30) days from the date of the CM's invoice shall bear interest at the annual rate of _____% from the due date, compounded monthly. In addition, since timely payment is an essential condition of this Agreement, the CM may, after giving seven (7) days written notice to the Owner, suspend services under this Agreement until the CM has been paid in full all amounts due for services, expenses and charges, including accrued interest.

8.2.5 Compensation for Additional Services: The CM shall be compensated and payments shall be made for performing Additional Services in the same amount and manner as provided in Article 8 for Basic Services. There shall be an increase in the fixed fee set out in Paragraph 8.2.1.1 in an amount, which is mutually agreeable between the Owner and CM.

8.3 Lump Sum

8.3.1 Compensation for Basic Services: The Owner shall compensate the CM for performing Basic Services described in Article 3, a total Lump Sum in the amount of_____

_____ dollars

($),   which amount shall be paid in

_____ monthly installments as

follows:

Installment No.	Installment Due Date	Installment Amount

Copyright Construction Management Association of America, Inc., 2005. All rights reserved. Reproduction or translation of any part of this Document without the permission of the copyright owner is unlawful.

CMAA Document A-1 (2005 Edition)

_____
_____
_____
_____
_____
_____
_____

8.3.2 <u>Direct Expenses:</u> The cost of direct expenses incurred shall be included in the Lump Sum.

8.3.3 <u>Payments:</u> Payments shall be made monthly, not later than fifteen (15) days after receipt of the CM's invoice by the Owner.

8.3.3.1 No deductions shall be made from the CM's compensation due to any claim of the Owner, Contractor or others not a party to this agreement or due to any liquidated damages, retainage or other sums withheld from payments to Contractor or others not a party to this Agreement.

8.3.3.2 Payments due the CM that are unpaid for more than thirty (30) days from the date of the CM's invoice shall bear interest at the annual rate of _____% from the due date, compounded monthly. In addition, since timely payment is an essential condition of this Agreement, the CM may, after giving seven (7) days written notice to the Owner, suspend services under this Agreement until the CM has been paid in full all amounts due for services, expenses and charges, including accrued interest.

8.3.4 <u>Compensation for Additional Services:</u> The CM shall be compensated and payments shall be made for performing Additional Services in an amount and on terms mutually agreeable between the Owner and CM.

8.4 <u>Fixed Billable Rates</u>

8.4.1 Compensation for Basic Services: The Owner shall compensate the CM for performing Basic Services described in Article 3 on the basis of fixed billable rates in accordance with the terms and conditions of this Agreement and specifically as follows:

8.4.1.1 The cost of employees working on the Project, in an amount which equals the billable rates of the employees as established in Paragraphs 8.4.1.1.1 and 8.4.1.1.2, multiplied by the hours for each such employee. These billable hourly rates are inclusive of all profit (fee), general administrative overhead costs, and personnel expense for each employee. Personnel expense includes the base hourly wage, payroll taxes, employee benefits and Workers' Compensation insurance. The specified rates shall remain constant for a twelve (12) month period following the date of this Agreement. Thereafter, the rates established in the referenced paragraphs shall be adjusted by a _____% escalation factor for each successive twelve (12) month period.

8.4.1.1.1 Employees assigned to the Project and working at the construction site, or employees for which the Owner provides all office facilities and services, the following hourly rates:

_____
_____
_____
_____
_____
_____
_____
_____

8.4.1.1.2 Employees assigned to the Project and working in the CM's administrative office, the following hourly rates:

_____
_____
_____
_____
_____
_____
_____
_____
_____

Copyright Construction Management Association of America, Inc., 2005. All rights reserved. Reproduction or translation of any part of this Document without the permission of the copyright owner is unlawful.

CMAA Document A-1 (2005 Edition)

8.4.1.2 Independent engineers, architects and other consultants employed by the CM and performing services related to the Project, a multiplier of _____(_____) times the amount of the invoice for such services.

8.4.2 Direct Expenses: In addition to the compensation for Basic and Additional Services stated herein, the CM shall be reimbursed for its direct expenses incurred in providing Basic and Additional Services. Direct expenses for those actual expenditures made by the CM, its principals, employees, independent engineers, architects and other consultants in the interest of the Project, including, without limitation:

8.4.2.1 Long distance telephone calls, telegrams and fees paid for securing approval of authorities having jurisdiction over the over the Project;

8.4.2.2 Handling, shipping, mailing and reproduction of materials and documents;

8.4.2.3 Transportation and living expenses when traveling in connection with the Project;

8.4.2.4 Computer equipment rental or service fees;

8.4.2.5 Computer software purchased;

8.4.2.6 Electronic data processing service and rental of electronic data processing equipment;

8.4.2.7 Word processing equipment rental;

8.4.2.8 Premiums for insurance beyond the limits normally carried by the CM that are required by the terms of this Agreement;

8.4.2.9 Relocation of employees and families;

8.4.2.10 Temporary living expenses of employees who are not relocated, but assigned to the Project;

8.4.2.11 Gross receipts taxes, sales or use taxes, service taxes and other similar taxes required to be paid as a result of this Agreement;

8.4.2.12 Field office expenses, including the cost of office rentals, telephones, utilities, furniture, equipment and supplies; and

8.4.2.13 Premium time work.

8.4.3 CM's Account Records: Records of the CM's personnel expense, independent engineers', architects' and other consultants' fees and direct expenses pertaining to the Project shall be maintained on the basis of generally accepted account practices and shall be available for inspection by the Owner or the Owner's representative at mutually convenient times for a period of two (2) years after completion of the construction phase Basic Services.

8.4.4 Payments: Payments to the CM shall be made monthly, not later than fifteen (15) days after receipt of the CM's invoice by the Owner, as follows:

8.4.4.1 Payment of the fixed hourly rate for employees shall be in amounts equal to the actual hours spent during the billing period on the Project multiplied by the rates stated in Paragraphs 8.4.1.1.1 and 8.4.1.1.2;

8.4.4.2 Payment of independent engineers', architects' and other consultants' services shall be in amounts equal to the invoice in receipt by the CM for the billing period times the multiplier stated in Paragraph 8.4.1.2;

8.4.4.3 Reimbursement for direct expense shall be in amounts equal to expenditures made during the billing period and during previous billing periods not yet invoiced;

8.4.4.4 No deductions shall be made from the CM's compensation due to any claim by the Owner, Contractor or others not a party to this Agreement or due to any liquidated damages, retainage or other sums withheld from payments to the Contractor or others not a party to this Agreement

8.4.4.5 Payments due the CM that are unpaid for more than thirty (30) days from the date of receipt by the Owner of the CM's invoice shall bear interest at the annual rate of _____% from the due date, compounded monthly. In additional, since timely payment is an essential condition of this Agreement, the CM may, after giving seven (7) days written notice to the Owner, suspend services under this

Copyright Construction Management Association of America, Inc., 2005. All rights reserved. Reproduction or translation of any part of this Document without the permission of the copyright owner is unlawful.

CMAA Document A-1 (2005 Edition)

Agreement until the CM has been paid in full all amounts due for services, expenses and charges, including accrued interest.

8.4.5 Compensation for Additional Services: The CM shall be compensated and payments shall be made for performing Additional Services in an amount and on terms mutually agreeable between the Owner and the CM.

ARTICLE 9
INSURANCE AND MUTUAL INDEMNITY

9.1   Construction Manager's Liability Insurance

9.1.1   General Liability: The CM shall procure and maintain insurance for protection from claims under Worker's Compensation Acts, from claims for damages because of bodily injury including personal injury, sickness or disease or death of any or all employees or of any person other than such employees, and from claims or damages because of injury to or destruction of property including loss of use resulting therefrom.

9.1.2   Commercial General Liability Insurance may be obtained under a single policy for the full limits required or by a combination of underlying policies with the balance provided by an excess or umbrella policy.

9.1.3   The foregoing policies shall contain a provision that coverages afforded under the policies shall not be cancelled or expire until at least thirty (30) days written notice has been given to the Owner and shall include either a liability endorsement covering this Agreement or an endorsement making the Owner an additional insured under the policies. Certificates of insurance showing such coverages to be in force shall be filed with the Owner prior to commencement of the CM's services.

9.1.4   Professional Liability: The CM shall procure and maintain professional liability insurance for protection from claims arising out of the performance of professional services caused by a negligent error, omission or act for which the insured is legally liable; such liability insurance will provide for coverage in such amounts, with such deductible provisions and for such period of time as required by the Owner and are

commercially available.   Certificates indicating that such insurance is in effect shall be delivered to the Owner. The CM shall also cause the independent engineers, architects and other consultants retained by the CM for the Project to procure and maintain professional liability insurance coverage, for at least such amounts, deductibles, and periods as determined by the Owner.

9.2   Owner's Insurance

9.2.1 The Owner shall be responsible for purchasing and maintaining its own liability insurance, and at the Owner's option, may purchase and maintain such additional insurance to protect the Owner against claims losses, or damages that may arise from the Project.

9.2.2 The CM, as agent of the Owner, shall be named as an additional insured in any insurance policy obtained by the Owner and the Contractor for the Project.

9.3   Notices and Recovery

9.3.1 The Owner and CM each shall provide the other with copies of all policies thus obtained for the Project. Each party shall provide the other thirty (30) days written notice of cancellation, non-renewal or endorsement reducing or restricting coverage.

9.4   Waiver of Subrogation

9.4.1 The Owner and the CM waive all rights against each other and against the Contractor, Designer, and other consultants, subcontractors, suppliers, agents and employees of the other for damages during construction covered by any property insurance as set forth in the Contract Documents. The Owner and the CM shall each require appropriate similar waivers from their contractors, designers, and other consultants, subcontractors, suppliers, agents and employees.

9.5   Indemnity

9.5.1 To the fullest extent permitted by law, the CM shall indemnify and hold harmless the Owner, its employees, agents, officers, directors and partners from and against any and all damages arising from bodily injury or property damage

Copyright Construction Management Association of America, Inc., 2005.   All rights reserved.   Reproduction or translation of any part of this Document without the permission of the copyright owner is unlawful.

CMAA Document A-1 (2005 Edition)

and reasonable attorneys' fees incurred by the Owner caused solely by the negligent act, error or omission of the CM, or the CM's consultants, or any other party for whom the CM is legally liable, in performance of services under this Agreement. The CM shall procure and maintain insurance as required by and set forth in this Agreement. Notwithstanding any other provision of this Agreement, and to the fullest extent permitted by law, the total liability, in the aggregate, of the CM and the CM's consultants and the officers, directors, partners, employees, and agents of any of them, to the Owner and anyone claiming by, through or under the Owner, for any and all claims, losses, costs or damages whatsoever arising out of, resulting from or in any way related to the Project or this Agreement from any cause or causes, including but limited to the negligence, professional errors or omissions, strict liability or breach of contract or warranty (express or implied) of the CM or the CM's consultants and the officers, directors, partners, employees, and agents of any of them, (hereafter "the Owner's claims"), shall not exceed the total insurance proceeds paid on behalf of or to the CM by the CM's insurers in settlement or satisfaction of the Owner's claims under the terms and conditions of the CM's insurance policies applicable thereto.

9.5.2 The Owner shall cause the Designer to indemnify and hold harmless the Owner, its employees, agents and representatives to the same extent and in the same manner that CM has provided indemnification for the Designer under Paragraph 9.5.1.

9.5.3 The Owner hereby indemnifies and holds harmless the CM and its employees, agents and representatives from and against any and all claims, demands, suits and damages for bodily injury and property damage for which the Owner is liable that arise out of or result from negligent acts or omissions of the Owner, its employees, agents, representatives, independent contractors, suppliers, the Contractor and the Designer.

9.5.4 The Owner shall cause the Contractor to indemnify and hold harmless the CM from and against any and all claims, demands, suits, damages, including consequential damages and damages resulting from personal injury or property damage, costs, and expenses and fees that are asserted against the CM and that arise out of or result from negligent acts or omissions by the Contractor, its employees, agents and representatives in performing the Work.

ARTICLE 10
TERMINATION AND SUSPENSION

10.1 Termination

10.1.1 This Agreement may be terminated by the Owner for convenience after seven (7) days written notice to the CM.

10.1.2 This Agreement may be terminated by either party hereto upon seven (7) days written notice should the other party fail substantially to perform in accordance with the terms hereof through no fault of the terminating party or if the Project in whole or substantial part is stopped for a period of sixty (60) days under an order of any court or other public authority having jurisdiction or as result of an act of government.

10.1.3 In the event of termination under Paragraph 10.1.1, the CM shall be paid its compensation for services performed to the date of termination, services of engineers, architects and consultants then due and all termination expenses. Termination expenses are defined as those expenses arising prior, during and subsequent to termination that are directly attributable to the termination, plus an amount computed as a percentage of the total compensation earned at the time of termination computed as follows:

10.1.3.1 Twenty (20) percent if the termination occurs during the Pre-Design Phase, Design Phase or Procurement Phase; or

10.1.3.2 Ten (10) percent if the termination occurs during the Construction Phase or Post-Construction Phase.

Copyright Construction Management Association of America, Inc., 2005. All rights reserved. Reproduction or translation of any part of this Document without the permission of the copyright owner is unlawful.

CMAA Document A-1 (2005 Edition)

10.1.4 In the event of termination under Paragraph 10.1.2, the CM shall be paid its compensation for services performed to the date of termination, services of professional consultants then due and all termination expenses. No amount computed as provided in Paragraphs 10.1.3.1 and 10.1.3.2 shall be paid in addition if the termination is due to the CM's failure to substantially perform in accordance with the terms of this Agreement.

10.2 Suspension

10.2.1 The Owner may, in writing, order the CM to suspend all or any part of the CM's services for the Project for the convenience of the Owner or for stoppage beyond the control of the Owner or the CM. If the performance of all or any part of the services for the Project is so suspended, an adjustment in the CM's compensation shall be made for the increase, if any, in the cost of the CM's performance of this Agreement caused by such suspension and this Agreement shall be modified in writing accordingly.

10.2.2 In the event the CM's services for the Project are suspended, the Owner shall reimburse the CM for all of the costs of its construction site staff, assigned Project home office staff and other costs as provided for by this Agreement for the first thirty (30) days of such suspension. The CM shall reduce the size of its project staff for the remainder of the suspension period as directed by the Owner and, during such period, the Owner shall reimburse the CM for all of the costs of its reduced staff. Upon cessation of the suspension, the CM shall restore the construction site staff and home office staff to its former size.

10.2.3 Persons assigned by the CM to another project during such suspension periods and not available to return to the Project upon cessation of the suspension shall be replaced. The Owner shall reimburse the CM for costs incurred for relocation of previous staff persons returning to the Project or for new persons assigned to the Project.

10.2.4 If the Project is suspended by the Owner for more than three (3) months, the CM shall be paid compensation for services performed prior to receipt of written notice from the Owner of the suspension, together with direct expenses then due and all expenses and costs directly resulting from the suspension. If the Project is to be resumed after being

suspended for more that six (6) months, the CM shall have the option of requiring that its compensation, including rates and fees, be renegotiated, that payments required hereunder shall have been made by the Owner, and that adjustments to this Agreement related to the suspension have been made by written amendment to this Agreement. Subject to the provisions off this Agreement relating to termination, a suspension of the Project does not void this Agreement.

ARTICLE 11
DISPUTE RESOLUTION

11.1 The Owner and the CM shall submit all unresolved claims, counterclaims, disputes, controversies, and other matters in question between them arising out of or relating to this Agreement or the breach thereof ("disputes"), to mediation prior to either party initiating against the other a demand for arbitration pursuant to Paragraph 11.2 below, unless delay in initiating or prosecuting a proceeding in an arbitration or judicial forum would prejudice the Owner or the CM. The Owner and the CM shall agree in writing as to the identity of the mediator and the rules and procedures of the mediation. If the Owner and CM cannot, agree, the dispute shall be submitted to mediation under the then current Construction Industry Mediation Rules of the American Arbitration Association.

11.2 All disputes that the Owner and CM are unable to resolve by mediation as aforesaid shall be decided by arbitration, subject to the limitations provided herein. The agreement to arbitrate, and any other agreement or consent to arbitrate entered into in accordance herewith shall be specifically enforceable under the prevailing law of any court having jurisdiction. The Owner and the CM shall agree in writing as to the identity of the arbitrator(s) and the rules and procedures of the arbitration. If the Owner and the CM do not so agree, then the Owner and the CM shall submit the dispute to arbitration under the then current Construction Industry Rules of the American Arbitration Association.

11.3 Notice of demand for arbitration must be filed in writing with the other party to this Agreement and with the arbitrator(s). The demand must be made within a reasonable time after the dispute has arisen, but not prior to or during the pendency of the mediation as agreed in Paragraph 11.1. In

Copyright Construction Management Association of America, Inc., 2005. All rights reserved. Reproduction or translation of any part of this Document without the permission of the copyright owner is unlawful.

CMAA Document A-1 (2005 Edition)

no event may the demand for arbitration be made after the date when institution of legal or equitable proceedings based on such dispute in question would be barred by the applicable statute of limitations or of repose.

11.4 No arbitration arising out of, or relating to, this Agreement may include, by consolidation, joinder or in any other manner, any person or entity who is not a party to this Agreement unless both parties agree otherwise in writing. No consent to arbitration in respect of a specifically described dispute will constitute consent to arbitrate any other dispute which is not specifically described in such consent or which is with any party not specifically described therein.

11.5 All demands for arbitration and answering statements thereto which include any monetary claim must contain a statement that the total sum or value in controversy as alleged by the party making such demand or answering statement is not more than $_____ (exclusive of interest and costs). The arbitrators will not have jurisdiction, power or authority to consider, or make findings (except in denial of their own jurisdiction) concerning any claim, counterclaim, dispute or other matter in question where the amount in controversy of any such claim, counterclaim, dispute or matter is more than $_____ (exclusive of interest and costs), or to render a monetary award in response thereto against any party which totals more than $_____ (exclusive of interest and costs).

11.6 The award rendered by the arbitrator(s) will be final, judgment may be entered upon it in any court having jurisdiction thereof, and the award will not be subject to modification or appeal. In any judicial proceeding to enforce this Agreement to arbitrate, the only issues to be determined shall be those set forth in 9 U.S.C. Section 4 Federal Arbitration Act, and such issues shall be determined by the Court without a jury. All other issues, such as, but not limited to, arbitrability, prerequisites to arbitration, compliance with contractual time limits, applicability of indemnity clauses, clauses limiting damages and statutes of limitations shall be for the arbitrator(s), whose decision thereon shall be final and binding. There shall be no interlocutory appeal of an order compelling arbitration.

11.7 Unless otherwise agreed in writing, the CM shall continue to carry out its responsibilities under this Agreement during any dispute, and the Owner shall continue to make payments in accordance with this Agreement.

ARTICLE 12
ADDITIONAL PROVISIONS

12.1 Confidentiality

12.1.1 The CM will keep all information concerning the Project confidential, except for communications incident to completion of the Project between the CM, Designer, and Contractor, and their independent professional engineers, architects and other consultants and subcontractors, and except for publicity approved by the Owner and communications in connection with filings with governmental bodies having jurisdiction over the design or construction of the Project.

12.2 Limitation and Assignment

12.2.1 The Owner and the CM each binds itself, its successors, assigns and legal representatives to the terms of this Agreement.

12.2.2 Neither the Owner nor the CM shall assign or transfer its interest in this Agreement without the written consent of the other, except that the CM may, without approval of the Owner, assign accounts receivable to a commercial bank for securing loans.

12.3 Governing Law

12.3.1 This Agreement shall, unless otherwise provided, be governed by the law of the state where the Project is located.

12.4 Extent of Agreement

12.4.1 This Agreement constitutes the entire agreement between the parties and incorporates all prior agreements and understandings in connection with the subject matter hereof. This Agreement may be amended only by a written amendment signed by the Owner and CM. Nothing contained

Copyright Construction Management Association of America, Inc., 2005. All rights reserved. Reproduction or translation of any part of this Document without the permission of the copyright owner is unlawful.

CMAA Document A-1 (2005 Edition)

in this Agreement is intended to benefit any person or party, other than the Designer, but for Designer only under the circumstances as specifically defined herein. In addition, it is expressly agreed that neither the Contractor, its subcontractors and suppliers, any other contractors or consultants of the Owner or CM, nor any other person or party providing any part of the design services or Work are intended beneficiaries of this Agreement.

12.5 Severability

12.5.1 If any portion of this Agreement is held as a matter of law to be unenforceable, the remainder of this Agreement shall be enforceable without such portion.

12.6 Meaning of Terms

12.6.1 References made in the singular shall include the plural and the masculine shall include the feminine or the neuter.

12.6.2 The meaning of terms used herein shall be consistent with the definitions expressed in the CMAA Standard Form Agreements, Contracts and General Conditions.

12.7 Notices

12.7.1 Whenever any provision of the Contract Documents requires the giving of written notice, it shall be deemed to have been validly given if delivered in person to the individual or to a member of the firm or to an officer of the corporation for whom it is intended or if delivered or sent by registered or certified mail, postage prepaid, addressed as follows:

To the Owner:

_____
_____
_____
_____

To the CM:

_____
_____
_____
_____

(Remainder of Page intentionally left blank.)

Copyright Construction Management Association of America, Inc., 2005. All rights reserved. Reproduction or translation of any part of this Document without the permission of the copyright owner is unlawful.

CMAA Document A-1 (2005 Edition)

IN WITNESS WHEREOF, the parties have duly executed this Contract as of the date set forth on page 1 hereof.

ATTEST: OWNER

Witness: _____   By: _____

                                                 Title: _____

                                                 CONSTRUCTION MANAGER

Witness: _____   By: _____

                                                 Title: _____

Copyright Construction Management Association of America, Inc., 2005. All rights reserved. Reproduction or translation of any part of this Document without the permission of the copyright owner is unlawful.

American Institute of Architects stipulated sum contract

 **Document A101™ – 2007**

Standard Form of Agreement Between Owner and Contractor *where the basis of payment is a Stipulated Sum*

AGREEMENT made as of the   day of
in the year
(In words, indicate day, month and year)

BETWEEN the Owner:
(Name, address and other information)

> This document has important legal consequences. Consultation with an attorney is encouraged with respect to its completion or modification.
>
> AIA Document A201™–2007, General Conditions of the Contract for Construction, is adopted in this document by reference. Do not use with other general conditions unless this document is modified.

and the Contractor:
(Name, address and other information)

for the following Project:
(Name, location, and detailed description)

The Architect:
(Name, address and other information)

The Owner ar d Contractor agree as follows.

Init.

/

AIA Document A101™ – 2007. Copyright © 1915, 1918, 1925, 1937, 1951, 1958, 1961, 1963, 1967, 1974, 1977, 1987, 1991, 1997 and 2007 by The American Institute of Architects. All rights reserved. WARNING: This AIA® Document is protected by U.S. Copyright Law and International Treaties. Unauthorized reproduction or distribution of this AIA® Document, or any portion of it, may result in severe civil and criminal penalties, and will be prosecuted to the maximum extent possible under the law. Purchasers are permitted to reproduce ten (10) copies of this document when completed. To report copyright violations of AIA Contract Documents, e-mail The American Institute of Architects' legal counsel, copyright@aia.org.

Reproduced with permission of The American Institute of Architects, 1735 New York Avenue, Washington, D.C. 20006

TABLE OF ARTICLES

ARTICLE 1 THE CONTRACT DOCUMENTS

The Contract Documents consist of this Agreement, Conditions of the Contract (General, Supplementary and other Conditions), Drawings, Specifications, Addenda issued prior to execution of this Agreement, other documents listed in this Agreement and Modifications issued after execution of this Agreement, all of which form the Contract, and are as fully a part of the Contract as if attached to this Agreement or repeated herein. The Contract represents the entire and integrated agreement between the parties hereto and supersedes prior negotiations, representations or agreements, either written or oral. An enumeration of the Contract Documents, other than a Modification, appears in Article 9.

ARTICLE 2 THE WORK OF THIS CONTRACT

The Contractor shall fully execute the Work described in the Contract Documents, except as specifically indicated in the Contract Documents to be the responsibility of others.

ARTICLE 3 DATE OF COMMENCEMENT AND SUBSTANTIAL COMPLETION

§ 3.1 The date of commencement of the Work shall be the date of this Agreement unless a different date is stated below or provision is made for the date to be fixed in a notice to proceed issued by the Owner.

(Insert the date of commencement if it differs from the date of this Agreement or, if applicable, state that the date will be fixed in a notice to proceed.)

If, prior to the commencement of the Work, the Owner requires time to file mortgages and other security interests, the Owner's time requirement shall be as follows:

§ 3.2 The Contract Time shall be measured from the date of commencement.

Init.

/

AIA Document A101™ – 2007. Copyright © 1915, 1918, 1925, 1937, 1951, 1958, 1961, 1963, 1967, 1974, 1977, 1980, 1987, 1991, 1997 and 2007 by The American Institute of Architects. All rights reserved. WARNING: This AIA® Document is protected by U.S. Copyright Law and International Treaties. Unauthorized reproduction or distribution of this AIA® Document, or any portion of it, may result in severe civil and criminal penalties, and will be prosecuted to the maximum extent possible under the law. Purchasers are permitted to reproduce ten (10) copies of this document when completed. To report copyright violations of AIA Contract Documents, e-mail The American Institute of Architects' legal counsel, copyright@aia.org.

§ 3.3 The Contractor shall achieve Substantial Completion of the entire Work not later than
(   ) days from the date of commencement, or as follows:
(Insert number of calendar days. Alternatively, a calendar date may be used when coordinated with the date of commencement. If appropriate, insert requirements for earlier Substantial Completion of certain portions of the Work.)

, subject to adjustments of this Contract Time as provided in the Contract Documents.
(Insert provisions, if any, for liquidated damages relating to failure to achieve Substantial Completion on time or for bonus payments for early completion of the Work.)

ARTICLE 4 CONTRACT SUM
§ 4.1 The Owner shall pay the Contractor the Contract Sum in current funds for the Contractor's performance of the Contract. The Contract Sum shall be
Dollars ($), subject to additions and deductions as provided in the Contract Documents.

§ 4.2 The Contract Sum is based upon the following alternates, if any, which are described in the Contract Documents and are hereby accepted by the Owner:
(State the numbers or other identification of accepted alternates. If the bidding or proposal documents permit the Owner to accept other alternates subsequent to the execution of this Agreement, attach a schedule of such other alternates showing the amount for each and the date when that amount expires.)

§ 4.3 Unit prices, if any:
(Identify and state the unit price; state quantity limitations, if any to which the unit price will be applicable.)

Item	Units and Limitations	Price Per Unit

§ 4.4 Allowances included in the Contract Sum, if any:
(Identify allowance and state exclusions, if any, from the allowance price.)

Item	Price

ARTICLE 5 PAYMENTS
§ 5.1 PROGRESS PAYMENTS
§ 5.1.1 Based upon Applications for Payment submitted to the Architect by the Contractor and Certificates for Payment issued by the Architect, the Owner shall make progress payments on account of the Contract Sum to the Contractor as provided below and elsewhere in the Contract Documents.

Init.

/

AIA Document A101™ – 2007. Copyright © 1915, 1918, 1925, 1937, 1951, 1958, 1961, 1963, 1967, 1974, 1977, 1980, 1987, 1991, 1997 and 2007 by The American Institute of Architects. All rights reserved. WARNING: This AIA® Document is protected by U.S. Copyright Law and International Treaties. Unauthorized reproduction or distribution of this AIA® Document, or any portion of it, may result in severe civil and criminal penalties, and will be prosecuted to the maximum extent possible under the law. Purchasers are permitted to reproduce ten (10) copies of this document when completed. To report copyright violations of AIA Contract Documents, e-mail The American Institute of Architects' legal counsel, copyright@aia.org.

§ 5.1.2 The period covered by each Application for Payment shall be one calendar month ending on the last day of the month, or as follows:

§ 5.1.3 Provided that an Application for Payment is received by the Architect not later than the
(　　　　　) day of a month, the Owner shall make payment of the certified amount to the Contractor not later than the
(　　　　　) day of the (　　　　　) month. If an Application for Payment
is received by the Architect after the application date fixed above, payment shall be made by the Owner not later than
(　　　　　) days after the Architect receives the Application for Payment.
(Federal, state or local laws may require payment within a certain period of time.)

§ 5.1.4 Each Application for Payment shall be based on the most recent schedule of values submitted by the Contractor in accordance with the Contract Documents. The schedule of values shall allocate the entire Contract Sum among the various portions of the Work. The schedule of values shall be prepared in such form and supported by such data to substantiate its accuracy as the Architect may require. This schedule, unless objected to by the Architect, shall be used as a basis for reviewing the Contractor's Applications for Payment.

§ 5.1.5 Applications for Payment shall show the percentage of completion of each portion of the Work as of the end of the period covered by the Application for Payment.

§ 5.1.6 Subject to other provisions of the Contract Documents, the amount of each progress payment shall be computed as follows:
  .1 Take that portion of the Contract Sum properly allocable to completed Work as determined by multiplying the percentage completion of each portion of the Work by the share of the Contract Sum allocated to that portion of the Work in the schedule of values, less retainage of
percent (　　　　　%). Pending final determination of cost to the Owner of changes in the Work, amounts not in dispute shall be included as provided in Section 7.3.9 of AIA Document A201™–2007, General Conditions of the Contract for Construction;
  .2 Add that portion of the Contract Sum properly allocable to materials and equipment delivered and suitably stored at the site for subsequent incorporation in the completed construction (or, if approved in advance by the Owner, suitably stored off the site at a location agreed upon in writing), less retainage of
percent (　　　　　%);
  .3 Subtract the aggregate of previous payments made by the Owner; and
  .4 Subtract amounts, if any, for which the Architect has withheld or nullified a Certificate for Payment as provided in Section 9.5 of AIA Document A201–2007.

§ 5.1.7 The progress payment amount determined in accordance with Section 5.1.6 shall be further modified under the following circumstances:
  .1 Add, upon Substantial Completion of the Work, a sum sufficient to increase the total payments to the full amount of the Contract Sum, less such amounts as the Architect shall determine for incomplete Work, retainage applicable to such work and unsettled claims; and
(Section 9.8.5 of AIA Document A201–2007 requires release of applicable retainage upon Substantial Completion of Work with consent of surety, if any.)
  .2 Add, if final completion of the Work is thereafter materially delayed through no fault of the Contractor, any additional amounts payable in accordance with Section 9.10.3 of AIA Document A201–2007.

§ 5.1.8 Reduction or limitation of retainage, if any, shall be as follows:
(If it is intended, prior to Substantial Completion of the entire Work, to reduce or limit the retainage resulting from the percentages inserted in Sections 5.1.6.1 and 5.1.6.2 above, and this is not explained elsewhere in the Contract Documents, insert here provisions for such reduction or limitation.)

Init.

/

AIA Document A101™ – 2007. Copyright © 1915, 1918, 1925, 1937, 1951, 1958, 1961, 1963, 1967, 1974, 1977, 1980, 1987, 1991, 1997 and 2007 by The American Institute of Architects. All rights reserved. WARNING: This AIA® Document is protected by U.S. Copyright Law and International Treaties. Unauthorized reproduction or distribution of this AIA® Document, or any portion of it, may result in severe civil and criminal penalties, and will be prosecuted to the maximum extent possible under the law. Purchasers are permitted to reproduce ten (10) copies of this document when completed. To report copyright violations of AIA Contract Documents, e-mail The American Institute of Architects' legal counsel, copyright@aia.org.

§ 5.1.9 Except with the Owner's prior approval, the Contractor shall not make advance payments to suppliers for materials or equipment which have not been delivered and stored at the site.

§ 5.2 FINAL PAYMENT
§ 5.2.1 Final payment, constituting the entire unpaid balance of the Contract Sum, shall be made by the Owner to the Contractor when
    .1   the Contractor has fully performed the Contract except for the Contractor's responsibility to correct Work as provided in Section 12.2.2 of AIA Document A201–2007, and to satisfy other requirements, if any, which extend beyond final payment; and
    .2   a final Certificate for Payment has been issued by the Architect.

§ 5.2.2 The Owner's final payment to the Contractor shall be made no later than 30 days after the issuance of the Architect's final Certificate for Payment, or as follows:

ARTICLE 6 DISPUTE RESOLUTION
§ 6.1 INITIAL DECISION MAKER
The Architect will serve as Initial Decision Maker pursuant to Section 15.2 of AIA Document A201–2007, unless the parties appoint below another individual, not a party to this Agreement, to serve as Initial Decision Maker.
(If the parties mutually agree, insert the name, address and other contact information of the Initial Decision Maker, if other than the Architect.)

§ 6.2 BINDING DISPUTE RESOLUTION
For any Claim subject to, but not resolved by, mediation pursuant to Section 15.3 of AIA Document A201–2007, the method of binding dispute resolution shall be as follows:
(Check the appropriate box. If the Owner and Contractor do not select a method of binding dispute resolution below, or do not subsequently agree in writing to a binding dispute resolution method other than litigation, Claims will be resolved by litigation in a court of competent jurisdiction.)

☐   Arbitration pursuant to Section 15.4 of AIA Document A201–2007

☐   Litigation in a court of competent jurisdiction

☐   Other *(Specify)*

ARTICLE 7 TERMINATION OR SUSPENSION
§ 7.1 The Contract may be terminated by the Owner or the Contractor as provided in Article 14 of AIA Document A201–2007.

§ 7.2 The Work may be suspended by the Owner as provided in Article 14 of AIA Document A201–2007.

Init.

/

AIA Document A101™ – 2007. Copyright © 1915, 1918, 1925, 1937, 1951, 1958, 1961, 1963, 1967, 1974, 1977, 1980, 1987, 1991, 1997 and 2007 by The American Institute of Architects. All rights reserved. WARNING: This AIA® Document is protected by U.S. Copyright Law and International Treaties. Unauthorized reproduction or distribution of this AIA® Document, or any portion of it, may result in severe civil and criminal penalties, and will be prosecuted to the maximum extent possible under the law. Purchasers are permitted to reproduce ten (10) copies of this document when completed. To report copyright violations of AIA Contract Documents, e-mail The American Institute of Architects' legal counsel, copyright@aia.org.

ARTICLE 8 MISCELLANEOUS PROVISIONS

§ 8.1 Where reference is made in this Agreement to a provision of AIA Document A201–2007 or another Contract Document, the reference refers to that provision as amended or supplemented by other provisions of the Contract Documents.

§ 8.2 Payments due and unpaid under the Contract shall bear interest from the date payment is due at the rate stated below, or in the absence thereof, at the legal rate prevailing from time to time at the place where the Project is located.
(Insert rate of interest agreed upon, if any.)

§ 8.3 The Owner's representative:
(Name, address and other information)

§ 8.4 The Contractor's representative:
(Name, address and other information)

§ 8.5 Neither the Owner's nor the Contractor's representative shall be changed without ten days written notice to the other party.

§ 8.6 Other provisions:

ARTICLE 9 ENUMERATION OF CONTRACT DOCUMENTS

§ 9.1 The Contract Documents, except for Modifications issued after execution of this Agreement, are enumerated in the sections below.

§ 9.1.1 The Agreement is this executed AIA Document A101–2007, Standard Form of Agreement Between Owner and Contractor.

§ 9.1.2 The General Conditions are AIA Document A201–2007, General Conditions of the Contract for Construction.

§ 9.1.3 The Supplementary and other Conditions of the Contract:

Document	Title	Date	Pages

§ 9.1.4 The Specifications:
(Either list the Specifications here or refer to an exhibit attached to this Agreement.)

Section	Title	Date	Pages

Init.

/

AIA Document A101™ – 2007. Copyright © 1915, 1918, 1925, 1937, 1951, 1958, 1961, 1963, 1967, 1974, 1977, 1980, 1987, 1991, 1997 and 2007 by The American Institute of Architects. All rights reserved. WARNING: This AIA® Document is protected by U.S. Copyright Law and International Treaties. Unauthorized reproduction or distribution of this AIA® Document, or any portion of it, may result in severe civil and criminal penalties, and will be prosecuted to the maximum extent possible under the law. Purchasers are permitted to reproduce ten (10) copies of this document when completed. To report copyright violations of AIA Contract Documents, e-mail The American Institute of Architects' legal counsel, copyright@aia.org.

§ 9.1.5 The Drawings:
(Either list the Drawings here or refer to an exhibit attached to this Agreement.)

Number	Title	Date

§ 9.1.6 The Addenda, if any:

Number	Date	Pages

Portions of Addenda relating to bidding requirements are not part of the Contract Documents unless the bidding requirements are also enumerated in this Article 9.

§ 9.1.7 Additional documents, if any, forming part of the Contract Documents:
    .1   AIA Document E201™–2007, Digital Data Protocol Exhibit, if completed by the parties, or the following:

    .2   Other documents, if any, listed below:
(List here any additional documents that are intended to form part of the Contract Documents. AIA Document A201–2007 provides that bidding requirements such as advertisement or invitation to bid, Instructions to Bidders, sample forms and the Contractor's bid are not part of the Contract Documents unless enumerated in this Agreement. They should be listed here only if intended to be part of the Contract Documents.)

ARTICLE 10   INSURANCE AND BONDS
The Contractor shall purchase and maintain insurance and provide bonds as set forth in Article 11 of AIA Document A201–2007.
(State bonding requirements, if any, and limits of liability for insurance required in Article 11 of AIA Document A201–2007.)

This Agreement entered into as of the day and year first written above.

_____   _____
OWNER *(Signature)*   **CONTRACTOR** *(Signature)*

_____   _____
(Printed name and title)   *(Printed name and title)*

CAUTION: You should sign an original AIA Contract Document, on which this text appears in RED. An original assures that changes will not be obscured.

Init.

/

AIA Document A101™ – 2007. Copyright © 1915, 1918, 1925, 1937, 1951, 1958, 1961, 1963, 1967, 1974, 1977, 1980, 1987, 1991, 1997 and 2007 by The American Institute of Architects. All rights reserved. WARNING: This AIA® Document is protected by U.S. Copyright Law and International Treaties. Unauthorized reproduction or distribution of this AIA® Document, or any portion of it, may result in severe civil and criminal penalties, and will be prosecuted to the maximum extent possible under the law. Purchasers are permitted to reproduce ten (10) copies of this document when completed. To report copyright violations of AIA Contract Documents, e-mail The American Institute of Architects' legal counsel, copyright@aia.org.

American Institute of Architects cost of work plus a fee with a guaranteed maximum price (GMP) contract

AIA® Document A102™ – 2007

Standard Form of Agreement Between Owner and Contractor where the basis of payment is the Cost of the Work Plus a Fee with a Guaranteed Maximum Price

AGREEMENT made as of the day of
in the year
(In words, indicate day, month and year)

BETWEEN the Owner:
(Name, address and other information)

This document has important legal consequences. Consultation with an attorney is encouraged with respect to its completion or modification.

This document is not intended for use in competitive bidding.

AIA Document A201™–2007, General Conditions of the Contract for Construction, is adopted in this document by reference. Do not use with other general conditions unless this document is modified.

and the Contractor:
(Name, address and other information)

for the following Project:
(Name, location, and detailed description)

The Architect:
(Name, address and other information)

The Owner and Contractor agree as follows.

Init.

/

AIA Document A102™ – 2007 (formerly A111™ – 1997). Copyright © 1920, 1925, 1951, 1958, 1961, 1963, 1967, 1974, 1978, 1987, 1997 and 2007 by The American Institute of Architects. All rights reserved. WARNING: This AIA® Document is protected by U.S. Copyright Law and International Treaties. Unauthorized reproduction or distribution of this AIA® Document, or any portion of it, may result in severe civil and criminal penalties, and will be prosecuted to the maximum extent possible under the law. Purchasers are permitted to reproduce ten (10) copies of this document when completed. To report copyright violations of AIA Contract Documents, e-mail The American Institute of Architects' legal counsel, copyright@aia.org.

Reproduced with permission of The American Institute of Architects, 1735 New York Avenue, Washington, D.C. 20006

TABLE OF ARTICLES

ARTICLE 1 THE CONTRACT DOCUMENTS
The Contract Documents consist of this Agreement, Conditions of the Contract (General, Supplementary and other Conditions), Drawings, Specifications, Addenda issued prior to execution of this Agreement, other documents listed in this Agreement and Modifications issued after execution of this Agreement, all of which form the Contract, and are as fully a part of the Contract as if attached to this Agreement or repeated herein. The Contract represents the entire and integrated agreement between the parties hereto and supersedes prior negotiations, representations or agreements, either written or oral. If anything in the other Contract Documents, other than a Modification, is inconsistent with this Agreement, this Agreement shall govern.

ARTICLE 2 THE WORK OF THIS CONTRACT
The Contractor shall fully execute the Work described in the Contract Documents, except as specifically indicated in the Contract Documents to be the responsibility of others.

ARTICLE 3 RELATIONSHIP OF THE PARTIES
The Contractor accepts the relationship of trust and confidence established by this Agreement and covenants with the Owner to cooperate with the Architect and exercise the Contractor's skill and judgment in furthering the interests of the Owner; to furnish efficient business administration and supervision; to furnish at all times an adequate supply of workers and materials; and to perform the Work in an expeditious and economical manner consistent with the Owner's interests. The Owner agrees to furnish and approve, in a timely manner, information required by the Contractor and to make payments to the Contractor in accordance with the requirements of the Contract Documents.

Init.

/

AIA Document A102™ – 2007 (formerly A111™ – 1997). Copyright © 1920, 1925, 1951, 1958, 1961, 1963, 1967, 1974, 1978, 1987, 1997 and 2007 by The American Institute of Architects. All rights reserved. WARNING: This AIA® Document is protected by U.S. Copyright Law and International Treaties. Unauthorized reproduction or distribution of this AIA® Document, or any portion of it, may result in severe civil and criminal penalties, and will be prosecuted to the maximum extent possible under the law. Purchasers are permitted to reproduce ten (10) copies of this document when completed. To report copyright violations of AIA Contract Documents, e-mail The American Institute of Architects' legal counsel, copyright@aia.org

ARTICLE 4 DATE OF COMMENCEMENT AND SUBSTANTIAL COMPLETION
§ 4.1 The date of commencement of the Work shall be the date of this Agreement unless a different date is stated below or provision is made for the date to be fixed in a notice to proceed issued by the Owner.
(Insert the date of commencement, if it differs from the date of this Agreement or, if applicable, state that the date will be fixed in a notice to proceed.)

If, prior to commencement of the Work, the Owner requires time to file mortgages and other security interests, the Owner's time requirement shall be as follows:

§ 4.2 The Contract Time shall be measured from the date of commencement.

§ 4.3 The Contractor shall achieve Substantial Completion of the entire Work not later than
( ) days from the date of commencement, or as follows:
(Insert number of calendar days. Alternatively, a calendar date may be used when coordinated with the date of commencement. If appropriate, insert requirements for earlier Substantial Completion of certain portions of the Work.)

, subject to adjustments of this Contract Time as provided in the Contract Documents.
(Insert provisions, if any, for liquidated damages relating to failure to achieve Substantial Completion on time, or for bonus payments for early completion of the Work.)

ARTICLE 5 CONTRACT SUM
§ 5.1. The Owner shall pay the Contractor the Contract Sum in current funds for the Contractor's performance of the Contract. The Contract Sum is the Cost of the Work as defined in Article 7 plus the Contractor's Fee.

§ 5.1.1 The Contractor's Fee:
(State a lump sum, percentage of Cost of the Work or other provision for determining the Contractor's Fee.)

§ 5.1.2 The method of adjustment of the Contractor's Fee for changes in the Work:

§ 5.1.3 Limitations, if any, on a Subcontractor's overhead and profit for increases in the cost of its portion of the Work:

Init.

/

AIA Document A102™ – 2007 (formerly A111™ – 1997). Copyright © 1920, 1925, 1951, 1958, 1961, 1963, 1967, 1974, 1978, 1987, 1997 and 2007 by The American Institute of Architects. All rights reserved. WARNING: This AIA® Document is protected by U.S. Copyright Law and International Treaties. Unauthorized reproduction or distribution of this AIA® Document, or any portion of it, may result in severe civil and criminal penalties, and will be prosecuted to the maximum extent possible under the law. Purchasers are permitted to reproduce ten (10) copies of this document when completed. To report copyright violations of AIA Contract Documents, e-mail The American Institute of Architects' legal counsel, copyright@aia.org

§ 5.1.4 Rental rates for Contractor-owned equipment shall not exceed   percent (%) of the standard rate paid at the place of the Project.

§ 5.1.5 Unit prices, if any:
(Identify and state the unit price; state the quantity limitations, if any, to which the unit price will be applicable.)

Item	**Units and Limitations**	**Price Per Unit**

§ 5.2 GUARANTEED MAXIMUM PRICE
§ 5.2.1 The Contract Sum is guaranteed by the Contractor not to exceed
                    Dollars ($), subject to additions and deductions by
Change Order as provided in the Contract Documents. Such maximum sum is referred to in the Contract Documents as the Guaranteed Maximum Price. Costs which would cause the Guaranteed Maximum Price to be exceeded shall be paid by the Contractor without reimbursement by the Owner.
(Insert specific provisions if the Contractor is to participate in any savings.)

§ 5.2.2 The Guaranteed Maximum Price is based on the following alternates, if any, which are described in the Contract Documents and are hereby accepted by the Owner:
(State the numbers or other identification of accepted alternates. If bidding or proposal documents permit the Owner to accept other alternates subsequent to the execution of this Agreement, attach a schedule of such other alternates showing the amount for each and the date when the amount expires.)

§ 5.2.3 Allowances included in the Guaranteed Maximum Price, if any:
(Identify allowance and state exclusions, if any, from the allowance price.)

Item	**Price**

§ 5.2.4 Assumptions, if any, on which the Guaranteed Maximum Price is based:

§ 5.2.5 To the extent that the Drawings and Specifications are anticipated to require further development by the Architect, the Contractor has provided in the Guaranteed Maximum Price for such further development consistent with the Contract Documents and reasonably inferable therefrom. Such further development does not include such things as changes in scope, systems, kinds and quality of materials, finishes or equipment, all of which, if required, shall be incorporated by Change Order.

Init.

/

AIA Document A102™ – 2007 (formerly A111™ – 1997). Copyright © 1920, 1925, 1951, 1958, 1961, 1963, 1967, 1974, 1978, 1987, 1997 and 2007 by The American Institute of Architects. All rights reserved. WARNING: This AIA® Document is protected by U.S. Copyright Law and International Treaties. Unauthorized reproduction or distribution of this AIA® Document, or any portion of it, may result in severe civil and criminal penalties, and will be prosecuted to the maximum extent possible under the law. Purchasers are permitted to reproduce ten (10) copies of this document when completed. To report copyright violations of AIA Contract Documents, e-mail The American Institute of Architects' legal counsel, copyright@aia.org

ARTICLE 6 CHANGES IN THE WORK

§ 6.1 Adjustments to the Guaranteed Maximum Price on account of changes in the Work may be determined by any of the methods listed in Section 7.3.3 of AIA Document A201–2007, General Conditions of the Contract for Construction.

§ 6.2 In calculating adjustments to subcontracts (except those awarded with the Owner's prior consent on the basis of cost plus a fee), the terms "cost" and "fee" as used in Section 7.3.3.3 of AIA Document A201–2007 and the term "costs" as used in Section 7.3.7 of AIA Document A201–2007 shall have the meanings assigned to them in AIA Document A201–2007 and shall not be modified by Articles 5, 7 and 8 of this Agreement. Adjustments to subcontracts awarded with the Owner's prior consent on the basis of cost plus a fee shall be calculated in accordance with the terms of those subcontracts.

§ 6.3 In calculating adjustments to the Guaranteed Maximum Price, the terms "cost" and "costs" as used in the above-referenced provisions of AIA Document A201–2007 shall mean the Cost of the Work as defined in Article 7 of this Agreement and the term "fee" shall mean the Contractor's Fee as defined in Section 5.1.1 of this Agreement.

§ 6.4 If no specific provision is made in Article 5 for adjustment of the Contractor's Fee in the case of changes in the Work, or if the extent of such changes is such, in the aggregate, that application of the adjustment provisions of Article 5 will cause substantial inequity to the Owner or Contractor, the Contractor's Fee shall be equitably adjusted on the same basis that was used to establish the Fee for the original Work, and the Guaranteed Maximum Price shall be adjusted accordingly.

ARTICLE 7 COSTS TO BE REIMBURSED
§ 7.1 COST OF THE WORK
§ 7.1.1 The term Cost of the Work shall mean costs necessarily incurred by the Contractor in the proper performance of the Work. Such costs shall be at rates not higher than the standard paid at the place of the Project except with prior consent of the Owner. The Cost of the Work shall include only the items set forth in this Article 7.

§ 7.1.2 Where any cost is subject to the Owner's prior approval, the Contractor shall obtain this approval prior to incurring the cost. The parties shall endeavor to identify any such costs prior to executing this Agreement.

§ 7.2 LABOR COSTS
§ 7.2.1 Wages of construction workers directly employed by the Contractor to perform the construction of the Work at the site or, with the Owner's prior approval, at off-site workshops.

§ 7.2.2 Wages or salaries of the Contractor's supervisory and administrative personnel when stationed at the site with the Owner's prior approval.
(If it is intended that the wages or salaries of certain personnel stationed at the Contractor's principal or other offices shall be included in the Cost of the Work, identify in Article 15, the personnel to be included, whether for all or only part of their time, and the rates at which their time will be charged to the Work.)

§ 7.2.3 Wages and salaries of the Contractor's supervisory or administrative personnel engaged at factories, workshops or on the road, in expediting the production or transportation of materials or equipment required for the Work, but only for that portion of their time required for the Work.

§ 7.2.4 Costs paid or incurred by the Contractor for taxes, insurance, contributions, assessments and benefits required by law or collective bargaining agreements and, for personnel not covered by such agreements, customary benefits such as sick leave, medical and health benefits, holidays, vacations and pensions, provided such costs are based on wages and salaries included in the Cost of the Work under Sections 7.2.1 through 7.2.3.

§ 7.2.5 Bonuses, profit sharing, incentive compensation and any other discretionary payments paid to anyone hired by the Contractor or paid to any Subcontractor or vendor, with the Owner's prior approval.

§ 7.3 SUBCONTRACT COSTS
Payments made by the Contractor to Subcontractors in accordance with the requirements of the subcontracts.

§ 7.4 COSTS OF MATERIALS AND EQUIPMENT INCORPORATED IN THE COMPLETED CONSTRUCTION
§ 7.4.1 Costs, including transportation and storage, of materials and equipment incorporated or to be incorporated in the completed construction.

Init.

/

AIA Document A102™ – 2007 (formerly A111™ – 1997). Copyright © 1920, 1925, 1951, 1958, 1961, 1963, 1967, 1974, 1978, 1987, 1997 and 2007 by The American Institute of Architects. All rights reserved. WARNING: This AIA® Document is protected by U.S. Copyright Law and International Treaties. Unauthorized reproduction or distribution of this AIA® Document, or any portion of it, may result in severe civil and criminal penalties, and will be prosecuted to the maximum extent possible under the law. Purchasers are permitted to reproduce ten (10) copies of this document when completed. To report copyright violations of AIA Contract Documents, e-mail The American Institute of Architects' legal counsel, copyright@aia.org

§ 7.4.2 Costs of materials described in the preceding Section 7.4.1 in excess of those actually installed to allow for reasonable waste and spoilage. Unused excess materials, if any, shall become the Owner's property at the completion of the Work or, at the Owner's option, shall be sold by the Contractor. Any amounts realized from such sales shall be credited to the Owner as a deduction from the Cost of the Work.

§ 7.5 COSTS OF OTHER MATERIALS AND EQUIPMENT, TEMPORARY FACILITIES AND RELATED ITEMS

§ 7.5.1 Costs of transportation, storage, installation, maintenance, dismantling and removal of materials, supplies, temporary facilities, machinery, equipment and hand tools not customarily owned by construction workers that are provided by the Contractor at the site and fully consumed in the performance of the Work. Costs of materials, supplies, temporary facilities, machinery, equipment and tools that are not fully consumed shall be based on the cost or value of the item at the time it is first used on the Project site less the value of the item when it is no longer used at the Project site. Costs for items not fully consumed by the Contractor shall mean fair market value.

§ 7.5.2 Rental charges for temporary facilities, machinery, equipment and hand tools not customarily owned by construction workers that are provided by the Contractor at the site and costs of transportation, installation, minor repairs, dismantling and removal. The total rental cost of any Contractor-owned item may not exceed the purchase price of any comparable item. Rates of Contractor-owned equipment and quantities of equipment shall be subject to the Owner's prior approval.

§ 7.5.3 Costs of removal of debris from the site of the Work and its proper and legal disposal.

§ 7.5.4 Costs of document reproductions, facsimile transmissions and long-distance telephone calls, postage and parcel delivery charges, telephone service at the site and reasonable petty cash expenses of the site office.

§ 7.5.5 Costs of materials and equipment suitably stored off the site at a mutually acceptable location, subject to the Owner's prior approval.

§ 7.6 MISCELLANEOUS COSTS

§ 7.6.1 Premiums for that portion of insurance and bonds required by the Contract Documents that can be directly attributed to this Contract. Self-insurance for either full or partial amounts of the coverages required by the Contract Documents, with the Owner's prior approval.

§ 7.6.2 Sales, use or similar taxes imposed by a governmental authority that are related to the Work and for which the Contractor is liable.

§ 7.6.3 Fees and assessments for the building permit and for other permits, licenses and inspections for which the Contractor is required by the Contract Documents to pay.

§ 7.6.4 Fees of laboratories for tests required by the Contract Documents, except those related to defective or nonconforming Work for which reimbursement is excluded by Section 13.5.3 of AIA Document A201–2007 or by other provisions of the Contract Documents, and which do not fall within the scope of Section 7.7.3.

§ 7.6.5 Royalties and license fees paid for the use of a particular design, process or product required by the Contract Documents; the cost of defending suits or claims for infringement of patent rights arising from such requirement of the Contract Documents; and payments made in accordance with legal judgments against the Contractor resulting from such suits or claims and payments of settlements made with the Owner's consent. However, such costs of legal defenses, judgments and settlements shall not be included in the calculation of the Contractor's Fee or subject to the Guaranteed Maximum Price. If such royalties, fees and costs are excluded by the last sentence of Section 3.17 of AIA Document A201–2007 or other provisions of the Contract Documents, then they shall not be included in the Cost of the Work.

§ 7.6.6 Costs for electronic equipment and software, directly related to the Work with the Owner's prior approval.

§ 7.6.7 Deposits lost for causes other than the Contractor's negligence or failure to fulfill a specific responsibility in the Contract Documents.

§ 7.6.8 Legal, mediation and arbitration costs, including attorneys' fees, other than those arising from disputes between the Owner and Contractor, reasonably incurred by the Contractor after the execution of this Agreement in the performance of the Work and with the Owner's prior approval, which shall not be unreasonably withheld.

Init.

/

AIA Document A102™ – 2007 (formerly A111™ – 1997). Copyright © 1920, 1925, 1951, 1958, 1961, 1963, 1967, 1974, 1978, 1987, 1997 and 2007 by The American Institute of Architects. All rights reserved. WARNING: This AIA® Document is protected by U.S. Copyright Law and International Treaties. Unauthorized reproduction or distribution of this AIA® Document, or any portion of it, may result in severe civil and criminal penalties, and will be prosecuted to the maximum extent possible under the law. Purchasers are permitted to reproduce ten (10) copies of this document when completed. To report copyright violations of AIA Contract Documents, e-mail The American Institute of Architects' legal counsel, copyright@aia.org

§ 7.6.9 Subject to the Owner's prior approval, expenses incurred in accordance with the Contractor's standard written personnel policy for relocation and temporary living allowances of the Contractor's personnel required for the Work.

§ 7.6.10 That portion of the reasonable expenses of the Contractor's supervisory or administrative personnel incurred while traveling in discharge of duties connected with the Work.

§ 7.7 OTHER COSTS AND EMERGENCIES
§ 7.7.1 Other costs incurred in the performance of the Work if, and to the extent, approved in advance in writing by the Owner.

§ 7.7.2 Costs incurred in taking action to prevent threatened damage, injury or loss in case of an emergency affecting the safety of persons and property, as provided in Section 10.4 of AIA Document A201–2007.

§ 7.7.3 Costs of repairing or correcting damaged or nonconforming Work executed by the Contractor, Subcontractors or suppliers, provided that such damaged or nonconforming Work was not caused by negligence or failure to fulfill a specific responsibility of the Contractor and only to the extent that the cost of repair or correction is not recovered by the Contractor from insurance, sureties, Subcontractors, suppliers, or others.

§ 7.8 RELATED PARTY TRANSACTIONS
§ 7.8.1 For purposes of Section 7.8, the term "related party" shall mean a parent, subsidiary, affiliate or other entity having common ownership or management with the Contractor; any entity in which any stockholder in, or management employee of, the Contractor owns any interest in excess of ten percent in the aggregate; or any person or entity which has the right to control the business or affairs of the Contractor. The term "related party" includes any member of the immediate family of any person identified above.

§ 7.8.2 If any of the costs to be reimbursed arise from a transaction between the Contractor and a related party, the Contractor shall notify the Owner of the specific nature of the contemplated transaction, including the identity of the related party and the anticipated cost to be incurred, before any such transaction is consummated or cost incurred. If the Owner, after such notification, authorizes the proposed transaction, then the cost incurred shall be included as a cost to be reimbursed, and the Contractor shall procure the Work, equipment, goods or service from the related party, as a Subcontractor, according to the terms of Article 10. If the Owner fails to authorize the transaction, the Contractor shall procure the Work, equipment, goods or service from some person or entity other than a related party according to the terms of Article 10.

ARTICLE 8 COSTS NOT TO BE REIMBURSED
§ 8.1 The Cost of the Work shall not include the items listed below:
- .1 Salaries and other compensation of the Contractor's personnel stationed at the Contractor's principal office or offices other than the site office, except as specifically provided in Section 7.2. or as may be provided in Article 15;
- .2 Expenses of the Contractor's principal office and offices other than the site office;
- .3 Overhead and general expenses, except as may be expressly included in Article 7;
- .4 The Contractor's capital expenses, including interest on the Contractor's capital employed for the Work;
- .5 Except as provided in Section 7.7.3 of this Agreement, costs due to the negligence or failure of the Contractor, Subcontractors and suppliers or anyone directly or indirectly employed by any of them or for whose acts any of them may be liable to fulfill a specific responsibility of the Contract;
- .6 Any cost not specifically and expressly described in Article 7; and
- .7 Costs, other than costs included in Change Orders approved by the Owner, that would cause the Guaranteed Maximum Price to be exceeded.

ARTICLE 9 DISCOUNTS, REBATES AND REFUNDS
§ 9.1 Cash discounts obtained on payments made by the Contractor shall accrue to the Owner if (1) before making the payment, the Contractor included them in an Application for Payment and received payment from the Owner, or (2) the Owner has deposited funds with the Contractor with which to make payments; otherwise, cash discounts shall accrue to the Contractor. Trade discounts, rebates, refunds and amounts received from sales of surplus materials and equipment shall accrue to the Owner, and the Contractor shall make provisions so that they can be obtained.

§ 9.2 Amounts that accrue to the Owner in accordance with the provisions of Section 9.1 shall be credited to the Owner as a deduction from the Cost of the Work.

Init.

/

AIA Document A102™ – 2007 (formerly A111™ – 1997). Copyright © 1920, 1925, 1951, 1958, 1961, 1963, 1967, 1974, 1978, 1987, 1997 and 2007 by The American Institute of Architects. All rights reserved. WARNING: This AIA® Document is protected by U.S. Copyright Law and International Treaties. Unauthorized reproduction or distribution of this AIA® Document, or any portion of it, may result in severe civil and criminal penalties, and will be prosecuted to the maximum extent possible under the law. Purchasers are permitted to reproduce ten (10) copies of this document when completed. To report copyright violations of AIA Contract Documents, e-mail The American Institute of Architects' legal counsel, copyright@aia.org

ARTICLE 10   SUBCONTRACTS AND OTHER AGREEMENTS

§ 10.1 Those portions of the Work that the Contractor does not customarily perform with the Contractor's own personnel shall be performed under subcontracts or by other appropriate agreements with the Contractor. The Owner may designate specific persons from whom, or entities from which, the Contractor shall obtain bids. The Contractor shall obtain bids from Subcontractors and from suppliers of materials or equipment fabricated especially for the Work and shall deliver such bids to the Architect. The Owner shall then determine, with the advice of the Contractor and the Architect, which bids will be accepted. The Contractor shall not be required to contract with anyone to whom the Contractor has reasonable objection.

§ 10.2 When a specific bidder (1) is recommended to the Owner by the Contractor; (2) is qualified to perform that portion of the Work; and (3) has submitted a bid that conforms to the requirements of the Contract Documents without reservations or exceptions, but the Owner requires that another bid be accepted, then the Contractor may require that a Change Order be issued to adjust the Guaranteed Maximum Price by the difference between the bid of the person or entity recommended to the Owner by the Contractor and the amount of the subcontract or other agreement actually signed with the person or entity designated by the Owner.

§ 10.3 Subcontracts or other agreements shall conform to the applicable payment provisions of this Agreement, and shall not be awarded on the basis of cost plus a fee without the prior consent of the Owner. If the Subcontract is awarded on a cost-plus a fee basis, the Contractor shall provide in the Subcontract for the Owner to receive the same audit rights with regard to the Subcontractor as the Owner receives with regard to the Contractor in Article 11, below.

ARTICLE 11   ACCOUNTING RECORDS

The Contractor shall keep full and detailed records and accounts related to the cost of the Work and exercise such controls as may be necessary for proper financial management under this Contract and to substantiate all costs incurred. The accounting and control systems shall be satisfactory to the Owner. The Owner and the Owner's auditors shall, during regular business hours and upon reasonable notice, be afforded access to, and shall be permitted to audit and copy, the Contractor's records and accounts, including complete documentation supporting accounting entries, books, correspondence, instructions, drawings, receipts, subcontracts, Subcontractor's proposals, purchase orders, vouchers, memoranda and other data relating to this Contract. The Contractor shall preserve these records for a period of three years after final payment, or for such longer period as may be required by law.

ARTICLE 12   PAYMENTS
§ 12.1 PROGRESS PAYMENTS

§ 12.1.1 Based upon Applications for Payment submitted to the Architect by the Contractor and Certificates for Payment issued by the Architect, the Owner shall make progress payments on account of the Contract Sum to the Contractor as provided below and elsewhere in the Contract Documents.

§ 12.1.2 The period covered by each Application for Payment shall be one calendar month ending on the last day of the month, or as follows:

§ 12.1.3 Provided that an Application for Payment is received by the Architect not later than the
(   ) day of a month, the Owner shall make payment of the certified amount to the Contractor not later than the   (   ) day of the   (   ) month. If an Application for Payment is received by the Architect after the application date fixed above, payment shall be made by the Owner not later than   (   ) days after the Architect receives the Application for Payment.
(Federal, state or local laws may require payment within a certain period of time.)

§ 12.1.4 With each Application for Payment, the Contractor shall submit payrolls, petty cash accounts, receipted invoices or invoices with check vouchers attached, and any other evidence required by the Owner or Architect to demonstrate that cash disbursements already made by the Contractor on account of the Cost of the Work equal or exceed (1) progress payments already received by the Contractor; less (2) that portion of those payments attributable to the Contractor's Fee; plus (3) payrolls for the period covered by the present Application for Payment.

§ 12.1.5 Each Application for Payment shall be based on the most recent schedule of values submitted by the Contractor in accordance with the Contract Documents. The schedule of values shall allocate the entire Guaranteed Maximum Price among the various portions of the Work, except that the Contractor's Fee shall be shown as a single separate item.

Init.

/

AIA Document A102™ – 2007 (formerly A111™ – 1997). Copyright © 1920, 1925, 1951, 1958, 1961, 1963, 1967, 1974, 1978, 1987, 1997 and 2007 by The American Institute of Architects. All rights reserved. WARNING: This AIA® Document is protected by U.S. Copyright Law and International Treaties. Unauthorized reproduction or distribution of this AIA® Document, or any portion of it, may result in severe civil and criminal penalties, and will be prosecuted to the maximum extent possible under the law. Purchasers are permitted to reproduce ten (10) copies of this document when completed. To report copyright violations of AIA Contract Documents, e-mail The American Institute of Architects' legal counsel, copyright@aia.org

The schedule of values shall be prepared in such form and supported by such data to substantiate its accuracy as the Architect may require. This schedule, unless objected to by the Architect, shall be used as a basis for reviewing the Contractor's Applications for Payment.

§ 12.1.6 Applications for Payment shall show the percentage of completion of each portion of the Work as of the end of the period covered by the Application for Payment. The percentage of completion shall be the lesser of (1) the percentage of that portion of the Work which has actually been completed; or (2) the percentage obtained by dividing (a) the expense that has actually been incurred by the Contractor on account of that portion of the Work for which the Contractor has made or intends to make actual payment prior to the next Application for Payment by (b) the share of the Guaranteed Maximum Price allocated to that portion of the Work in the schedule of values.

§ 12.1.7 Subject to other provisions of the Contract Documents, the amount of each progress payment shall be computed as follows:

.1 Take that portion of the Guaranteed Maximum Price properly allocable to completed Work as determined by multiplying the percentage of completion of each portion of the Work by the share of the Guaranteed Maximum Price allocated to that portion of the Work in the schedule of values. Pending final determination of cost to the Owner of changes in the Work, amounts not in dispute shall be included as provided in Section 7.3.9 of AIA Document A201–2007;

.2 Add that portion of the Guaranteed Maximum Price properly allocable to materials and equipment delivered and suitably stored at the site for subsequent incorporation in the Work, or if approved in advance by the Owner, suitably stored off the site at a location agreed upon in writing;

.3 Add the Contractor's Fee, less retainage of   percent (%). The Contractor's Fee shall be computed upon the Cost of the Work at the rate stated in Section 5.1.1 or, if the Contractor's Fee is stated as a fixed sum in that Section, shall be an amount that bears the same ratio to that fixed-sum fee as the Cost of the Work bears to a reasonable estimate of the probable Cost of the Work upon its completion;

.4 Subtract retainage of   percent (%) from that portion of the Work that the Contractor self-performs;

.5 Subtract the aggregate of previous payments made by the Owner;

.6 Subtract the shortfall, if any, indicated by the Contractor in the documentation required by Section 12.1.4 to substantiate prior Applications for Payment, or resulting from errors subsequently discovered by the Owner's auditors in such documentation; and

.7 Subtract amounts, if any, for which the Architect has withheld or nullified a Certificate for Payment as provided in Section 9.5 of AIA Document A201–2007.

§ 12.1.8 The Owner and the Contractor shall agree upon a (1) mutually acceptable procedure for review and approval of payments to Subcontractors and (2) the percentage of retainage held on Subcontracts, and the Contractor shall execute subcontracts in accordance with those agreements.

§ 12.1.9 In taking action on the Contractor's Applications for Payment, the Architect shall be entitled to rely on the accuracy and completeness of the information furnished by the Contractor and shall not be deemed to represent that the Architect has made a detailed examination, audit or arithmetic verification of the documentation submitted in accordance with Section 12.1.4 or other supporting data; that the Architect has made exhaustive or continuous on-site inspections; or that the Architect has made examinations to ascertain how or for what purposes the Contractor has used amounts previously paid on account of the Contract. Such examinations, audits and verifications, if required by the Owner, will be performed by the Owner's auditors acting in the sole interest of the Owner.

§ 12.2 FINAL PAYMENT
§ 12.2.1 Final payment, constituting the entire unpaid balance of the Contract Sum, shall be made by the Owner to the Contractor when

.1 the Contractor has fully performed the Contract except for the Contractor's responsibility to correct Work as provided in Section 12.2.2 of AIA Document A201–2007, and to satisfy other requirements, if any, which extend beyond final payment;

.2 the Contractor has submitted a final accounting for the Cost of the Work and a final Application for Payment; and

.3 a final Certificate for Payment has been issued by the Architect.

§ 12.2.2 The Owner's auditors will review and report in writing on the Contractor's final accounting within 30 days after delivery of the final accounting to the Architect by the Contractor. Based upon such Cost of the Work as the Owner's

Init.

/

AIA Document A102™ – 2007 (formerly A111™ – 1997). Copyright © 1920, 1925, 1951, 1958, 1961, 1963, 1967, 1974, 1978, 1987, 1997 and 2007 by The American Institute of Architects. All rights reserved. WARNING: This AIA® Document is protected by U.S. Copyright Law and International Treaties. Unauthorized reproduction or distribution of this AIA® Document, or any portion of it, may result in severe civil and criminal penalties, and will be prosecuted to the maximum extent possible under the law. Purchasers are permitted to reproduce ten (10) copies of this document when completed. To report copyright violations of AIA Contract Documents, e-mail The American Institute of Architects' legal counsel, copyright@aia.org

auditors report to be substantiated by the Contractor's final accounting, and provided the other conditions of Section 12.2.1 have been met, the Architect will, within seven days after receipt of the written report of the Owner's auditors, either issue to the Owner a final Certificate for Payment with a copy to the Contractor, or notify the Contractor and Owner in writing of the Architect's reasons for withholding a certificate as provided in Section 9.5.1 of the AIA Document A201–2007. The time periods stated in this Section 12.2.2 supersede those stated in Section 9.4.1 of the AIA Document A201–2007. The Architect is not responsible for verifying the accuracy of the Contractor's final accounting.

§ 12.2.3 If the Owner's auditors report the Cost of the Work as substantiated by the Contractor's final accounting to be less than claimed by the Contractor, the Contractor shall be entitled to request mediation of the disputed amount without seeking an initial decision pursuant to Section 15.2 of A201–2007. A request for mediation shall be made by the Contractor within 30 days after the Contractor's receipt of a copy of the Architect's final Certificate for Payment. Failure to request mediation within this 30-day period shall result in the substantiated amount reported by the Owner's auditors becoming binding on the Contractor. Pending a final resolution of the disputed amount, the Owner shall pay the Contractor the amount certified in the Architect's final Certificate for Payment.

§ 12.2.4 The Owner's final payment to the Contractor shall be made no later than 30 days after the issuance of the Architect's final Certificate for Payment, or as follows:

§ 12.2.5 If, subsequent to final payment and at the Owner's request, the Contractor incurs costs described in Article 7 and not excluded by Article 8 to correct defective or nonconforming Work, the Owner shall reimburse the Contractor such costs and the Contractor's Fee applicable thereto on the same basis as if such costs had been incurred prior to final payment, but not in excess of the Guaranteed Maximum Price. If the Contractor has participated in savings as provided in Section 5.2, the amount of such savings shall be recalculated and appropriate credit given to the Owner in determining the net amount to be paid by the Owner to the Contractor.

ARTICLE 13 DISPUTE RESOLUTION
§ 13.1 INITIAL DECISION MAKER
The Architect will serve as Initial Decision Maker pursuant to Section 15.2 of AIA Document A201–2007, unless the parties appoint below another individual, not a party to the Agreement, to serve as Initial Decision Maker.
(If the parties mutually agree, insert the name, address and other contact information of the Initial Decision Maker, if other than the Architect.)

§ 13.2 BINDING DISPUTE RESOLUTION
For any Claim subject to, but not resolved by mediation pursuant to Section 15.3 of AIA Document A201–2007, the method of binding dispute resolution shall be as follows:
(Check the appropriate box. If the Owner and Contractor do not select a method of binding dispute resolution below, or do not subsequently agree in writing to a binding dispute resolution method other than litigation, Claims will be resolved by litigation in a court of competent jurisdiction.)

☐ Arbitration pursuant to Section 15.4 of AIA Document A201–2007

☐ Litigation in a court of competent jurisdiction

☐ Other *(Specify)*

Init.

/

AIA Document A102™ – 2007 (formerly A111™ – 1997). Copyright © 1920, 1925, 1951, 1958, 1961, 1963, 1967, 1974, 1978, 1987, 1997 and 2007 by The American Institute of Architects. All rights reserved. WARNING: This AIA® Document is protected by U.S. Copyright Law and International Treaties. Unauthorized reproduction or distribution of this AIA® Document, or any portion of it, may result in severe civil and criminal penalties, and will be prosecuted to the maximum extent possible under the law. Purchasers are permitted to reproduce ten (10) copies of this document when completed. To report copyright violations of AIA Contract Documents, e-mail The American Institute of Architects' legal counsel, copyright@aia.org

ARTICLE 14   TERMINATION OR SUSPENSION

§ **14.1** Subject to the provisions of Section 14.2 below, the Contract may be terminated by the Owner or the Contractor as provided in Article 14 of AIA Document A201–2007.

§ **14.2** If the Owner terminates the Contract for cause as provided in Article 14 of AIA Document A201–2007, the amount, if any, to be paid to the Contractor under Section 14.2.4 of AIA Document A201–2007 shall not cause the Guaranteed Maximum Price to be exceeded, nor shall it exceed an amount calculated as follows:

   .1   Take the Cost of the Work incurred by the Contractor to the date of termination;

   .2   Add the Contractor's Fee computed upon the Cost of the Work to the date of termination at the rate stated in Section 5.1.1 or, if the Contractor's Fee is stated as a fixed sum in that Section, an amount that bears the same ratio to that fixed-sum Fee as the Cost of the Work at the time of termination bears to a reasonable estimate of the probable Cost of the Work upon its completion; and

   .3   Subtract the aggregate of previous payments made by the Owner.

§ **14.3** The Owner shall also pay the Contractor fair compensation, either by purchase or rental at the election of the Owner, for any equipment owned by the Contractor that the Owner elects to retain and that is not otherwise included in the Cost of the Work under Section 14.2.1. To the extent that the Owner elects to take legal assignment of subcontracts and purchase orders (including rental agreements), the Contractor shall, as a condition of receiving the payments referred to in this Article 14, execute and deliver all such papers and take all such steps, including the legal assignment of such subcontracts and other contractual rights of the Contractor, as the Owner may require for the purpose of fully vesting in the Owner the rights and benefits of the Contractor under such subcontracts or purchase orders.

§ **14.4** The Work may be suspended by the Owner as provided in Article 14 of AIA Document A201–2007; in such case, the Guaranteed Maximum Price and Contract Time shall be increased as provided in Section 14.3.2 of AIA Document A201–2007, except that the term "profit" shall be understood to mean the Contractor's Fee as described in Sections 5.1.1 and Section 6.4 of this Agreement.

ARTICLE 15   MISCELLANEOUS PROVISIONS

§ **15.1** Where reference is made in this Agreement to a provision of AIA Document A201–2007 or another Contract Document, the reference refers to that provision as amended or supplemented by other provisions of the Contract Documents.

§ **15.2** Payments due and unpaid under the Contract shall bear interest from the date payment is due at the rate stated below, or in the absence thereof, at the legal rate prevailing from time to time at the place where the Project is located. *(Insert rate of interest agreed upon, if any.)*

§ **15.3** The Owner's representative:
(Name, address and other information)

§ **15.4** The Contractor's representative:
(Name, address and other information)

§ **15.5** Neither the Owner's nor the Contractor's representative shall be changed without ten days' written notice to the other party.

Init.

/

AIA Document A102™ – 2007 (formerly A111™ – 1997). Copyright © 1920, 1925, 1951, 1958, 1961, 1963, 1967, 1974, 1978, 1987, 1997 and 2007 by The American Institute of Architects. All rights reserved. WARNING: This AIA® Document is protected by U.S. Copyright Law and International Treaties. Unauthorized reproduction or distribution of this AIA® Document, or any portion of it, may result in severe civil and criminal penalties, and will be prosecuted to the maximum extent possible under the law. Purchasers are permitted to reproduce ten (10) copies of this document when completed. To report copyright violations of AIA Contract Documents, e-mail The American Institute of Architects' legal counsel, copyright@aia.org

§ 15.6 Other provisions:

ARTICLE 16 ENUMERATION OF CONTRACT DOCUMENTS
§ 16.1 The Contract Documents, except for Modifications issued after execution of this Agreement, are enumerated in the sections below.

§ 16.1.1 The Agreement is this executed AIA Document A102–2007, Standard Form of Agreement Between Owner and Contractor.

§ 16.1.2 The General Conditions are AIA Document A201–2007, General Conditions of the Contract for Construction.

§ 16.1.3 The Supplementary and other Conditions of the Contract:

Document	Title	Date	Pages

§ 16.1.4 The Specifications:
(Either list the Specifications here or refer to an exhibit attached to this Agreement.)

Section	Title	Date	Pages

§ 16.1.5 The Drawings:
(Either list the Drawings here or refer to an exhibit attached to this Agreement.)

Number	Title	Date

§ 16.1.6 The Addenda, if any:

Number	Date	Pages

Portions of Addenda relating to bidding requirements are not part of the Contract Documents unless the bidding requirements are also enumerated in this Article 16.

Init.

/

AIA Document A102™ – 2007 (formerly A111™ – 1997). Copyright © 1920, 1925, 1951, 1958, 1961, 1963, 1967, 1974, 1978, 1987, 1997 and 2007 by The American Institute of Architects. All rights reserved. WARNING: This AIA® Document is protected by U.S. Copyright Law and International Treaties. Unauthorized reproduction or distribution of this AIA® Document, or any portion of it, may result in severe civil and criminal penalties, and will be prosecuted to the maximum extent possible under the law. Purchasers are permitted to reproduce ten (10) copies of this document when completed. To report copyright violations of AIA Contract Documents, e-mail The American Institute of Architects' legal counsel, copyright@aia.org

§ 16.1.7 Additional documents, if any, forming part of the Contract Documents:

    .1   AIA Document E201™–2007, Digital Data Protocol Exhibit, if completed by the parties, or the following:

    .2   Other documents, if any, listed below:

(List here any additional documents that are intended to form part of the Contract Documents. AIA Document A201–2007 provides that bidding requirements such as advertisement or invitation to bid, Instructions to Bidders, sample forms and the Contractor's bid are not part of the Contract Documents unless enumerated in this Agreement. They should be listed here only if intended to be part of the Contract Documents.)

ARTICLE 17 INSURANCE AND BONDS

The Contractor shall purchase and maintain insurance and provide bonds as set forth in Article 11 of AIA Document A201–2007.

(State bonding requirements, if any, and limits of liability for insurance required in Article 11 of AIA Document A201–2007.)

This Agreement entered into as of the day and year first written above.

_____   _____
OWNER *(Signature)*   **CONTRACTOR** *(Signature)*

_____   _____
(Printed name and title)   *(Printed name and title)*

CAUTION: You should sign an original AIA Contract Document, on which this text appears in RED. An original assures that changes will not be obscured.

Init.

/

AIA Document A102™ – 2007 (formerly A111™ – 1997). Copyright © 1920, 1925, 1951, 1958, 1961, 1963, 1967, 1974, 1978, 1987, 1997 and 2007 by The American Institute of Architects. All rights reserved. WARNING: This AIA® Document is protected by U.S. Copyright Law and International Treaties. Unauthorized reproduction or distribution of this AIA® Document, or any portion of it, may result in severe civil and criminal penalties, and will be prosecuted to the maximum extent possible under the law. Purchasers are permitted to reproduce ten (10) copies of this document when completed. To report copyright violations of AIA Contract Documents, e-mail The American Institute of Architects' legal counsel, copyright@aia.org

Index

377

Printed in the United States
by Bookmasters

Printed in the United States
By Bookmasters